Written for the NEW **AQA** *Specificati*

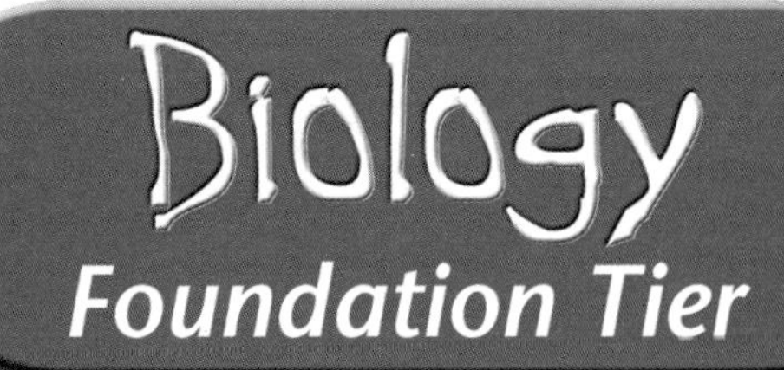

AQA Modular Science

David Coppock

OXFORD
UNIVERSITY PRESS

OXFORD
UNIVERSITY PRESS

Great Clarendon Street, Oxford OX2 6DP

Oxford University Press is a department of the University of Oxford.
It furthers the University's objective of excellence in research, scholarship, and education by publishing worldwide in

Oxford New York

Athens Auckland Bangkok Bogotá Buenos Aires Cape Town Chennai Dar es Salaam Delhi Florence Hong Kong Istanbul Karachi Kolkata Kuala Lumpur Madrid Melbourne Mexico City Mumbai Nairobi Paris São Paulo Shanghai Singapore Taipei Tokyo Toronto Warsaw with associated companies in Berlin Ibadan

Oxford is a registered trade mark of Oxford University Press in the UK and in certain other countries

First published 2001

Reprinted 2001

British Library Cataloguing in Publication Data

Data available

ISBN 0-19-914827-9

Typeset in Stone Serif
by Ian Foulis & Associates, Plymouth UK

Printed by
Gráficas Estella, S.A. Spain

Acknowledgements

The publisher would like to thank the following for their kind permission to reproduce copyright material (l = left, r = right, t = top, c = centre, b = bottom):

Cover Stone; 8/9 Corel Professional Pictures; 10tl SPL/Jeff Lepore; 10tr B. Coleman/Christer Fredriksson; 10bl Holt Studios/Phil McLean; 10br OSF/Animals Animals/John Pontier; 10t OSF/Satoshi Kuribayashi; 10c SPL/J. Krasemann; 10b B. Coleman/Sven Halling; 12l SPL/Eye of Science; 12c SPL/Quest; 12r SPL/Quest; 13t Biophoto Associates; 13tc SPL/Dr Gopal Murti; 13bc Biophoto Associates; 13b Biophoto Associates; 14 Holt Studios/Nigel Cattlin; 15 Holt Studios/Nigel Cattlin; 21 Peter Gould; 22b Corel Professional Pictures; 23t Corel Professional Pictures; 23b Stockbyte; 26 Biophoto Associates; 31tl SPL/Quest; 31tr Stock Market/Michael Keller; 31bl Photofusion/Mark Campbell; 31br Biophoto Associates; 32l OSF/Mark Hamblin; 32r Stock Market/David Stoecklein; 33t John Walmsley; 33bl John Walmsley; 33bc John Walmsley; 33br John Walmsley; 34 SPL; 38 SPL; 40l SPL/CNRI; 40r SPL/Juergen Berger, Max-Planck Institute; 41 SPL/Ricardo Arias, Latin Stock; 43 SPL/Matt Meadows; 45 SPL/Saturn Stills; 48c Corbis; 48bl Ann Ronan Picture Library; 48br Science and Society Picture Library; 49t Corbis; 49b Science and Society Picture Library; 56/57 Robert Harding Picture Library; 58l Biophoto Associates; 58c OSF/Richard Kirby; 58r SPL/Andrew Syred; 60 Natural Visions/Heather Angel; 61l B. Coleman/Luiz Claudio Marigo; 61c FLPA/Silvestris; 61r Premaphotos Wildlife/Ken Preston-Mafham; 62 SPL/Microfield Scientific Ltd; 63 OSF/Harold Taylor; 64l Holt Studios/Nigel Cattlin; 64c Holt Studios/Gordon Roberts; 64r Holt Studios/Nigel Cattlin; 65 Holt Studios/Nigel Cattlin; 68l Worm's Way; 68c Still Pictures/Mark Edwards; 68r Holt Studios/Nigel Cattlin; 69t Holt Studios/Rosemary Mayer; 69b Holt Studios/Nigel Cattlin; 70tl Biophoto Associates; 70tr Ian Parsons; 70bl Biophoto Associates; 70br Holt Studios/Nigel Cattlin; 71 Holt Studios/Nigel Cattlin; 73tl Holt Studios/Nigel Cattlin; 73tr OSF/ Philip Sharpe; 74 Holt Studios/Inga Spence; 75t Holt Studios/Phil McLean; 75b Holt Studios/Nigel Cattlin; 77t Microscopix/Andrew Syred; 77b Microscopix/Andrew Syred; 78 Holt Studios/Phil McLean; 81l Holt Studios/Nigel Cattlin; 81c Holt Studios/Nigel Cattlin; 81r Natural Visions/Soames Summerhays; 82l Holt Studios/Nigel Cattlin; 82c OSF/David M. Dennis; 82r Harry Smith Collection; 84t Holt Studios/Rosemary Mayer; 84b Holt Studios/Nigel Cattlin; 85l Holt Studios/David Burton; 85r Ian Parsons; 86 SPL/Oscar Burriel; 89t SPL/Adam Jones; 89b Holt Studios/Mike Amphlett; 95 John Walmsley; 96t Ian Parsons; 96b Stock Market/Mark Lawrence; 97t Network Photographers/Gideon Mendel; 97b Ian Parsons; 98 SPL/Eye of Science; 99tl Photofusion/Peter Olive; 99bl Corel Professional Pictures; 99tc Corbis; 99bc Hulton Deutsch Collection Ltd; 99tr The Hutchison Library/John Wright; 100l Wellcome Trust Photo Library; 100ct Biophoto Associates; 100cb Biophoto Associates; 100r SPL/Quest; 101l Photofusion/Clarissa Leahy; 101r Photofusion/Paul Doyle; 105 SPL/Mark Clarke; 108 OSF/Konrad Wothe; 110/111 Corel Professional Pictures; 110t Corel Professional Pictures; 110b Corel Professional Pictures; 111t Corbis/Paul A Souders; 111c Corbis/David Spear; 111b Holt Studios/Nigel Cattlin; 118/119 Corel Professional Pictures; 122tl B. Coleman/Rita Meyer; 122tr B. Coleman/Kim Taylor; 122b SPL/J. Krasemann; 123t FLPA/L. & O. Bahat; 123b Premaphotos Wildlife/Ken Preston-Mafham; 124t Natural Visions/Heather Angel; 124b Holt Studios/Chris & Tilde Stuart; 125tl OSF/Daniel J. Cox; 125tr B. Coleman/Luiz Claudio Marigo; 125b Holt Studios/Nigel Cattlin; 126 OSF/Leonard Lee Rue III; 127t Holt Studios/Wayne Hutchinson; 127ct OSF/Christian Grzimek/Okapia; 127cb OSF/Tim Shepherd; 127b OSF/John McCammon; 128tl Holt Studios/Phil McLean; 128tc SPL/Renee Lynn; 128tr Holt Studios/Nigel Cattlin; 128bl OSF/Steve Turner; 128bc FLPA/Mark Newman; 128br Holt Studios/Phil McLean; 129 OSF/Steve Turner; 130tl Natural Visions/Heather Angel; 130tr OSF/Alastair Shay; 130bl OSF/Mark Hamblin; 130bc Holt Studios/Mike Lane; 130br Holt Studios/Paul Hobson; 131 Holt Studios/Rosemary Mayer; 133t Natural Visions/Heather Angel; 133b Holt Studios/Richard Anthony; 135 Holt Studios/Inga Spence; 137t B. Coleman/Nigel Blake; 137b Holt Studios/Nigel Cattlin; 139 Natural Visions/Heather Angel; 140t John Walmsley; 140bl Still Pictures/Jean-Leo Dugast; 140br OSF/Frithjof Skibbe; 142tl Ardea/Ron & Valerie Taylor; 142tc Photofusion/Sam Tanner; 142tr OSF/Terry Heathcote; 142b Holt Studios/Willem Harinck; 143t Holt Studios/Mary Cherry; 143b Holt Studios/Inga Spence; 144t Holt Studios/Richard Anthony; 144b Holt Studios/Nigel Cattlin; 145 Still Pictures/Mark Edwards; 146l Stock Market/Lester Lefkowitz; 146c OSF/Richard Packwood; 146r Still Pictures/Paul Glendell; 147t Holt Studios/Nigel Cattlin; 147b Format/Brenda Prince; 148 Holt Studios/Nigel Cattlin; 149t FLPA/Celtic P. A.; 149c OSF/Niall Benvie; 149b Format/Paula Solloway; 150 Holt Studios/Nigel Cattlin; 151tl Holt Studios/Nigel Cattlin; 151tr Holt Studios/Nigel Cattlin; 151b United States Department of Agriculture; 152l OSF/John McCammon; 152r Holt Studios/Mary Cherry; 153t Holt Studios/Nigel Cattlin; 153b Premaphotos Wildlife/Ken Preston-Mafham; 154tl Holt Studios/Gordon Roberts; 54tr Press Association; 154bl Holt Studios/Nigel Cattlin; 154br Network Photographers/John Sturrock; 155t OSF/Terry Heathcote; 155bl Premaphotos Wildlife/Rod Preston-Mafham; 155br Holt Studios/Duncan Smith; 156bl Heather Angel; 156t K. G. Preston-Matham/PremaPhotos Wildlife; 156cl Heather Angel; 156cr Biophoto Associates; 156br Heather Angel; 157t Bruce Coleman Collection; 157cr Heather Angel; 157cl Heather Angel; 157bl Heather Angel; 158t SPL/Eye of Science; 158bl SPL/Cristina Pedrazzi; 158br SPL/Eye of Science; 158t SPL/Simon Fraser; 166/167 Francesc Muntada/Corbis; 168tl Photofusion/Liam Bailey; 168tr John Walmsley; 168b John Walmsley; 169l SPL/Mauro Fermariello; 169r Agripicture/Peter Dean; 170 Biophoto Associates; 171t Stock Market; 171b Stock Market/Norbert Schafer; 174 Biophoto Associates; 175t Natural Visions/Heather Angel; 175b SPL/M. I. Walker; 183tl SPL/Professors P. M. Motta & S. Makabe; 183tc SPL/Garry Watson; 183tr SPL/Corey Meitchik/Custom Medical Stock Photo; 183b SPL/Gaillard, Jerrican; 184 Ian Parsons; 185tl SPL/Saturn Stills; 185tr SPL/Cordelia Molloy; 185bl SPL/Gary Parker; 185br SPL/Gary Parker; 186 SPL/Eye of Science; 187t Format/Jacky Chapman; 187b SPL/Dept. Medical Photography, St Stephen's Hospital, London; 188 Holt Studios/Nigel Cattlin; 189tl Holt Studios/Nigel Cattlin; 189tr Holt Studios/Nigel Cattlin; 189bl FLPA/Mark Newman; 189br Natural Visions/Heather Angel; 190 OSF/Richard Packwood; 191t Holt Studios/Nigel Cattlin; 191b OSF/Sean Morris; 192tl OSF/Buttner/Okapia; 192tc Holt Studios/Nigel Cattlin; 192tr Holt Studios/Nigel Cattlin; 192bl Holt Studios/Nigel Cattlin; 192bc Holt Studios/Nigel Cattlin; 192br OSF/G. A. Maclean; 193tl OSF/Breck P. Kent; 193tc Holt Studios/Mike Amplett; 193tr Holt Studios/Bob Gibbons; 193bl Holt Studios/Nigel Cattlin; 193bc Natural Visions/Heather Angel; 193br Premaphotos Wildlife/Ken Preston-Mafham; 195 Holt Studios/Nigel Cattlin; 196tl Holt Studios/Rosemary Mayer; 196tc Holt Studios/Nigel Cattlin; 196tr Holt Studios/Nigel Cattlin; 196bl OSF/Geoff Kidd; 196bc Harry Smith Collection; 196br Harry Smith Collection; 197 Holt Studios/Nigel Cattlin; 199tl SPL/John Heseltine; 199tr Holt Studios/Nigel Cattlin; 199bl Holt Studios/Nigel Cattlin; 199bc Holt Studios/Nigel Cattlin; 199br Holt Studios/Wayne Hutchinson; 200 SPL/Mark Clarke; 201 PPL Therapeutics/John Chadwick; 202/203 Corbis; 202 Popperfoto/Reuters; 204tl Holt Studios/Bob Gibbons; 204tr Holt Studios/Nigel Cattlin; 204c Holt Studios/Nigel Cattlin; 204bl FLPA/Gerard Lacy; 204br OSF/Derek Bromhall; 207t SPL; 207bl SPL/Michael W. Tweedie; 207br SPL/Michael W. Tweedie; 208t FLPA/Philip Perry; 208c OSF/Sinclair Stammers; 208b FLPA/Peter Reynolds; 209 Holt Studios/Willem Harinck; 210l John Walmsley; 210r Photofusion/Roderick Smith; 211l John Walmsley; 211r John Walmsley; 212/213 Garden Picture Library; 213 Corbis; 215t Wellcome Trust Photo Library; 215b Wellcome Trust Photo Library/Wessex Regional Genetic Lab; 216l SPL/Hattie Young; 216r SPL/Biophoto Associates; 217 SPL/Conor Caffrey; 218t SPL/Eye of Science; 218b SPL/Dr Gopal Murti; 219t SPL/Simon Fraser/RVI, Newcastle-upon-Tyne; 219bl SPL/Alexander Tsiaras; 219bc SPL/Pekka Parviainen; 219br SPL/Sheila Terry; 220/221 SPL/Alfred Pasieka; 220 SPL/Dr Linda Stannard, UCT; 221t SPL/Tek Image; 221b FLPA/J. C. Allen; 238l Holt Studios/Richard Anthony; 238r Premaphotos Wildlife/Rod Preston-Mafham; 240tl SPL/Secchi-Lecaque/Roussel-UCLAF/CNRI; 240tr SPL/Microfield Scientific; 240cl SPL/Eye of Science; 240cr SPL/Claude Nuridsany & Marie Perennou; 240bl Biophoto Associates; 240br Holt Studios/Peter Wilson; 241tl SPL/Andrew Syred; 241tc Biophoto Associates; 241tr OSF/Barrie E. Watts; 241bl SPL/Sinclair Stammers; 241bc Holt Studios/Nigel Cattlin; 241br Holt Studios/Inga Spence; 242tl SPL/Gilbert S. Grant; 242tcl SPL/Eric Grave; 242tcr Holt Studios/Nigel Cattlin; 242tr Holt Studios/Nigel Cattlin; 242bl SPL/Dr John Brackenbury; 242bcl SPL/Dr Tony Brain; 242bcr B. Coleman/Kim Taylor; 242br SPL/Sinclair Stammers; 243tl Holt Studios/Nigel Cattlin; 243tc Holt Studios/Phil McLean; 243tr OSF/Paul Franklin; 243bl SPL/Tom McHugh; 243br OSF/Sean Morris; 244tl Holt Studios/Bob Gibbons; 244tc OSF/Larry Crowhurst; 244tr Holt Studios/Jean Hall; 244bl OSF/Deni Bown; 244bc Holt Studios/Nigel Cattlin; 244br SPL/Simon Fraser; 245tl OSF/Bob Gibbons; 245tr OSF/Bob Gibbons; 245bl OSF/Deni Bown; 245bc Holt Studios/Nigel Cattlin; 245br OSF/Warwick Johnson

Illustrations are by Ian Foulis & Associates, John Haslam, Clive Goodyer

Introduction

Science is about asking questions. Biology is the science that asks questions about the biological world around you, its practical uses, and some of the social issues it raises.

You will find this book useful if you are studying biology as part of the AQA Modular Science Single or Double Award GCSE science course.

Everything in this book has been organized to help you find out things quickly and easily. It is written in two-page units called spreads.

Use the contents page

If you are looking for information on a large topic, look it up in the contents page.

Use the index

If there is something small you want to check on, look up the most likely word in the index. The index gives the page number(s) where you'll find information about that word.

Use the questions

Asking questions and answering them is a very good way of learning. There are questions at the end of every Module. At the end of the book there is a set of further exam-style questions and a selection of multiple-choice questions. Answers to numerical questions and some pointers to those requiring short answers are provided.

Use the key words glossary

At the end of each Module there are a selection of key words and their meanings to help you understand the main ideas given in the Module.

Helping you revise

To help you revise, in addition to the questions and the end-of-Module glossaries of important terms, there are some revision notes and some further exam-style questions.

Biology is an important and exciting subject. It doesn't just happen in laboratories. It is all around you: it is taking place deep in the Earth and (probably!) far out in space. You'll find biology everywhere.

I hope that this book helps you with your studies, that you enjoy using it, and that at the end of your course, you agree with me that biology is exciting!

David Coppock
July 2001

Contents

Although you will probably follow the AQA specification module by module, these 'routemaps' show you different ways to work through the material. They are designed to help you understand the subject in small, manageable chunks by suggesting groupings of spreads relevant to particular topics and highlighting connections between them. They are especially useful when you are revising because they help you to identify and revise logical sections of material at a time, and if you have missed any work, they can also help you to catch up.

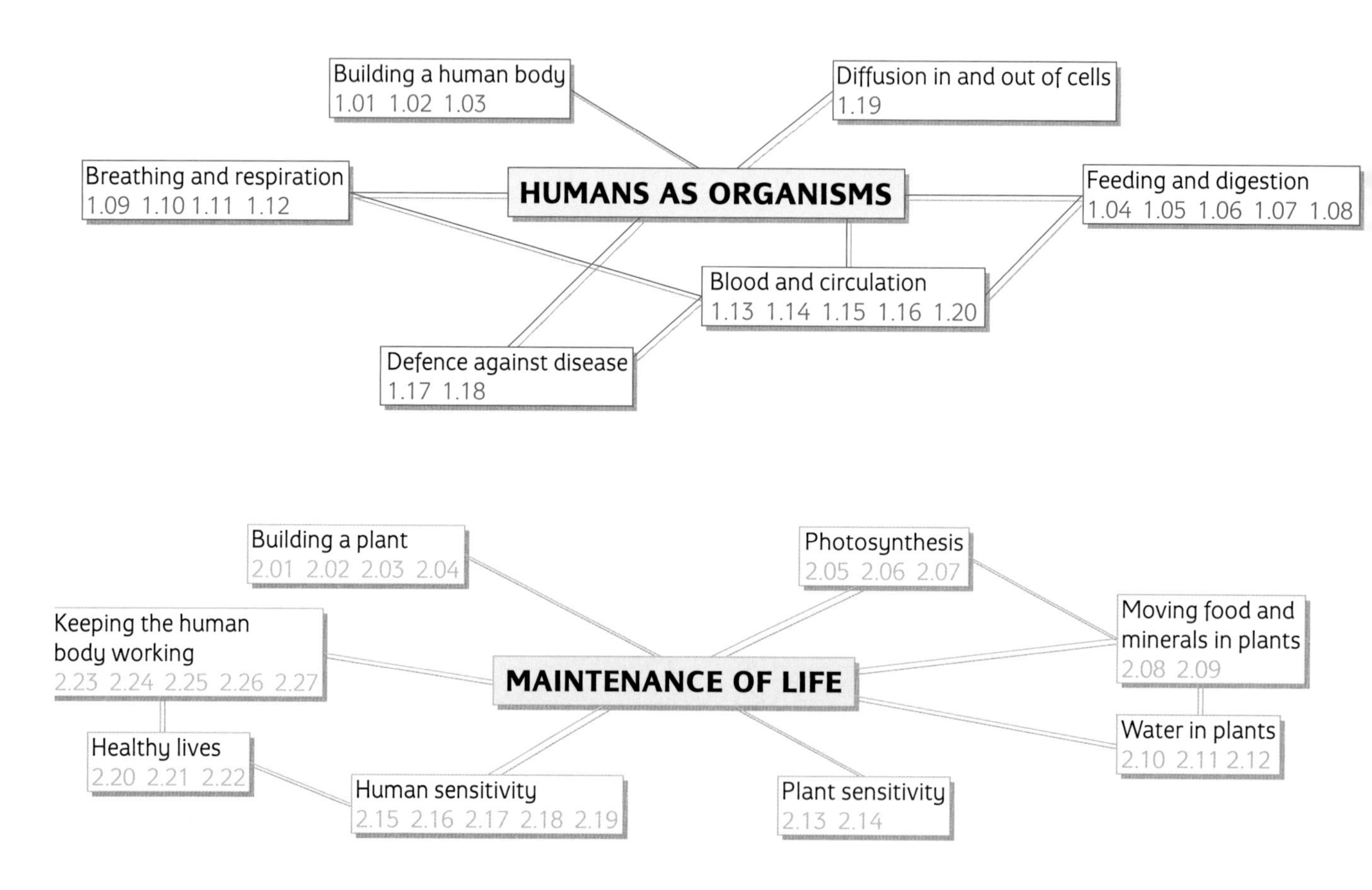

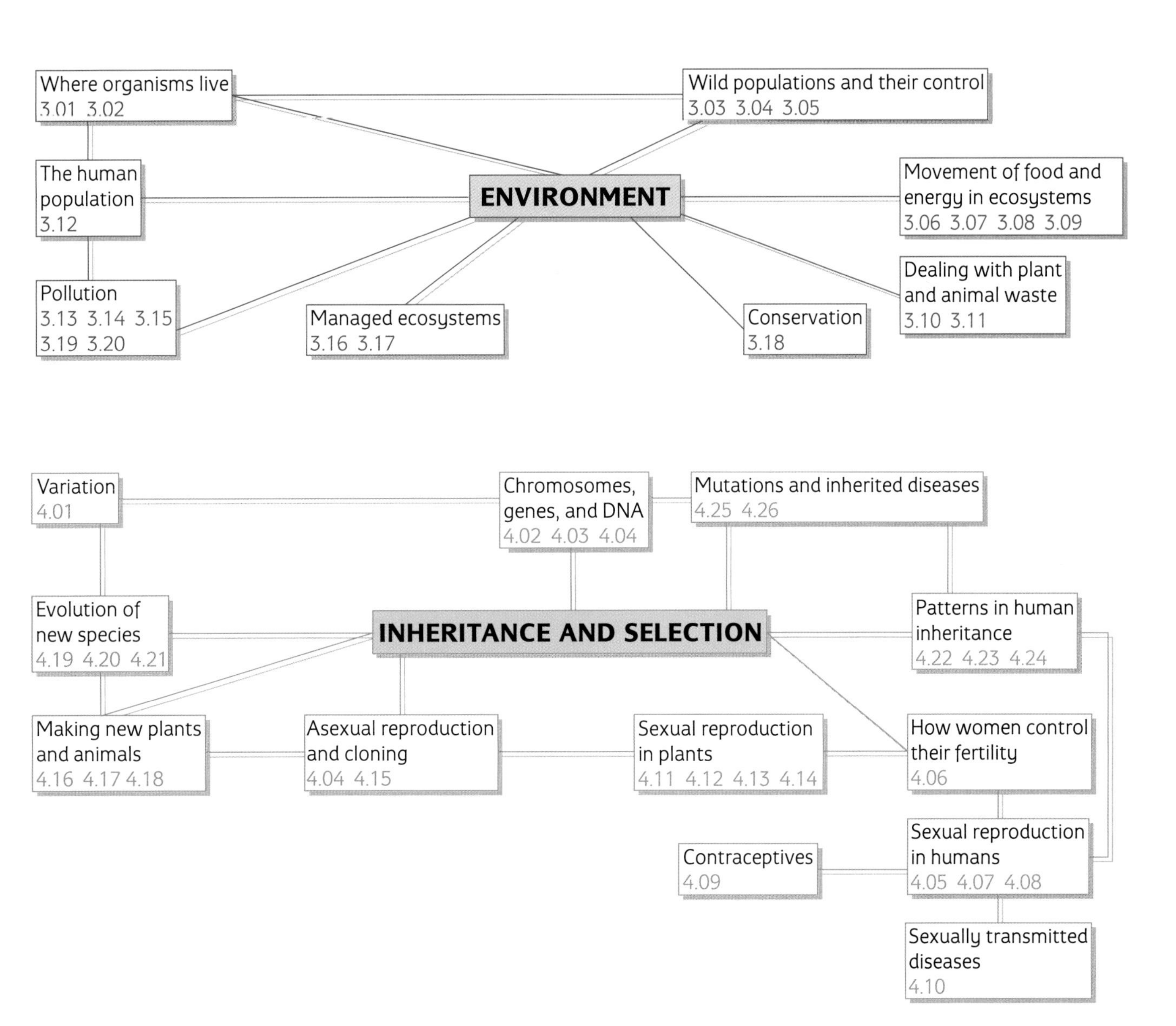

Where organisms live
3.01 3.02
Wild populations and their control
3.03 3.04 3.05
The human population
3.12
ENVIRONMENT
Movement of food and energy in ecosystems
3.06 3.07 3.08 3.09
Dealing with plant and animal waste
3.10 3.11
Pollution
3.13 3.14 3.15
3.19 3.20
Managed ecosystems
3.16 3.17
Conservation
3.18
Variation
4.01
Chromosomes, genes, and DNA
4.02 4.03 4.04
Mutations and inherited diseases
4.25 4.26
Evolution of new species
4.19 4.20 4.21
INHERITANCE AND SELECTION
Patterns in human inheritance
4.22 4.23 4.24
Making new plants and animals
4.16 4.17 4.18
Asexual reproduction and cloning
4.04 4.15
Sexual reproduction in plants
4.11 4.12 4.13 4.14
How women control their fertility
4.06
Sexual reproduction in humans
4.05 4.07 4.08
Contraceptives
4.09
Sexually transmitted diseases
4.10

Humans as organisms

All living things are made of cells. Different types of cell do different jobs. Substances get in and out of cells through the cell membrane. The movement of molecules through the membrane is carefully controlled.

Humans and other animals have to eat food to stay alive. Food supplies the materials we need for the chemical reactions in our bodies. It gives you energy and the building materials for us to grow. It is the job of the digestive system to change food into soluble substances and then absorb it into the bloodstream.

The cells in our bodies need oxygen to be able to release energy from food during respiration. Respiration takes place inside the cell cytoplasm of all living things. Carbon dioxide is a waste product of respiration. Humans and other animals get their oxygen from the air around them. It is the job of the breathing system to take oxygen into the body and get rid of carbon dioxide.

All of our cells need a regular supply of food, water, and oxygen. We have a transport system in which blood is pumped by the heart through blood vessels. Our blood is one of our defence systems that protect us from diseases.

Module 1

1.01 What is life?

Objectives

This spread should help you to

- describe the seven life processes

Life processes

All living things have seven **life processes**. These are things that all animals and plants do.

The processes are:

1 movement
2 respiration
3 sensitivity
4 feeding
5 excretion
6 reproduction
7 growth.

Movement *– animals can walk, run, hop, crawl, swim, or fly from place to place. Plants are fixed in one place, but some parts do move. Stomata open and close to let gases in and out of leaves. Daises have flowers that open in the daytime and close at night.*

Respiration *– this is how energy is released from food. Animals and plants need energy to grow, to move, and to help the body work properly.*

Sensitivity *– animals use sense organs such as eyes and ears to detect what is going on around them and respond to it. Plants don't have sense organs but they still respond, for example, by growing towards light and water.*

Feeding *– living things feed to provide them with energy and to grow and repair damaged parts. Animals eat other living things. Plants make their own food by photosynthesis.*

Excretion *– this is how living things get rid of waste. All living things produce waste, such as carbon dioxide and urea from the chemical reactions that happen in their cells.*

Reproduction *– living things must reproduce to replace those that die. Asexual reproduction does not involve sex. Most animals and plants have sex organs and use sexual reproduction to have young.*

Growth *– most animals grow until they reach a certain size, then stop growing. Plants, however, grow continuously all through their lives.*

Questions

1. Suggest why an animal moves from place to place.
2. During respiration, where does the energy come from?
3. Animals have these to help them respond. Plants don't. What are they?
4. Why do we eat food?
5. Name two things that are excreted by living things.
6. **a** Name the two kinds of reproduction.
   b What is the difference between them?
7. What is different about the way animals and plants grow?

1.02 Human cells

Objectives

This spread should help you to

- describe the structure of a human cell
- link the shape of a cell with its job
- describe the differences between a light microscope and an electron microscope

Made up of cells

All living things are made up of cells. Plants and animals are **multicellular**. This means they are made up of lots of cells. Humans are multicellular and have millions of cells.

Cells carry out many different jobs. They take in food, release energy, get rid of waste, grow, and reproduce.

All human cells have these parts.

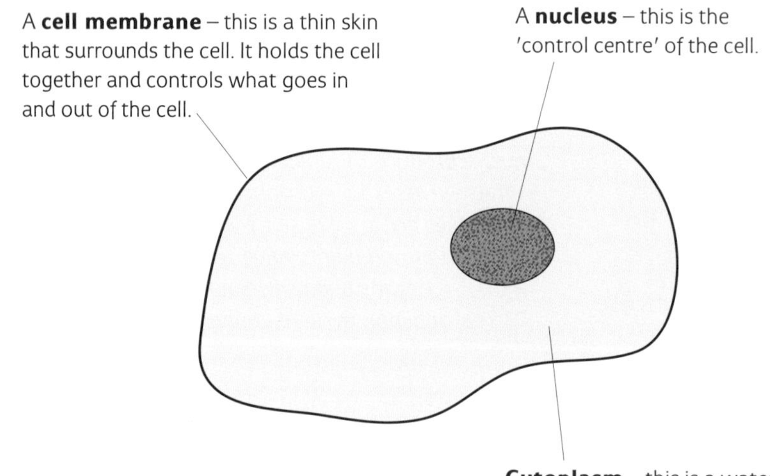

There are lots of different kinds of human cell. Each one is suited to the job it has to do.

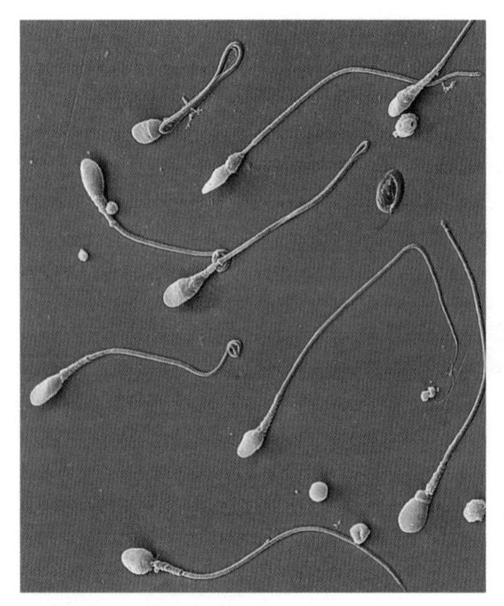

These sperm cells have a streamlined shape and a tail. This to helps them swim through the fluid in the female reproductive system to the egg to fertilize it.

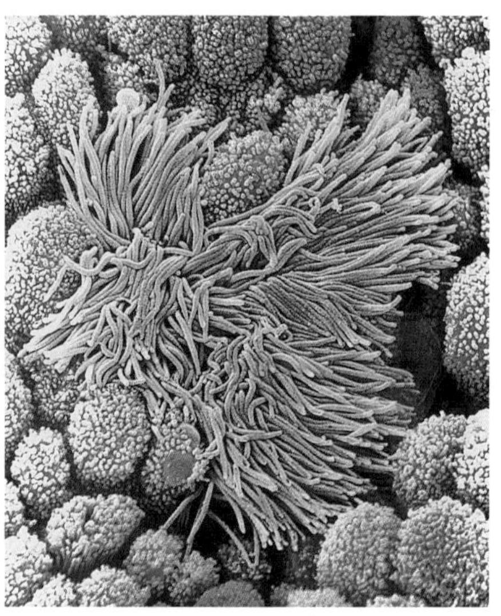

These are ciliated epithelial cells that keep our lungs clean. Notice the hair-like cilia on the surface, shown in green. These beat in rhythm, moving mucus away from the lungs.

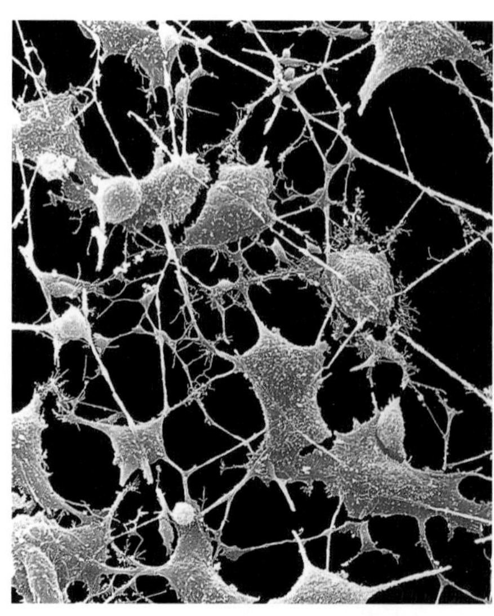

These nerve cells have long thin fibres which carry messages around the body.

In a light microscope, light passes from the object on the stage through two lenses into your eye. The knobs on the side are for focusing.

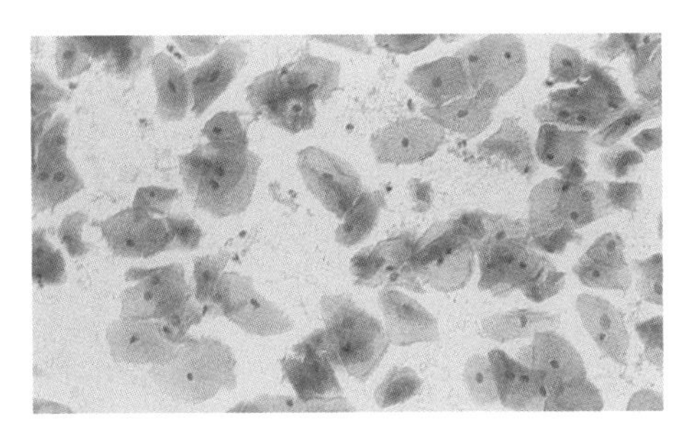

This is what human cells look like when seen through a light microscope. Cells are colourless and transparent, so stains are used to make different parts show up clearly.

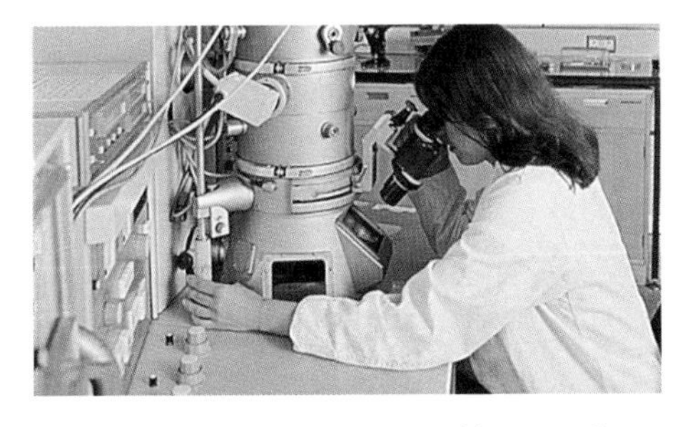

An electron microscope in use. You can alter the magnification, the focusing, and the brightness of the image. A beam of electrons passes down the tube through the object.

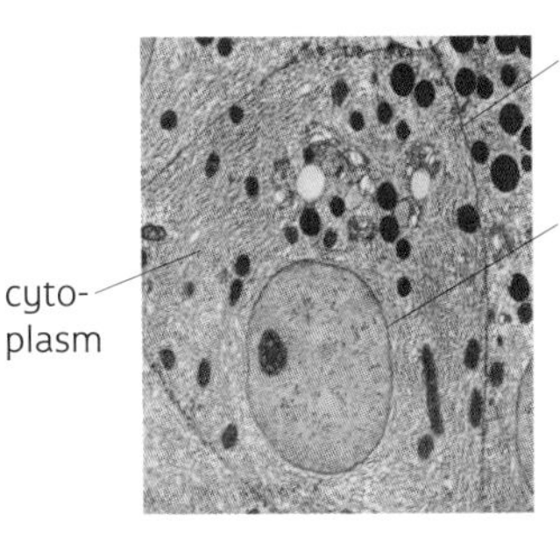

This is what a human cell looks like when seen using an electron microscope. You can see a lot more detail now.

Looking at human cells

You can't see an individual human cell without using a microscope. When you look at something through a microscope, it appears to get larger – the microscope magnifies what you see.

The photograph alongside shows the sort of microscope used in schools. You will probably have used one like it. Some school microscopes have a built-in light, while others use a mirror for directing light through the object and the lenses.

This sort of microscope is called a **light microscope** because it uses light to carry the image of the object though the lenses into your eye. A light microscope can only magnify a cell large enough for you to see the main parts.

Even the very best light microscopes can only magnify something by about 1500 times.

To get higher magnifications and see more detail we would need to use an **electron microscope**. Electron microscopes can magnify something by more than 500 000 times. This is because they use a beam of electrons instead of a beam of light. Focusing is done by electromagnets rather than lenses. The object has to be put in a vacuum so air molecules don't interfere with the electron beam. Because we can't see electrons, the beam hits a screen (like a television screen) where the image is shown.

Questions

1 What is a multicellular organism?

2 How are these cells suited to their job?

- **a** sperm cells
- **b** ciliated epithelial cells
- **c** nerve cells

3 What is the highest magnification of a light microscope?

4 Why are cells sometimes stained before looking at them through a microscope?

5 How did the electron microscope get its name?

6 Why is the object put in a vacuum when looked at through an electron microscope?

7 What is the difference in the way the light microscope and electron microscope are focused?

1.03 Cells, tissues, and organs

Objectives

This spread should help you to
- describe how an organism is built

Cells working together

Humans and other multicellular organisms are made up of different types of cell. Each type carries out a different job. Cells that do the same job join together to make a **tissue**.

Muscle cells, for example, form tissue that is able to contract and relax.

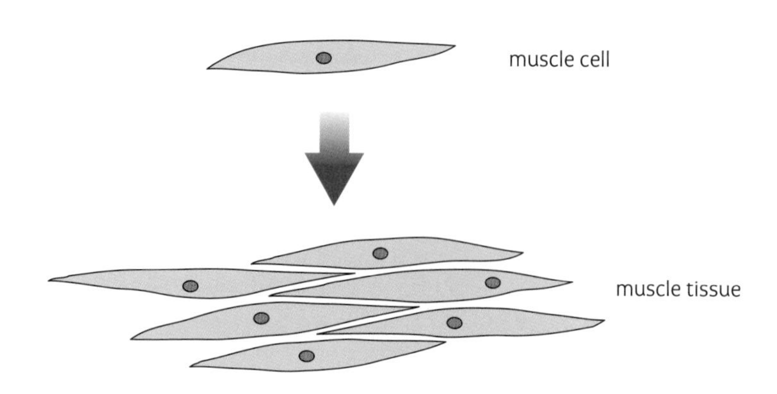

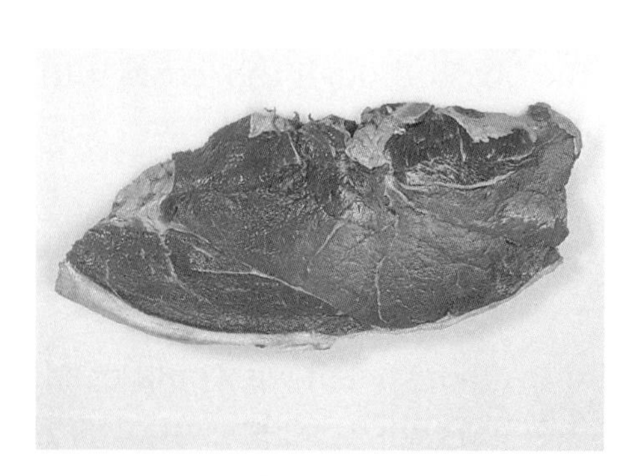

Meat is mainly muscle tissue.

Different tissues combine to make an **organ**.

The stomach is an organ that digests food. Muscle tissue, nerve tissue, and blood tissue are just some of the types of tissue that make up the stomach.

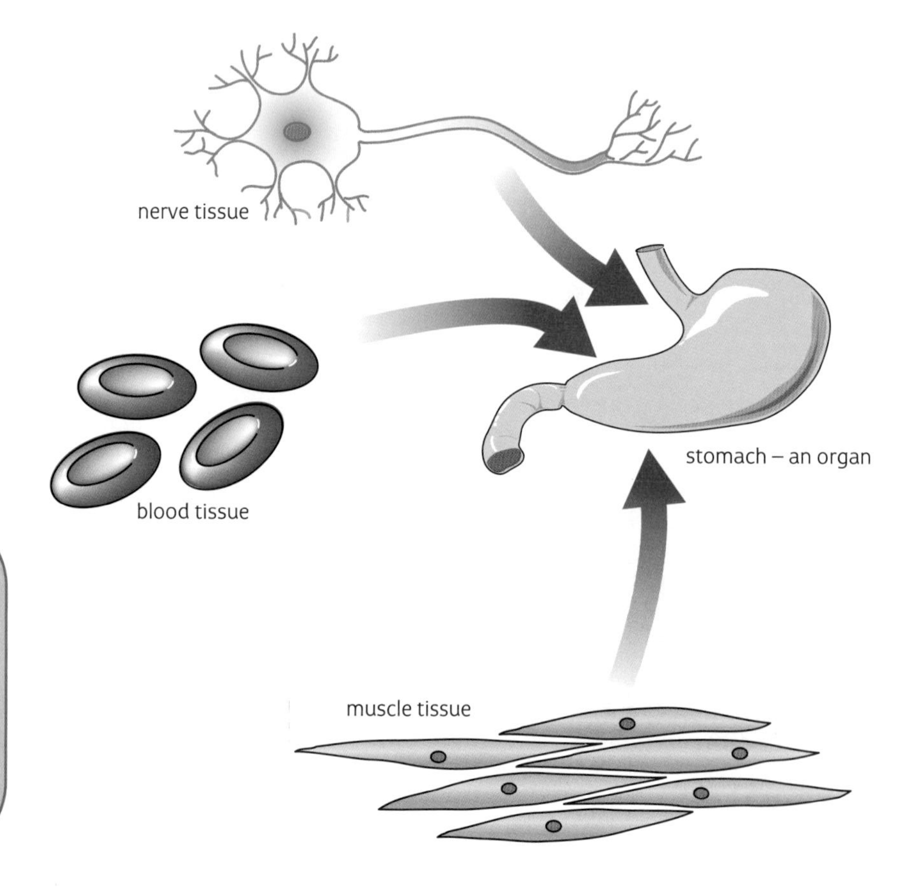

Questions

1. What is a tissue?
2. Name three types of human tissue.
3. Name an organ found in a human.
4. List three tissues which make up this organ.

Organ systems and organisms

Organs work together to form **organ systems**. Organ systems carry out much larger jobs than a single organ is able to.

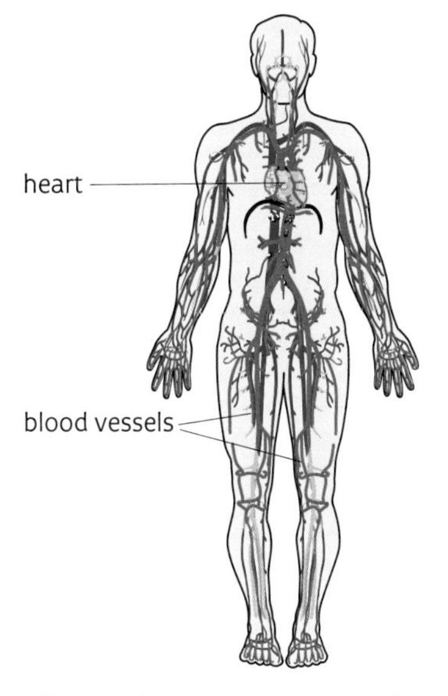

The circulatory system is made up of the heart, blood vessels, and blood.

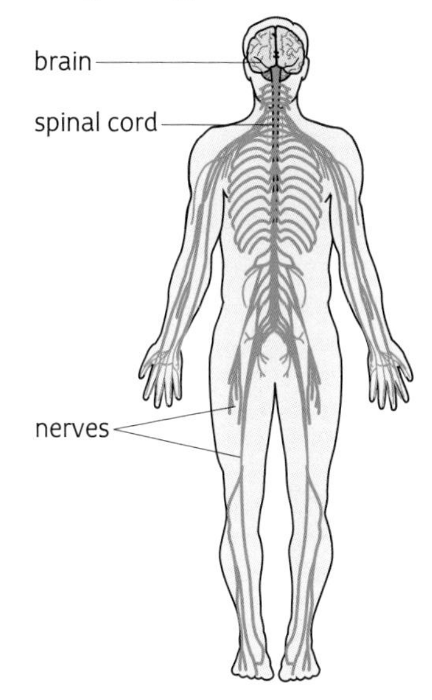

The nervous system is made up of the brain, spinal cord, and nerves.

An **organism** is made up of organ systems which work together to carry out all the jobs in a living thing.

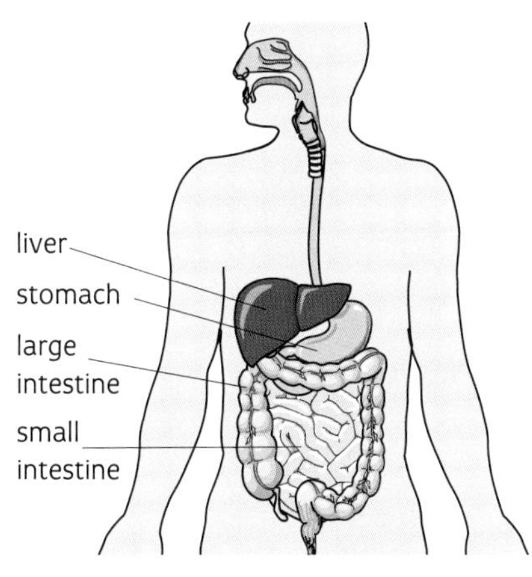

The digestive system is made up of various organs such as the stomach, liver, and intestines.

A human is an organism.

Questions

5 What is an organ system? Give one example.

6 Name three organs which make up this organ system.

7 Put these words in the right order, starting with the simplest.

organ *cell* *organism* *tissue*

8 Explain the difference between an organ and an organism.

1.04 Food

Objectives

This spread should help you to

- describe a balanced diet
- name some foods that contain the six food types
- carry out food tests for starch, glucose, protein, and fat

Balancing types of food

Food gives us all the things needed to keep the body working properly. The things we eat contain different types of food. The body needs these in the right amounts, otherwise we can become ill. For a healthy body we need a **balanced diet**. This means there must be the right balance between:

- foods that give us energy
- foods that give us building materials
- foods that help control chemical reactions
- foods that give us dietary fibre (roughage).

There are six food types.

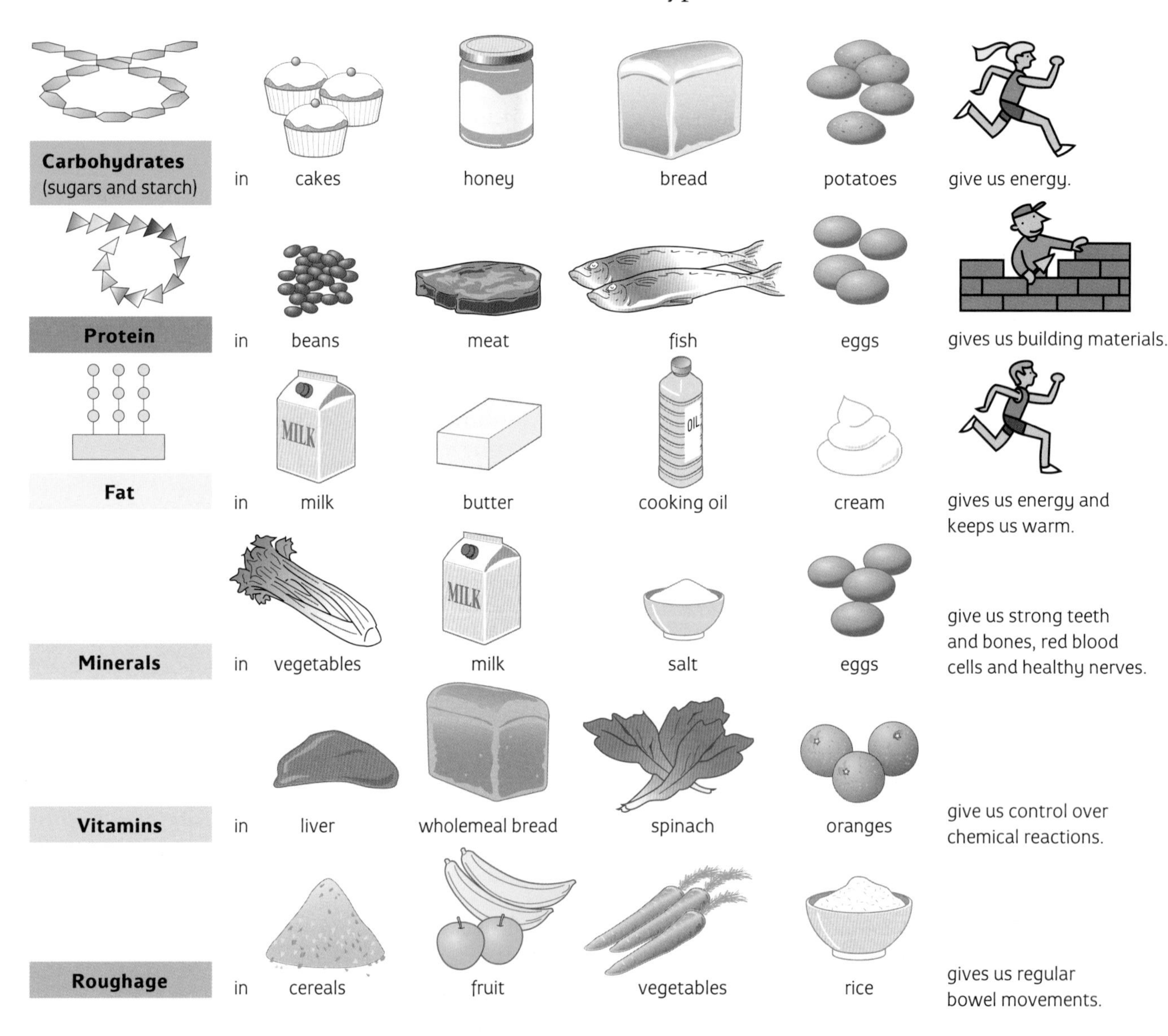

We also need lots of **water** because all the chemical reactions in the body take place in solution.

Food tests

You can find out what food contains by doing these simple tests in the laboratory. *Remember to wear your safety glasses when doing these experiments.*

Testing for starch
Add a few drops of iodine solution to the food.
If starch is present the iodine will change from brown to blue/black.

Testing for glucose (a simple sugar)
Add Benedict's solution to the food in a test tube. Put the test tube in a beaker of water and heat gently. If glucose is present the Benedict's solution will change from blue to orange.

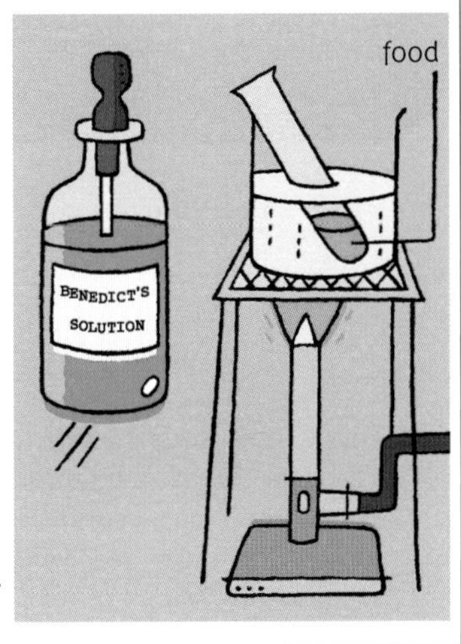

Testing for protein
Add some Biuret solution to the food in a test tube and shake it carefully. If protein is present the Biuret solution will change from light blue to purple.

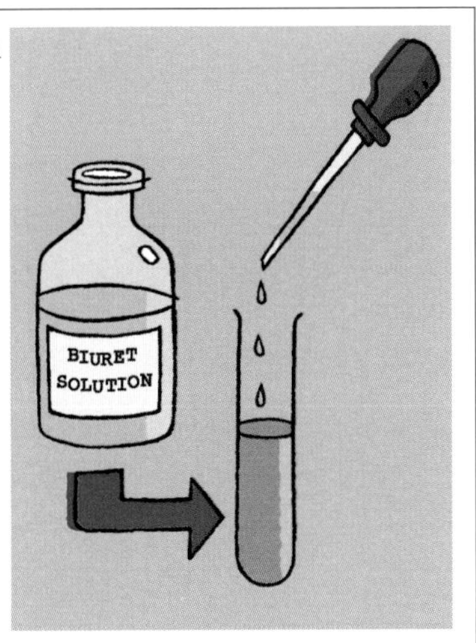

Testing for fat
Add some ethanol to the food in a test tube and shake it carefully.
Filter the mixture then add some clean water. If fat is present a white cloudy emulsion will appear.

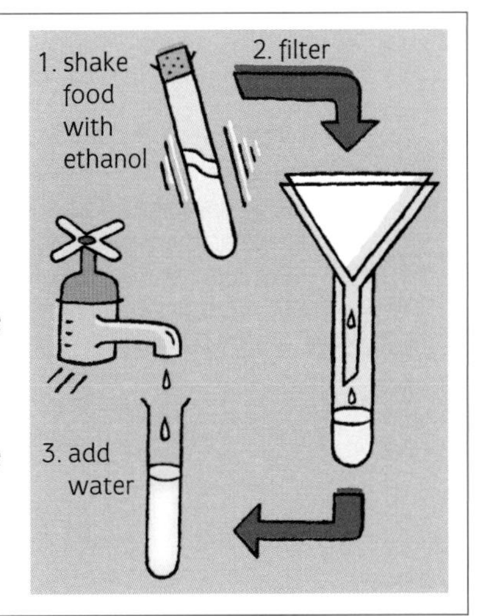

Questions

1 Why is a balanced diet important?
2 What does a balanced diet contain?
3 Name two foods that contain:
 a carbohydrates **b** fat **c** protein
 d minerals **e** vitamins.
4 Name a food that gives us:
 a building materials **b** energy
 c insulation **d** control over chemical reactions.
5 What is another name for roughage?
6 Why is it important to eat lots of roughage?
7 Name four 'foods' that contain water.
8 Describe how you would find out whether:
 a milk contains protein
 b cooking oil contains fat
 c biscuits contain glucose
 d potato contains starch.
9 What safety precautions should you take when doing food tests?

1.05 Digestion

Objectives

This spread should help you to

- describe what digestion is
- describe what happens when food is digested
- describe the human digestive system

Breaking it down – digestion

Only very small molecules of food can be absorbed by the body and pass into the bloodstream. Even after a lot of chewing, food still consists of large pieces which cannot be dissolved. These must be broken down. The breaking down of food into soluble bits is called **digestion**. Digestion takes place in the **digestive system**.

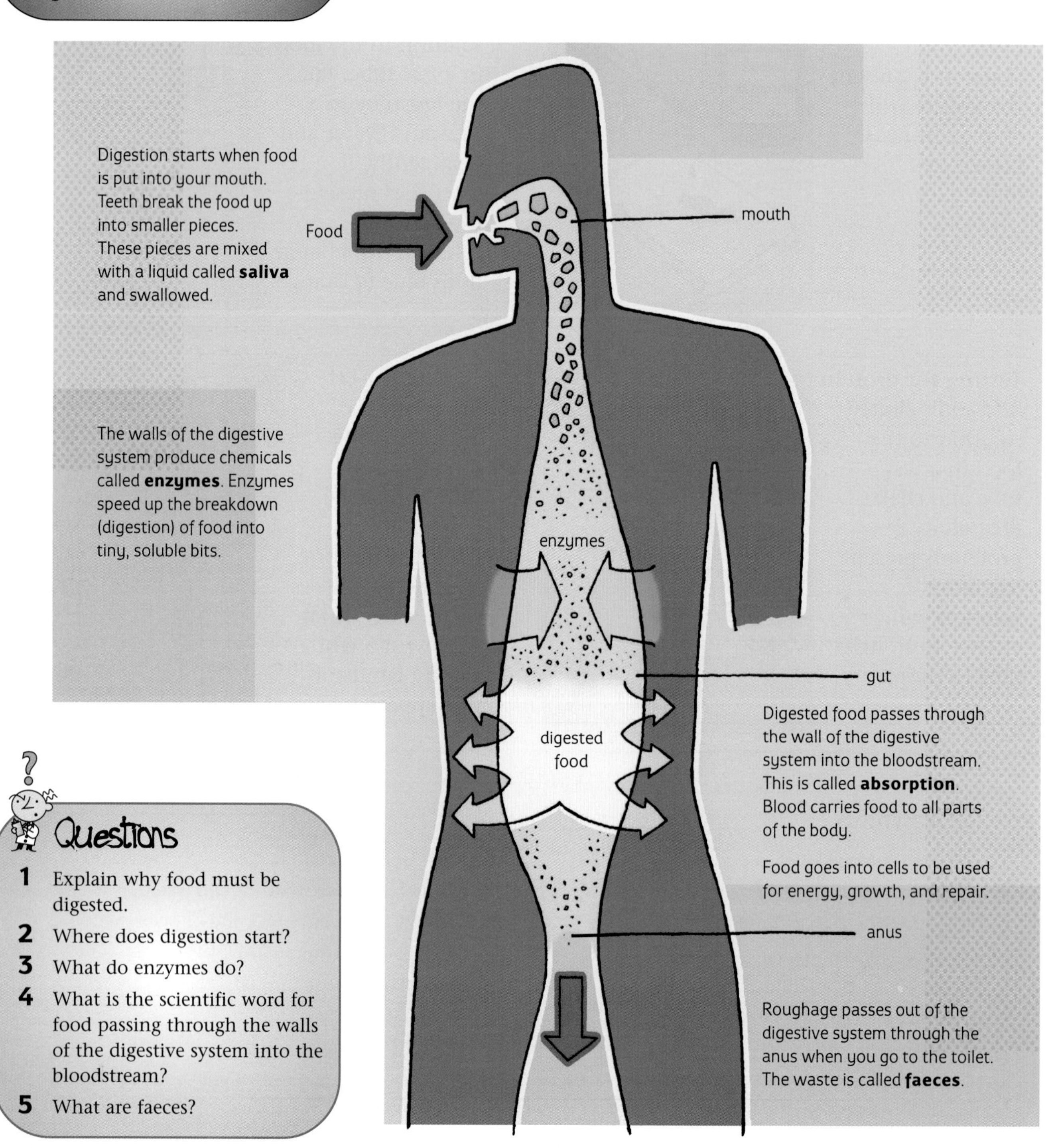

Questions

1. Explain why food must be digested.
2. Where does digestion start?
3. What do enzymes do?
4. What is the scientific word for food passing through the walls of the digestive system into the bloodstream?
5. What are faeces?

The human digestive system

The digestive system is a tube running from mouth to anus. It is about 6 m long and food takes about 24 hours to pass from one end to the other.

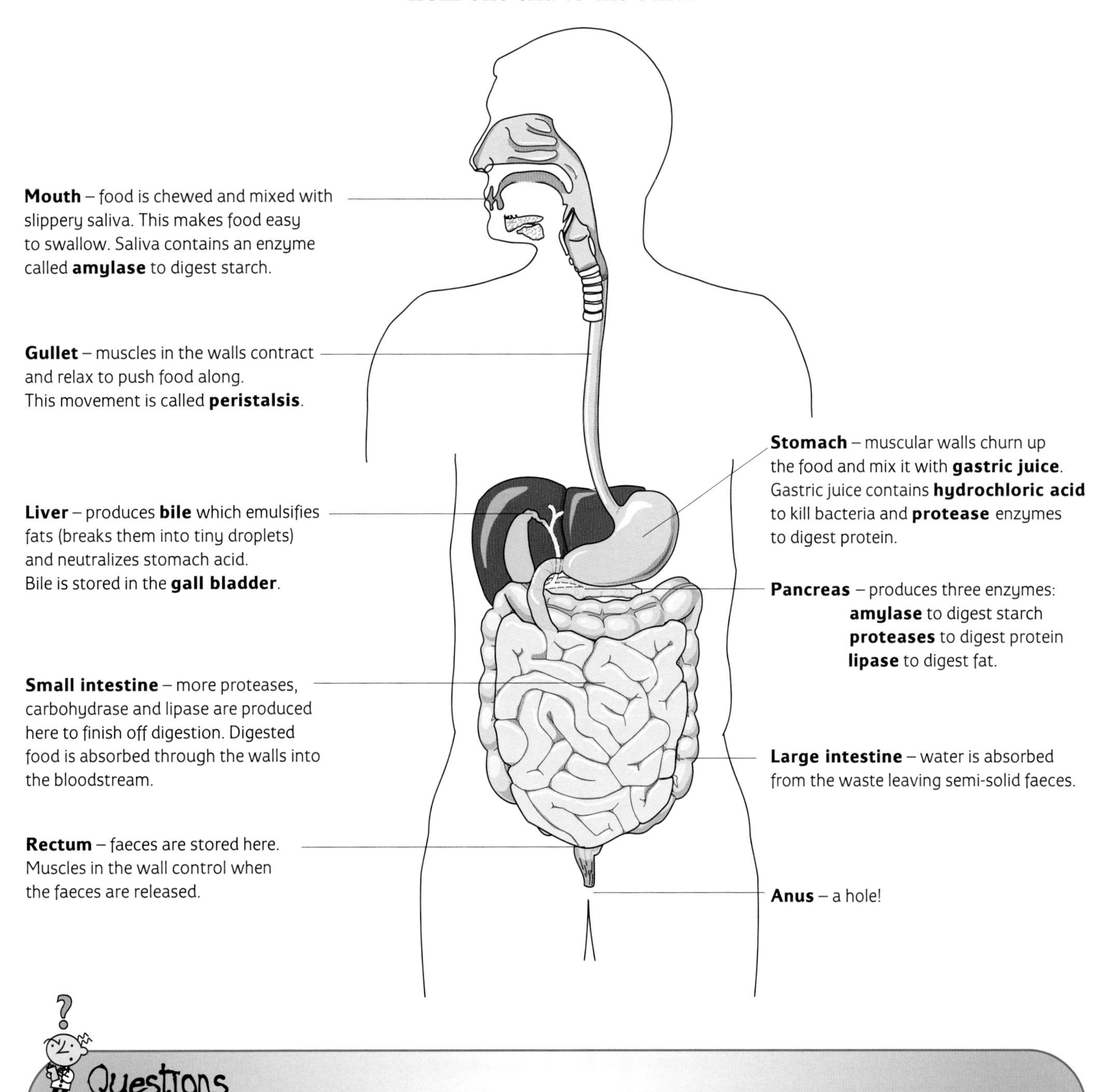

Questions

6 How long does it take for food to pass through the digestive system?

7 What is peristalsis?

8 What does emulsify mean?

9 What food types are broken down by these enzymes?
a protease **b** amylase **c** lipase

10 Describe what happens:
a in the large intestine **b** in the rectum.

1.06 Digestive enzymes

Objectives

This spread should help you to

- describe what enzymes do
- name the enzymes that digest starch, protein, and fat
- name the end products of digestion

Enzymes break it down

Enzymes break down large food molecules into small ones. Starch, protein, and fat are big molecules. Sugars, amino acids, fatty acids, and glycerol are small molecules. These can get into the bloodstream easily.

Enzymes have strange names but usually an enzyme's name gives you a clue about what it does.

Carbohydrases (such as amylase) break down starch into simple sugars

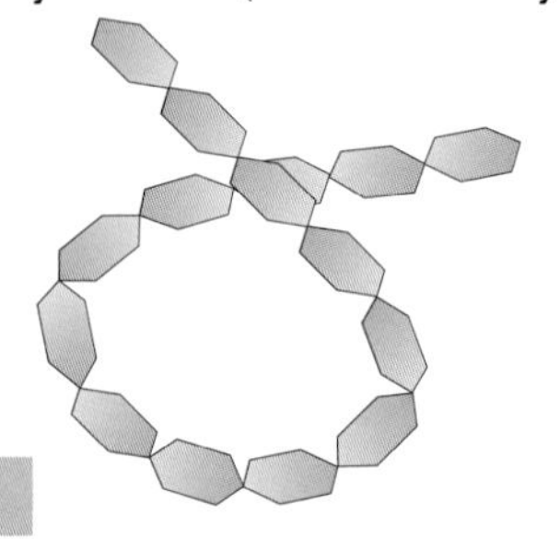

starch

Starch molecules are made up of **glucose** molecules. The glucose molecules are joined together like a string of beads.

Carbohydrase enzyme cuts the links that hold the beads together.

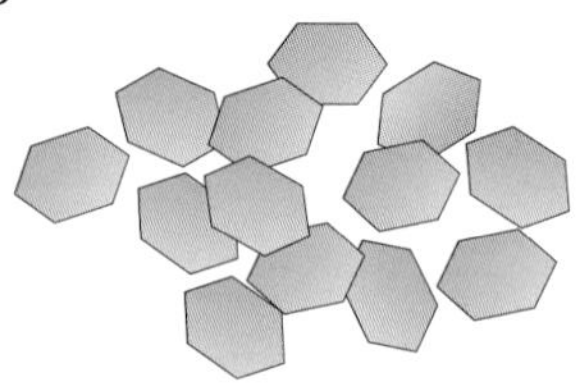

glucose

The glucose molecules are now free.

Proteases break down protein into amino acids

protein

Protein molecules are made up of long chains of **amino acids**. These amino acids are not all the same. Different proteins have different amino acids in the chains.

Protease enzyme cuts the links in the chain.

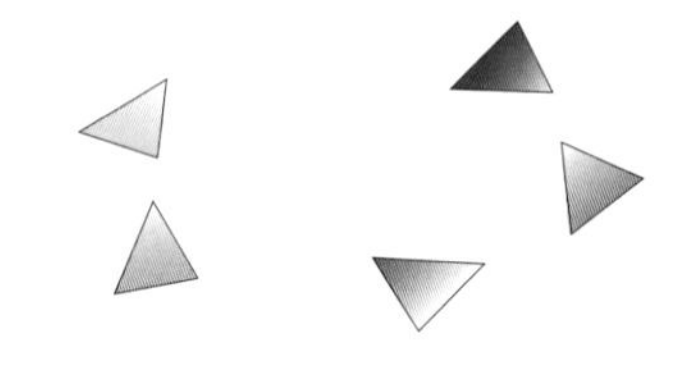

amino acids

The amino acids are released.

Lipase breaks down fats into fatty acids and glycerol

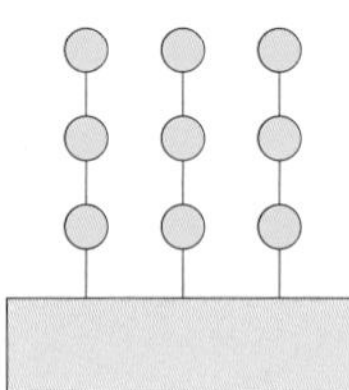

fat

Fat molecules are not chains. They are a bit like a fork without the handle. Small **fatty acid** chains are joined to a **glycerol** bar.

Lipase enzyme cuts the links between the fatty acids and the glycerol.

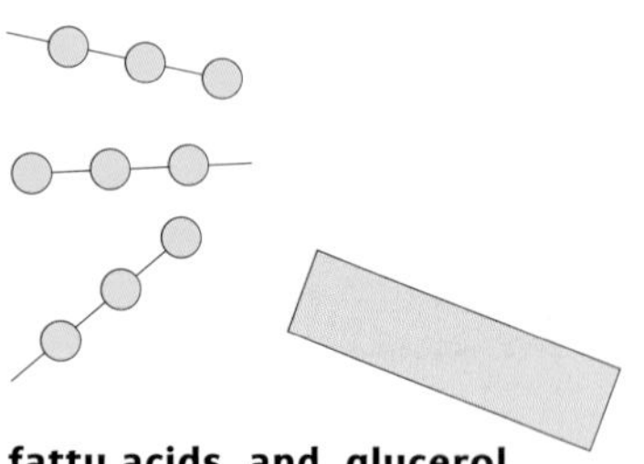

fatty acids and glycerol

The fatty acid chains and glycerol bar are now separated.

Did you know?

Enzymes aren't only involved in digestion. They also control many of the chemical reactions that happen in the body.

Each enzyme works best at a particular pH

The hydrochloric acid in the stomach is there to protect your body against infection by bacteria in your food. The liquid in the stomach is strongly acidic, about pH 2. You have probably noticed how your throat burns after you have been sick! The sick is very acidic. The stomach protease enzyme works best in strongly acidic conditions.

When the food moves from the stomach into the small intestine, the acid must be neutralized otherwise the rest of the digestive system will be damaged. Bile contains an alkali which neutralizes the acid. The enzymes that break down food in the small intestine work best in neutral conditions.

The gastric juice in the stomach is strongly acidic.

Questions

1 What do enzymes do?

2 Name three large food molecules.

3 Name three small food molecules.

4 Describe the structure of a starch molecule.

5 What is the difference between a starch molecule and a protein molecule?

6 Draw diagrams to show the digestion of fat.

7 What is the job of the hydrochloric acid in the stomach?

8 How could you prove that the liquid in the stomach is acidic?

9 Why is the hydrochloric acid neutralized when it leaves the stomach?

10 How is the hydrochloric acid neutralized?

1.07

Enzymes in the food industry

Enzymes from microorganisms

Enzyme technology is a new industry. The enzymes come from microorganisms such as bacteria and fungi, so they are called **microbial enzymes**. Microbial enzymes have lots of uses in the food industry.

Amylase

Amylase breaks down starch into sugar. One use of amylase is to make sugar from the starch in potatoes or corn. This is used as a sweetener in the food industry. The sweetener in your favourite canned or bottled drink has probably been made by amylase.

Cellulase

Cellulase breaks down cellulose into sugar. The walls of plant cells are made of cellulose. Cellulase is used to break down the cell walls in green beans to make them more tender to eat.

Lactase

Lactase breaks down the sugar in milk (called lactose) to glucose. Lactose makes the crystals in frozen ice-cream which make the ice-cream 'gritty'. Without lactose the ice-cream is smoother and more easily scooped.

Pectinase

Pectinase breaks down pectin into sugars. Pectin makes fruit juice cloudy. Pectinases are added to crushed fruit such as apples and grapes to get more, clear juice out and to get more colour out of the skins.

Proteases

Proteases break down protein to amino acids. The cheese-making industry uses proteases to clot milk. This makes curds and whey. Cheese is made from the curd, and the whey is thrown away.

Talking points

1. What is the most important use for amylase in the food industry?
2. Explain how lactase makes ice-cream easier to scoop.
3. Milk clots when it goes sour. It separates into curds and whey. How are curds used in the food industry?

1.08 Absorption

Objectives

This spread should help you to

- describe how digested food is absorbed
- describe how the small intestine is adapted to absorb digested food
- describe what happens to digested food when it gets to the liver

Taking it in – the small intestine

The food that has been broken down during digestion has to be absorbed into the bloodstream before the body can use it.

The digested food molecules are so small that they pass through the wall of the small intestine into the blood. This happens by diffusion and active transport. The small intestine has a special shape so that enough food gets into the bloodstream:

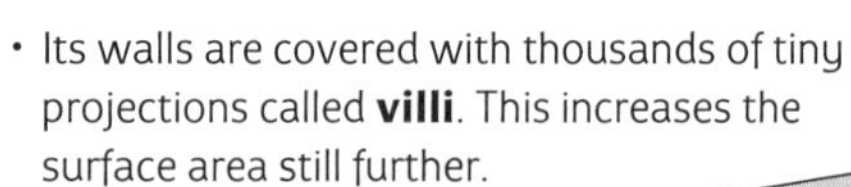

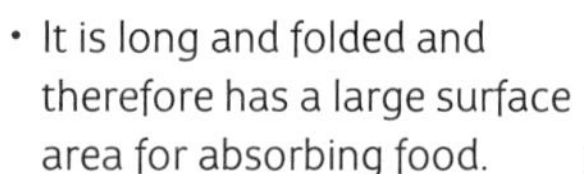

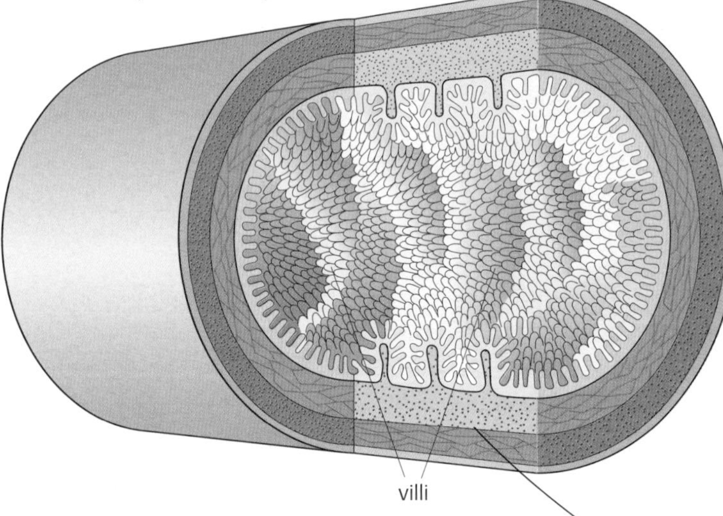

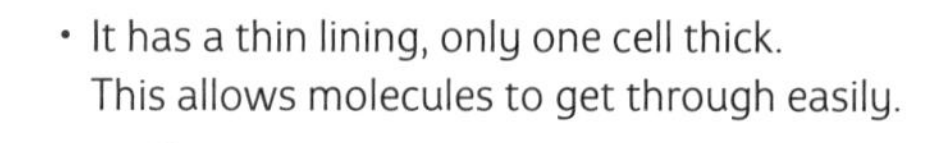

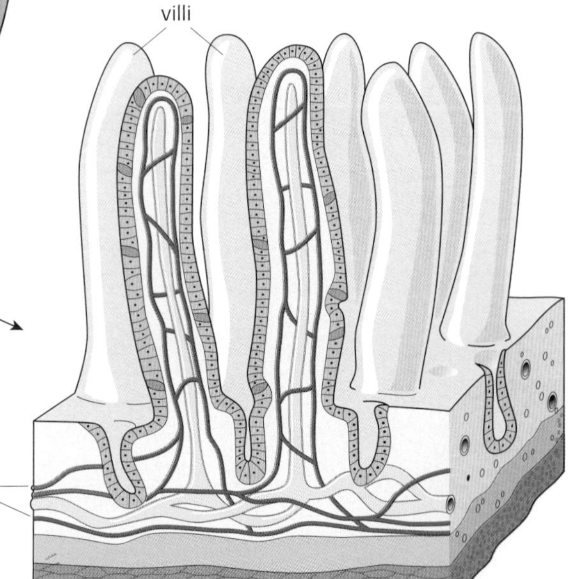

• It has a good supply of blood to take the food away.

Questions

1. Why does it take so long for digested food molecules to get into the bloodstream?
2. Give four ways in which the small intestine is shaped to do its job.

Did you know?

The human body can give off as much heat as a 100 W light bulb. Most of this heat comes from the liver.

Where does it go?

Glucose, amino acids, fatty acids, glycerol, minerals, and vitamins go through the wall of the villi into the bloodstream.

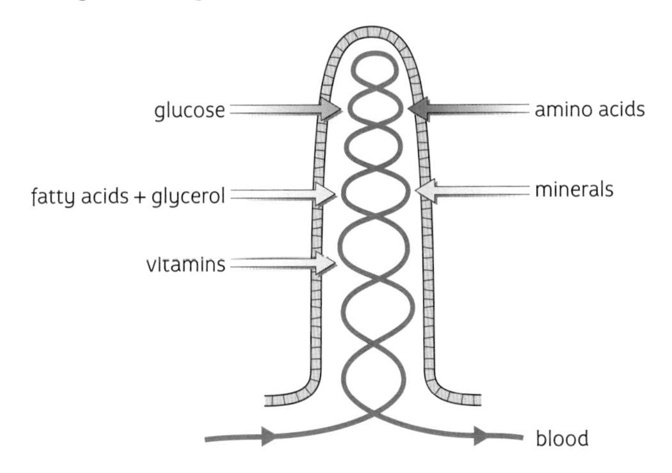

Blood full of digested food goes to the liver.

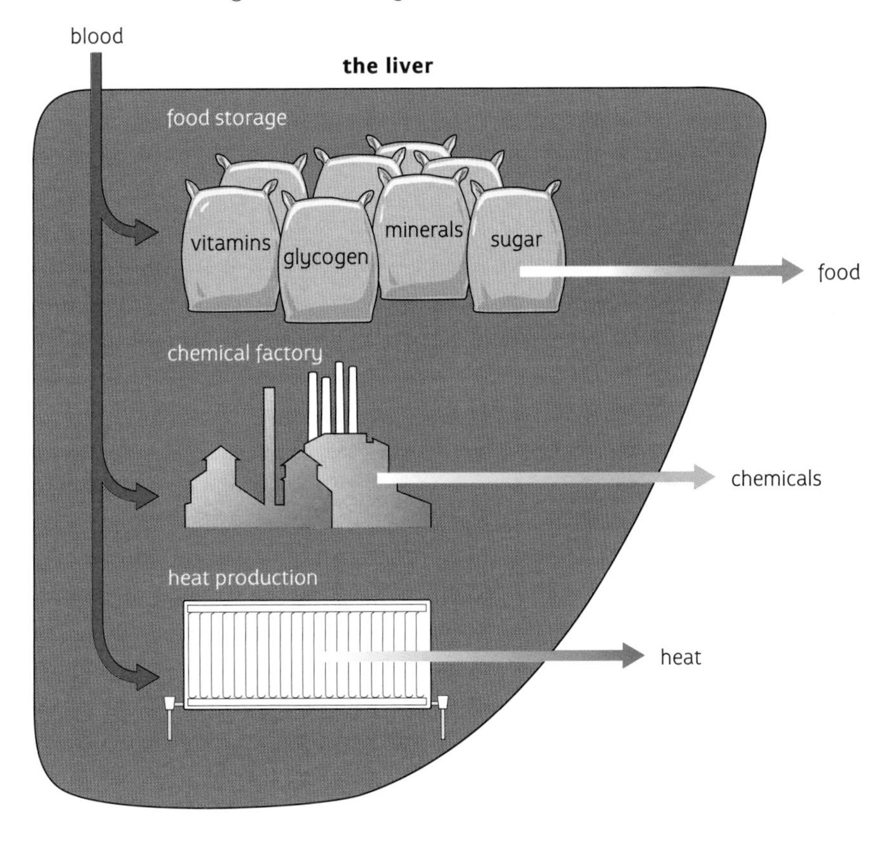

Questions

3 What are villi and where are they found?

4 Where does the blood go to when it leaves the intestine?

5 Name three things stored in the liver.

6 How does the liver help keep you warm?

1.09 Lungs

Objectives

This spread should help you to

- describe how the air you breathe in is different from the air you breathe out
- describe the structure of the lungs
- describe how gases are exchanged in the alveoli

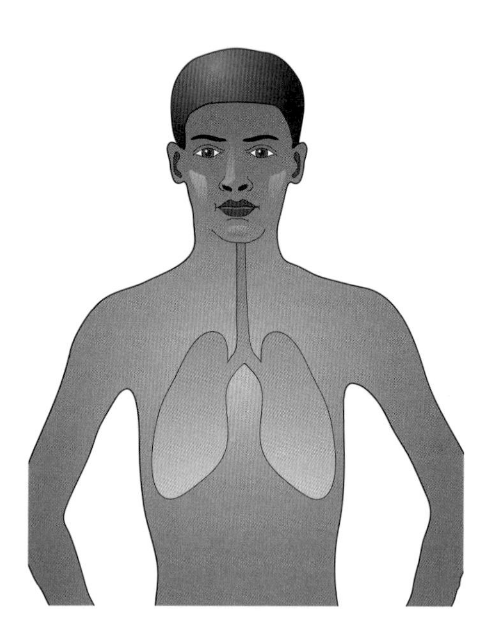

Exchanging gases

The body needs oxygen for respiration. Carbon dioxide is a waste gas produced in respiration.

This is how the air changes when you breathe in and out.

Gas	Amount in the air breathed in	Amount in the air breathed out
oxygen	21%	17%
carbon dioxide	0.04%	4%
nitrogen	79%	79%
water vapour	varies	saturated

It is the job of the lungs to get oxygen in and carbon dioxide out of the body. This **gas exchange** takes place in the lungs.

The lungs are in the chest cavity. This area is called the **thorax**. The lungs are like two pink sponges because they contain lots of air sacs and blood.

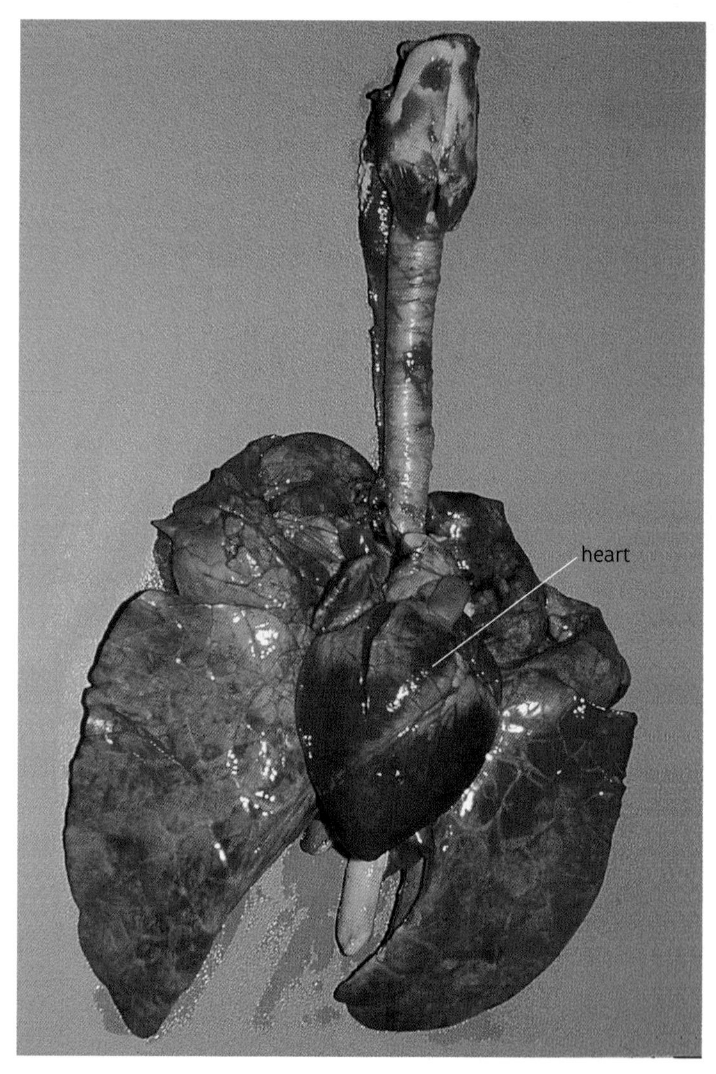

Questions

1. Why does the body need oxygen?
2. What is the job of the lungs?
3. Explain what 'gas exchange' means.
4. Describe what a lung looks like.

The structure of the lungs

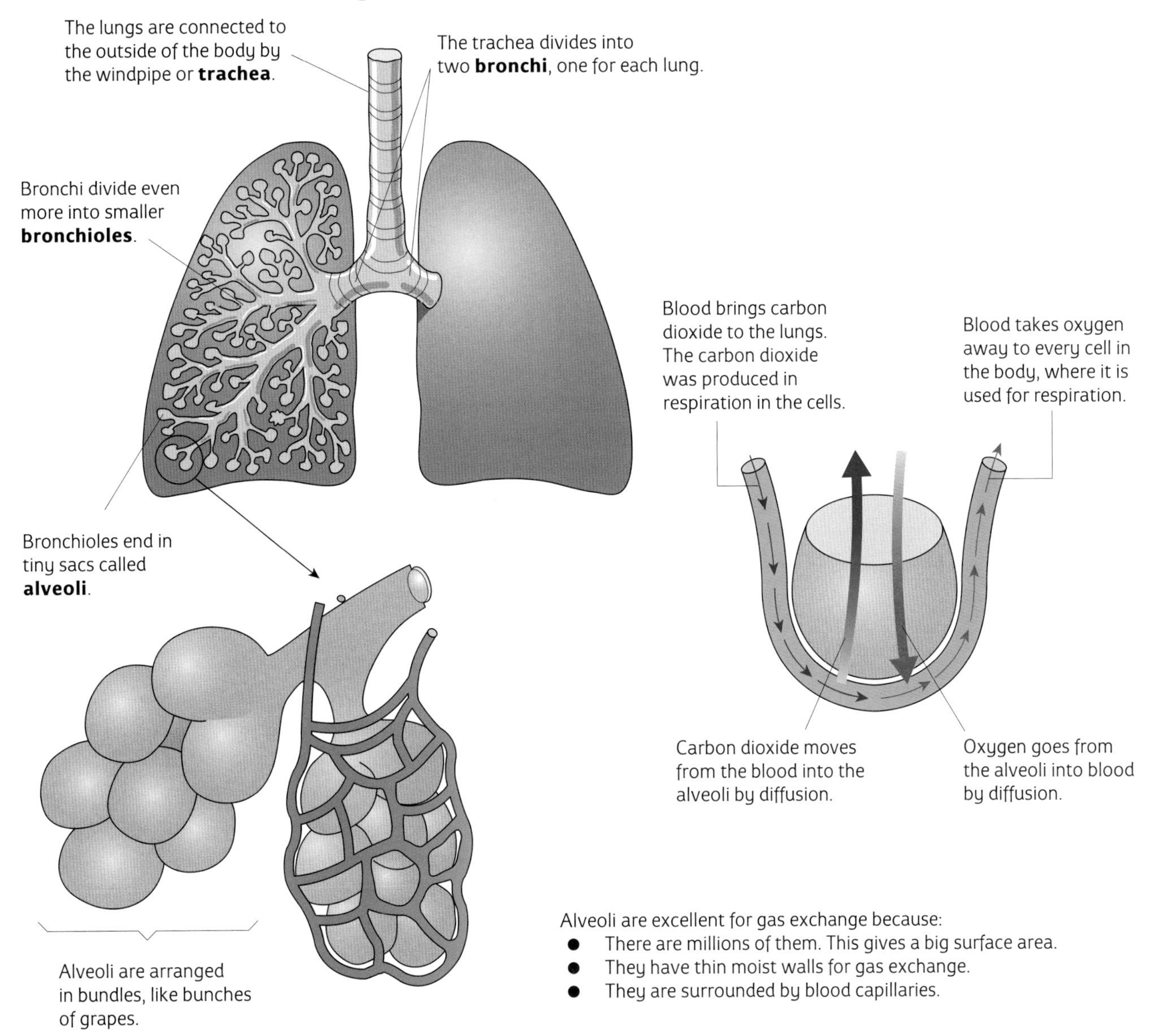

Alveoli are excellent for gas exchange because:

- There are millions of them. This gives a big surface area.
- They have thin moist walls for gas exchange.
- They are surrounded by blood capillaries.

Gas exchange in alveoli.

Questions

5 What is another name for the windpipe?

6 Why are there two bronchi?

7 Give three reasons why alveoli are good for gas exchange.

8 Draw a diagram showing how gases are exchanged in the alveoli.

9 Explain why oxygen moves from the alveoli to the blood. (Hint: Look at spread 1.19.)

1.10 Breathing

Objectives

This spread should help you to

- describe the structure of the thorax
- describe how you breathe in and out

Take a breath...

Breathing is the way we get air in and out of our lungs.

The thorax is separated from the lower part of the body (the **abdomen**) by a sheet of muscle called the **diaphragm**. The ribcage surrounds and protects the lungs. Muscles covering the ribs move the ribcage up and down. This, together with movements of the diaphragm, changes the volume inside the ribcage.

The thorax

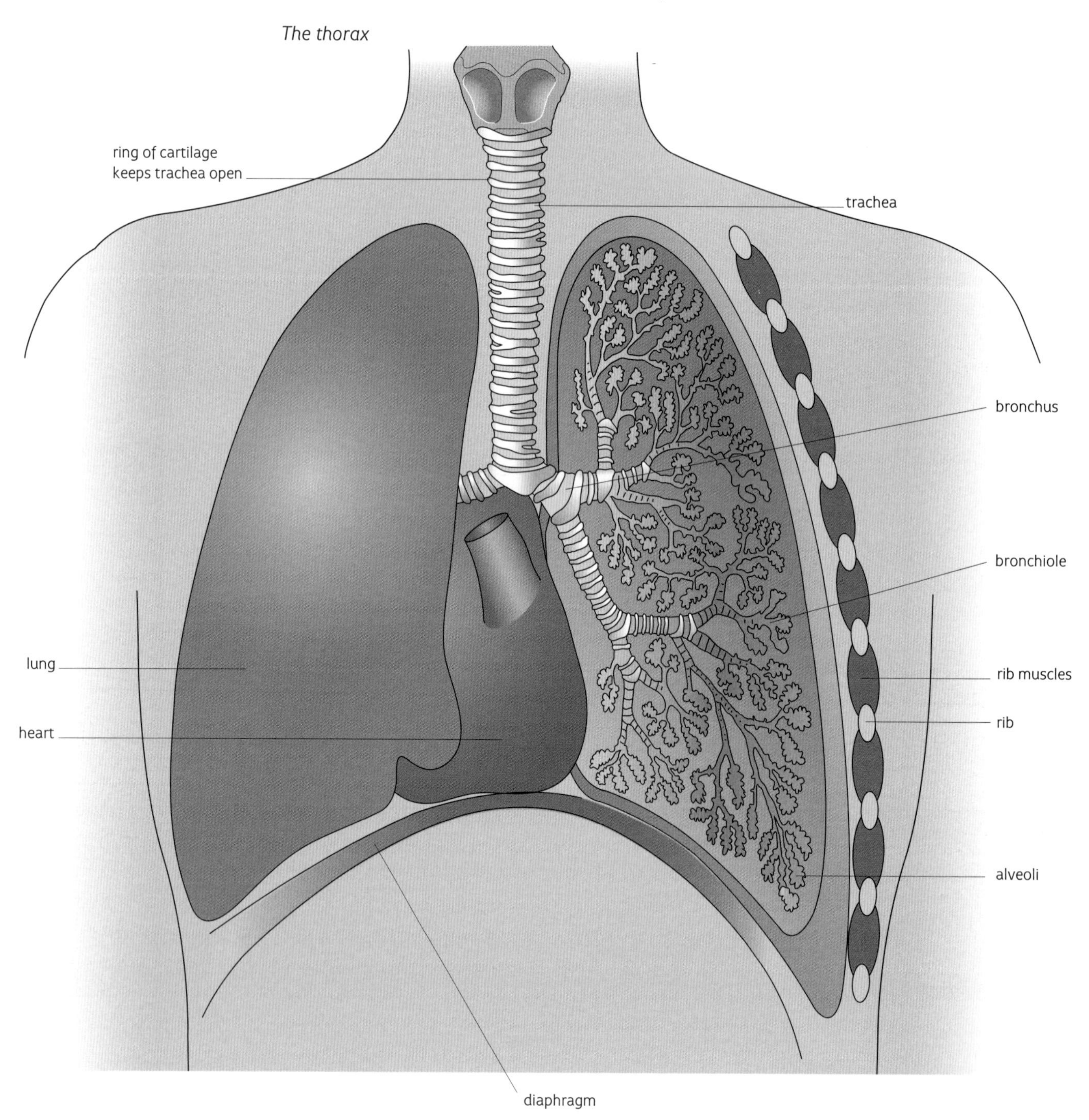

Breathing in...

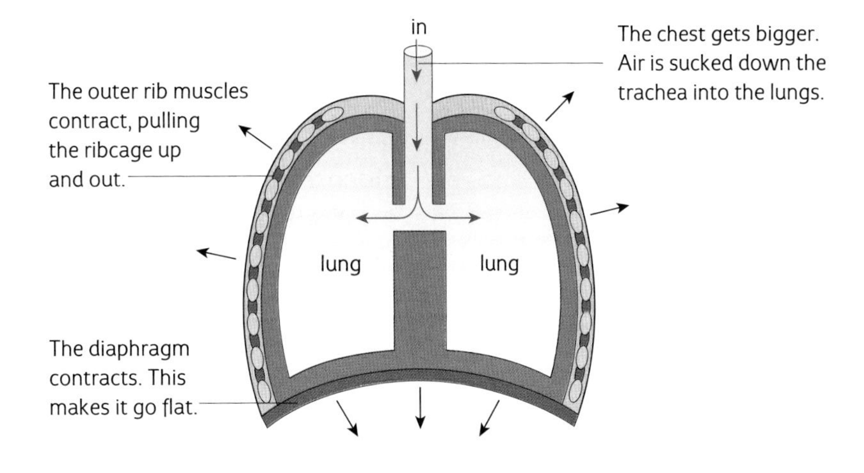

plunger

syringe

plastic tubing

flask

balloon

***Model lungs**: You can show how we breathe in and out by using this simple piece of equipment.*

...and out

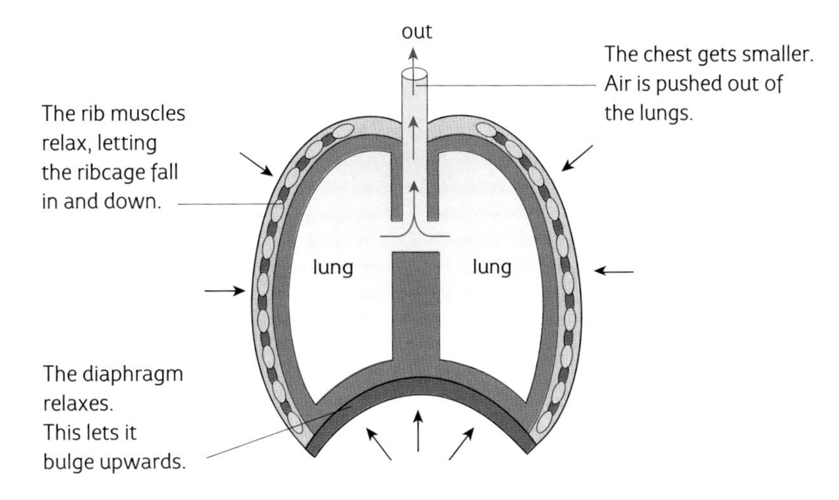

Questions

1. What is breathing?
2. What separates the thorax and abdomen?
3. What causes the ribcage to move up and down?
4. Describe how the chest gets bigger when you breathe in.
5. Describe how the chest gets smaller as you breathe out.
6. What shape is your diaphragm when you:
   a breathe in **b** breathe out?
7. The breathing rate (number of breaths per minute) can change depending upon what you are doing. Give two examples of when your breathing rate will increase.
8. In the model lungs shown on this page, which part represents:
   a the trachea **b** the lungs
   c the ribcage **d** the diaphragm?

1.11 Respiration

Objectives

This spread should help you to

- describe what happens in aerobic respiration
- describe an experiment you could do to find out how much energy is in foods
- list some things that energy is used for

Respiration is not breathing

Respiration is all about getting energy from glucose. It is not breathing!

All living cells respire. Glucose (a sugar) is broken down to release the energy in it. Oxygen is needed to break down the glucose completely. This is called **aerobic respiration**. Aerobic means 'with air'.

Aerobic respiration can be written like this:

glucose + oxygen ⟶ carbon dioxide + water + ENERGY

Energy in food

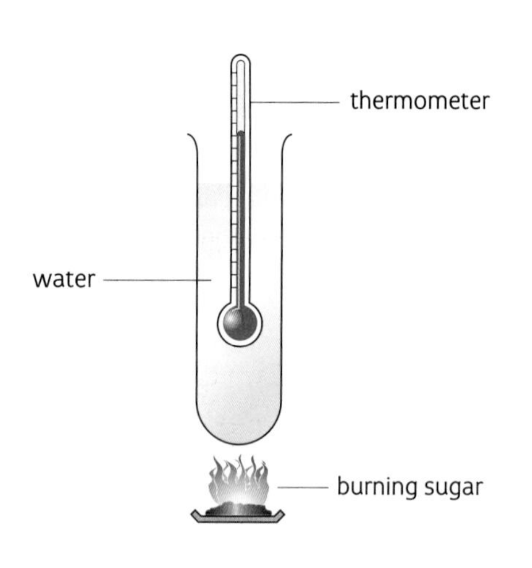

We can do this simple experiment to measure how much energy is in different foods. As the sugar burns it gives off heat energy. This energy goes into the water and makes the temperature rise. The more energy is in the food, the higher the temperature will go. The table shows the results for sugar, a peanut, and some dried bread.

	Sugar	Peanut	Dried bread
volume of water in test tube	50 ml	50 ml	50 ml
mass of water	50 g	50 g	50 g
mass of food	1 g	1 g	1 g
temperature of water at start	20 °C	20 °C	20 °C
temperature of water at finish	70 °C	90 °C	50 °C
temperature rise	50 °C	70 °C	30 °C

This formula can be used to work out the energy in each food:

$$\text{energy} = \frac{\text{mass of water (g)} \times 4.2 \times \text{temperature rise (°C)}}{\text{mass of food (g)}}$$

For example, the energy in 1 g of sugar is:

$$\text{energy in sugar} = \frac{50\text{ g} \times 4.2 \times 50\text{ °C}}{1\text{ g}}$$

= 10 500 joules (10.5 kilojoules)

Questions

1. What is respiration?
2. Write a word equation for respiration.
3. What does aerobic mean?
4. Use the information in the table to work out how much energy there is in the peanut and the dried bread.

What is energy used for?

The energy released during respiration is needed for many things.

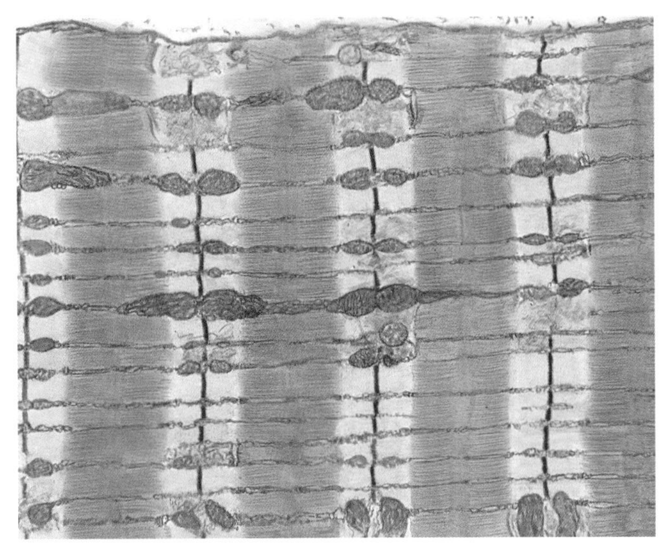

Muscle cells need energy to help you move quickly.

As we grow, energy is needed to make more proteins from amino acids.

Keeping warm in cold weather uses up energy.

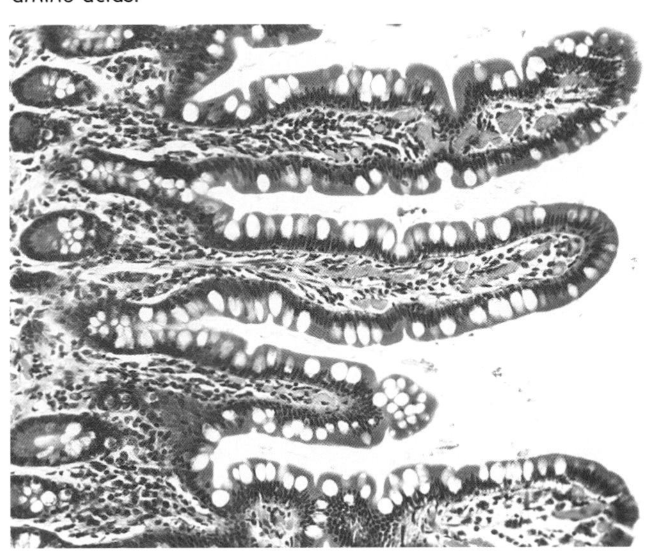

Intestinal wall cells need energy to absorb food by active transport.

Questions

5 Look at the experiment to find the energy in food on the opposite page. Suggest three reasons why the experiment is not very accurate.

6 Give four uses for the energy released in respiration.

7 People sometimes take glucose tablets when playing sport. Why?

8 Suggest why people feel hungry after they have done lots of physical activity.

1.12 Anaerobic respiration in humans

Objectives

This spread should help you to

- describe the difference between aerobic and anaerobic respiration
- describe what happens in anaerobic respiration in humans
- explain what an oxygen debt is

Breathing without air

Most living cells in animals and plants respire aerobically. However, sometimes your breathing rate can't get oxygen to the cells fast enough. You don't die, because respiration can also happen without oxygen. This is called **anaerobic respiration**. Anaerobic means 'without air'.

Walking is an aerobic exercise. The body gets enough oxygen for aerobic respiration to supply all the energy needed. Carbon dioxide and water are breathed out instead of building up in the body. This is why you can walk for a long time without getting tired.

Walking is an aerobic exercise.

Sprinting is an anaerobic exercise.

Running fast is an anaerobic exercise. No matter how fast you breathe or how fast your heart beats, the body can't get enough oxygen for aerobic respiration. This time anaerobic respiration supplies the energy needed. Unfortunately there is a time limit on anaerobic exercise.

How does anaerobic respiration work?

Instead of being broken down completely into carbon dioxide and water, glucose breaks down into **lactic acid**. Lactic acid is a sort of halfway stage between glucose and carbon dioxide and water.

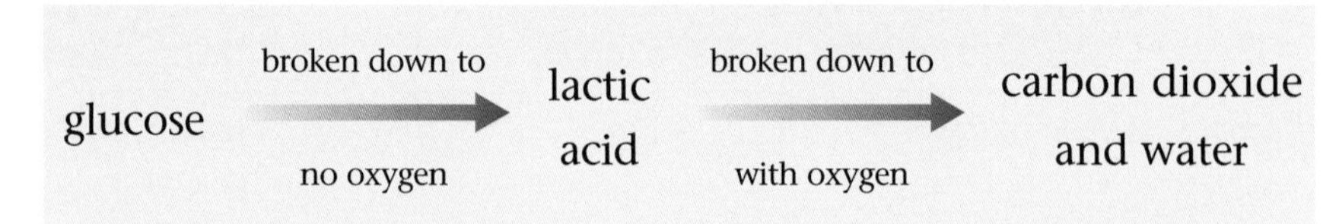

Anaerobic respiration is not the best way to get energy from glucose because the glucose is not completely broken down and much less energy is produced. However, anaerobic respiration is very useful when energy is needed in a hurry.

Anaerobic respiration in animals can be written like this:

glucose ⟶ lactic acid + ENERGY (small amount)

Questions

1. What does anaerobic mean?
2. Give one example of an:
   - **a** aerobic exercise
   - **b** anaerobic exercise.
3. Write a word equation for anaerobic respiration in animals.
4. What is the main difference between aerobic and anaerobic respiration?

The oxygen debt

You can't breathe lactic acid out like carbon dioxide! During vigorous exercise the body 'borrows' some energy from glucose in anaerobic respiration. Lactic acid builds up in your muscles causing them to ache. This quickly changes to painful **cramp**. Cramp makes you stop what you are doing and rest. You keep breathing hard and your heart keeps pumping fast even when you have stopped. This is to get oxygen to the muscles so the lactic acid can be broken down into harmless carbon dioxide and water.

During the anaerobic exercise you build up an **'oxygen debt'**. Getting oxygen to the muscles repays this debt and stops the cramp.

How fit are you?

The fitter you are the quicker you recover after vigorous exercise. This is called your **'recovery time'**. You can measure your own recovery time. (Check before trying this activity.)

1 *Take your pulse rate when you are resting.*

2 *Do some vigorous exercise such as going for a run.*

3 *Take your pulse rate after the exercise. Time how long it takes for your pulse rate to get back to normal. This is your recovery time.*

Questions

5 What is lactic acid?

6 **a** What is cramp? **b** Explain how it is caused.

7 **a** What is an oxygen debt?
b How do you repay it?

8 Footballers sometimes get cramp during a long hard game. The trainer comes on the field and stretches the footballer's leg and rubs it vigorously. Suggest how this rubbing helps stop cramp.

1.13 The heart

Objectives

This spread should help you to
- describe the structure of the heart
- describe how the heart works

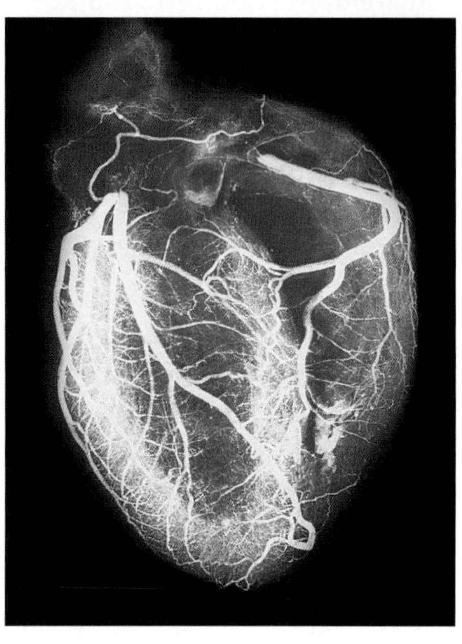

A coloured X-ray image of the heart showing the coronary arteries.

The **heart** is a pump made of muscle. It is about the size of a clenched fist. The heart has the job of pumping the blood round the body.

Outside the heart

This is a **vein**. It brings blood from the body, except the lungs.

These are **arteries**. They carry blood all over the body.

The heart has four chambers (see below). The top two are called **atria**.

The heart has its own blood supply carried by the **coronary artery** and **coronary vein**.

The bottom two chambers are called **ventricles**.

Inside the heart

This artery carries blood to the lungs.

This artery carries blood to the body.

This **valve** stops blood going back into the heart.

This vein carries blood into the heart from the lungs.

This vein carries blood into the heart from the body.

The **right atrium** has thin walls.

The **left atrium** has thin walls.

This valve stops blood going back into the right atrium.

This valve stops blood going back into the left atrium.

These **valve strings** hold the valve flaps in place.

The **right ventricle** has thick walls. It pumps blood to the lungs.

The **left ventricle** has very thick walls. It pumps blood to the body *except* the lungs.

Questions

1. What is the heart made of?
2. What are the:
   a upper **b** lower
   chambers of the heart called?
3. Which ventricle has the thickest walls?

The pumping cycle

The heart pumps by contracting and relaxing. The contractions can be heard as **heartbeats**.

Did you know?

Some athletes have a heart beat of more than 200 beats per minute.

Blood flows into the two atria.

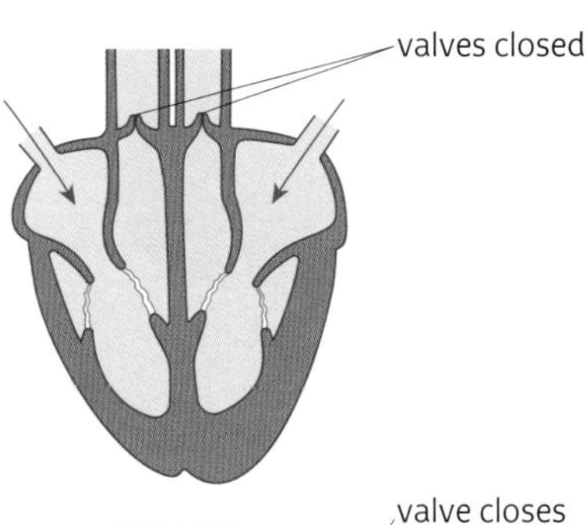

The atria contract and push blood through the valves into the ventricles.

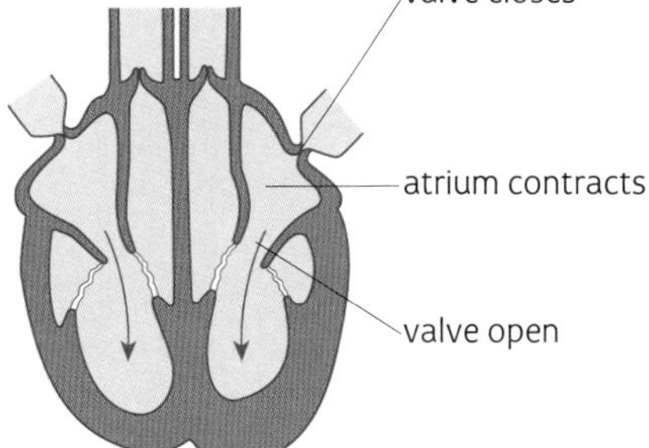

A split second later, the ventricles contract forcing blood out into the arteries.

Blood flows along the arteries, the atria fill up with blood, and the whole cycle starts again.

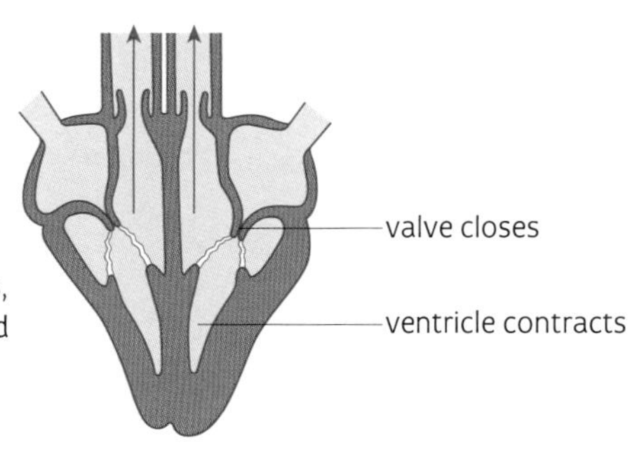

Your heart beats about 70 times a minute when you are resting but this can increase to well over 100 times a minute during physical activity or excitement. You can easily measure your heartbeat by taking your **pulse**. This is a surge of blood produced in the arteries every time the ventricles contract. You can feel a pulse with your fingertips at your wrist, where an artery passes between a bone and the surface of the skin.

Questions

4 What do the valves do in the heart?

5 The heart is a part of the living body. How does it get its blood supply?

6 Describe the pumping cycle of the heart.

7 If you listen carefully to someone else's heartbeat through a stethoscope you can hear a double beat – a faint one followed rapidly by a louder one. Suggest a reason for this.

8 How fast does the heart usually beat when you are resting?

1.14 Blood vessels and circulation

Objectives

This spread should help you to

- describe three types of blood vessel
- describe how blood circulates round the body
- explain what a double circulation is

The heart pumps blood round the body in tubes called **blood vessels**. The heart and blood vessels together make up the **circulatory system**.

Blood vessels

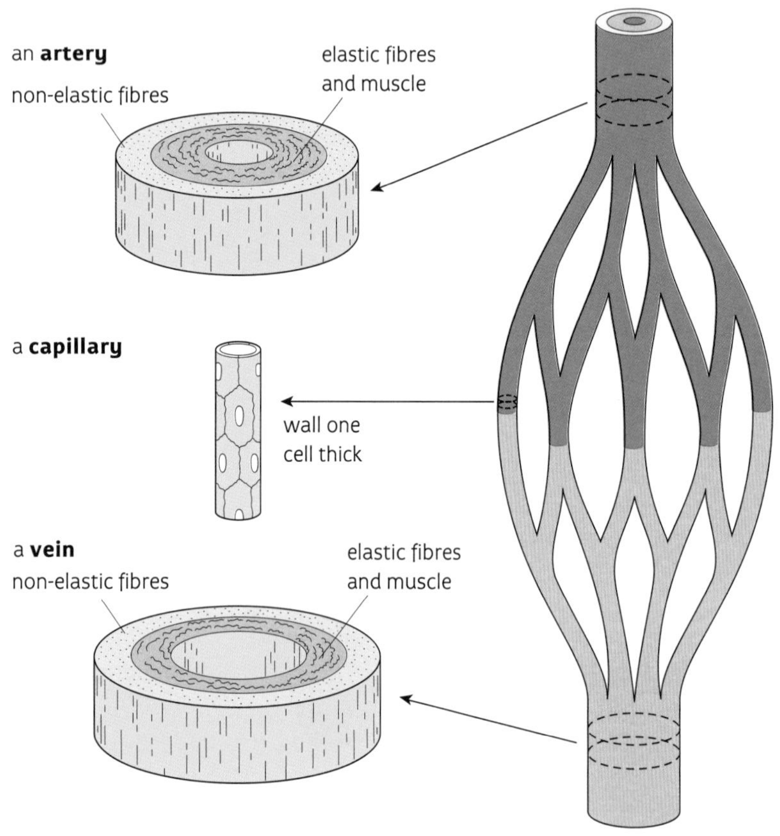

Arteries carry blood away from the heart. They have thick muscular walls to cope with the high pressure of blood as it leaves the heart. Arteries divide into smaller tubes called **capillaries**.

Capillaries spread through the body tissues so that the blood supply gets to every cell in the body. The walls of a capillary are very thin. This means that liquid (tissue fluid) carrying oxygen, food, carbon dioxide, and waste can pass between the blood and the cells. Capillaries join up to form **veins**.

Veins carry blood back to the heart. They are wider than arteries and have thinner walls. This is because the blood pressure is low and the blood flows slowly. When you move around, your body muscles help push the blood along veins back to your heart. Veins have valves to keep the blood flowing in the right direction.

Vein valves

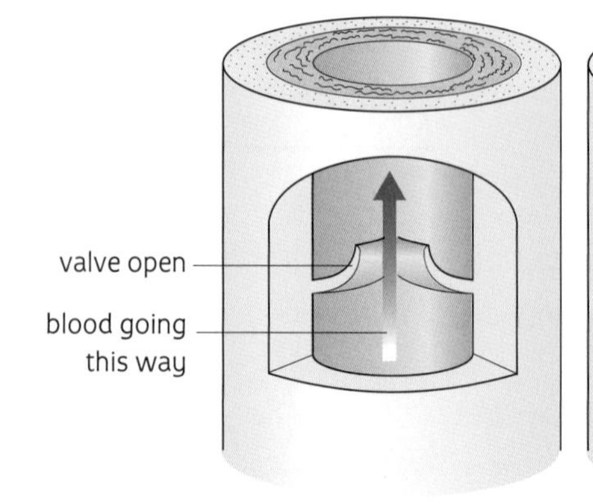

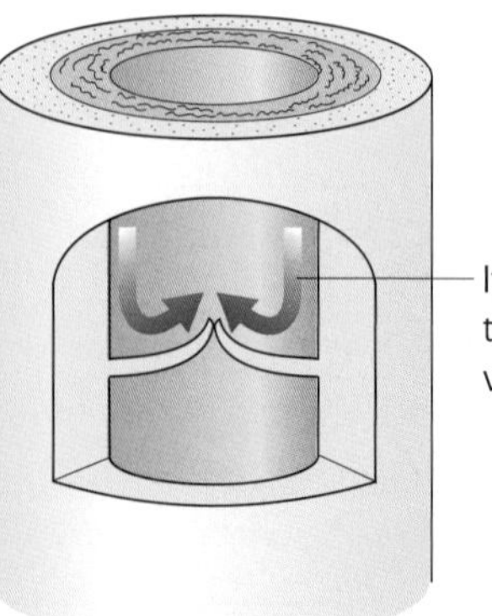

Questions

1. What two things make the circulatory system?
2. Give three differences between an artery and a vein.
3. Explain why capillary walls are so thin.
4. Suggest why the blood pressure in arteries is high but the blood pressure in veins is low.

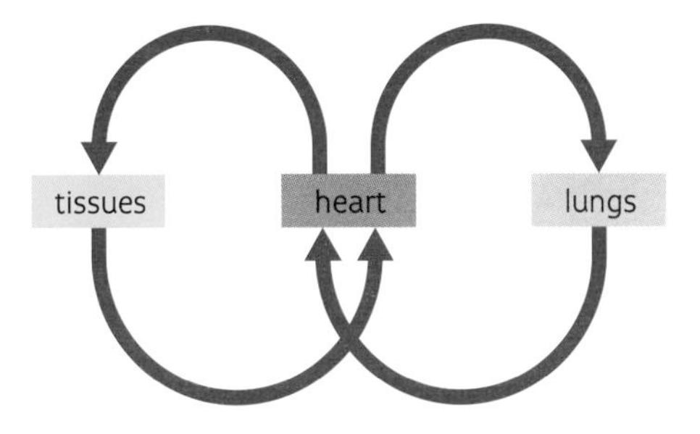

The circulatory system

The circulatory system is a transport system that carries things to and from different places in the body. We have a **double circulation**. This means that blood passes through the heart twice as it goes once round the body. It works like this:

The right side of the heart pumps blood to the lungs to collect oxygen and get rid of carbon dioxide.

The left side of the heart pumps blood to the rest of the body to supply cells with oxygen and pick up carbon dioxide and other waste.

When the blood gets to the cells, liquid called tissue fluid passes across the capillary wall. Substances move in both directions by diffusion.

Did you know?

It takes about 30 seconds for blood to go once round the body. An adult human has about 100 000 km of blood vessels.

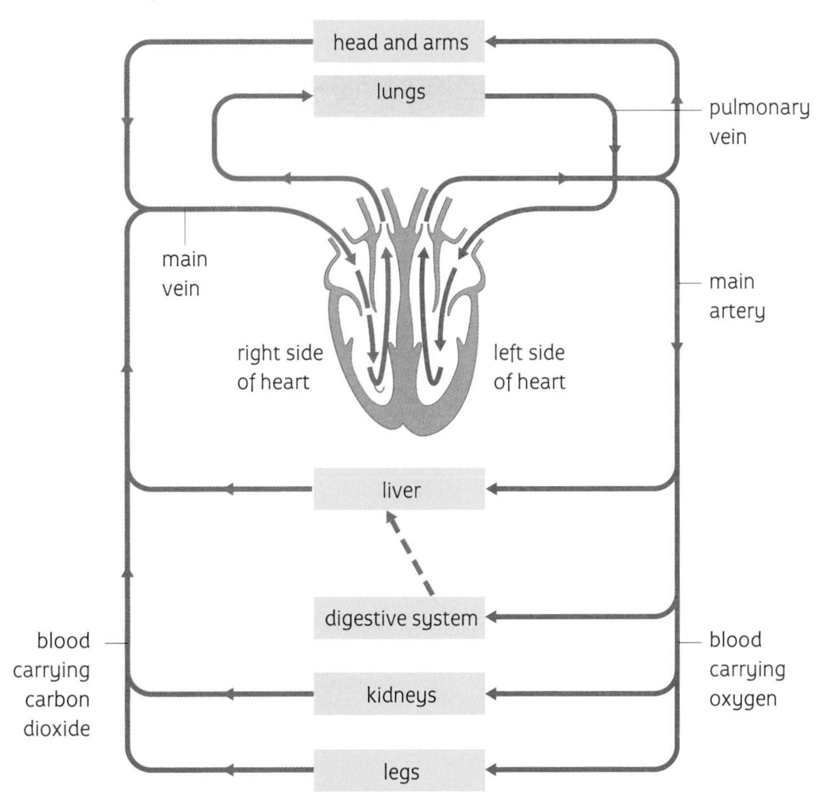

Questions

5 What is the job of the circulatory system?

6 Explain what is meant by 'double circulation'.

7 Name two things that diffuse:

- **a** from blood to cells
- **b** from cells to blood.

8 Name:

- **a** an artery with blood full of oxygen
- **b** a vein with blood full of oxygen
- **c** an artery with blood full of carbon dioxide
- **d** a vein with blood full of carbon dioxide.

1.15 Blood

Objectives

This spread should help you to

- describe what blood is

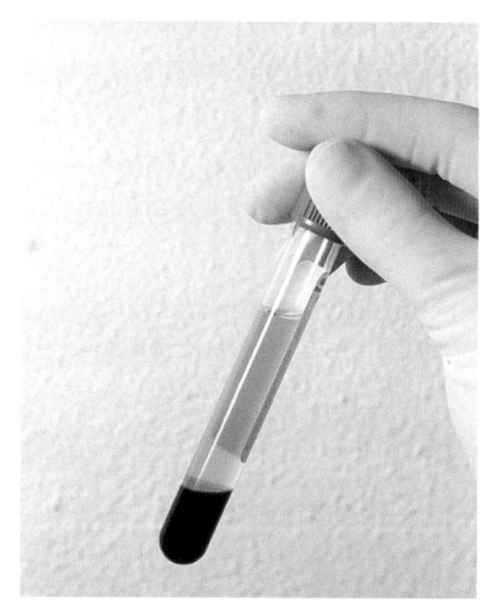

When blood is left for a while, the cells sink to the bottom of the plasma.

What's in blood?

Blood is liquid called **plasma** with **red cells**, **white cells**, and **platelets** floating in it. Blood looks red because there are lots more red cells than white cells. There are about 5.5 litres of blood in your body.

Plasma

Plasma is a straw-coloured liquid. It is mainly water with things like digested food, carbon dioxide, and urea dissolved in it.

Red cells

Red cells are made in the bone marrow. They are round with a dent on both sides, giving a large surface area. They have no nucleus. Red cells are red because they contain a chemical called **haemoglobin**. Their job is to carry oxygen from the lungs to all the cells in the body.

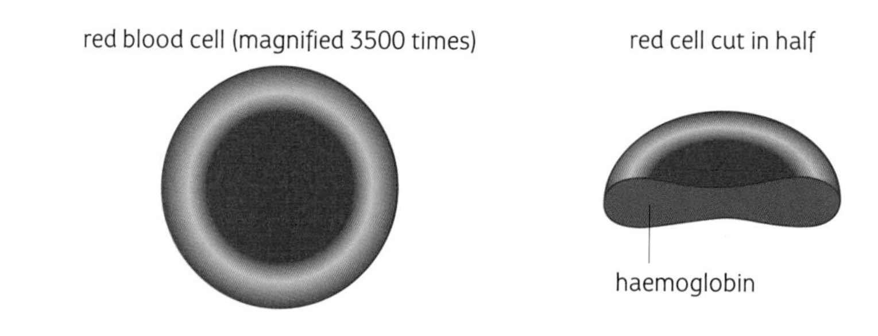

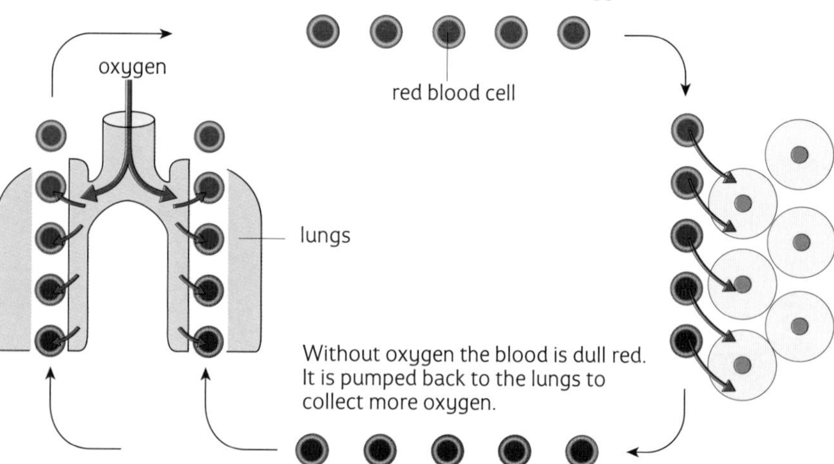

White cells

White cells have a big nucleus and can change their shape. Their job is to protect the body against disease. Some white cells eat the microorganisms that cause disease.

Questions

1. How much blood is there in your body?
2. Name four things that make up the blood.
3. What is the job of a red cell?
4. Describe how the blood carries oxygen round the body.

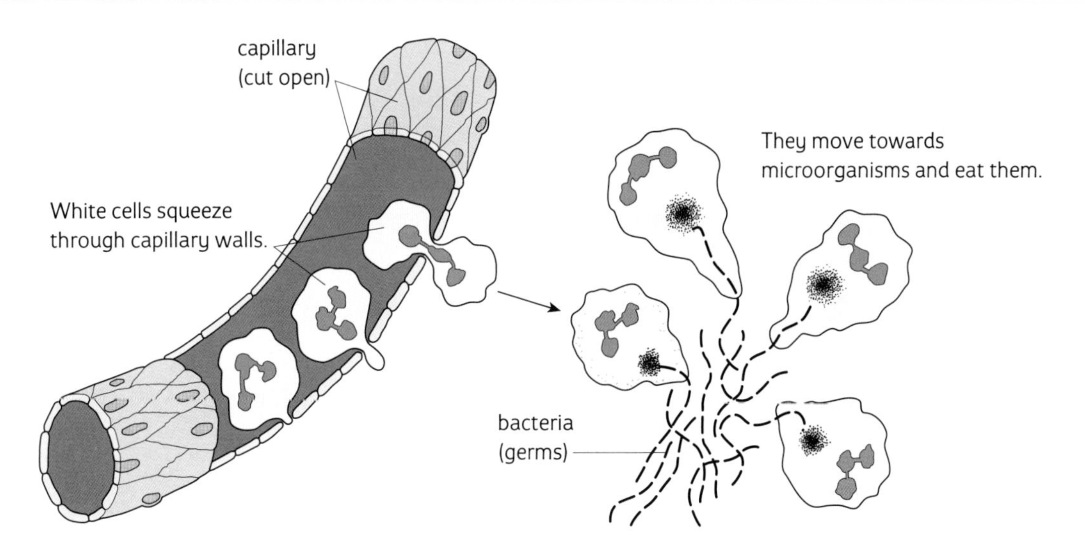

Other white cells make **antibodies** or **antitoxins**. Antibodies are chemicals that destroy microorganisms by dissolving them. There is a different antibody for each kind of microorganism. Antitoxins break down **toxins** (poisons) that microorganisms make.

Did you know?

There are about 5 000 000 000 red cells and 5 000 000 white cells in 1 cm^3 of blood.

Platelets

Platelets are bits of cell broken off larger cells in the bone marrow. Their job is to help the blood to clot and stop bleeding at cuts.

1 Platelets cause tiny fibres to form a net. Red cells get caught in this net forming a **blood clot**.
2 The surface of the clot hardens to form a scab. This keeps the cut clean while new skin grows.

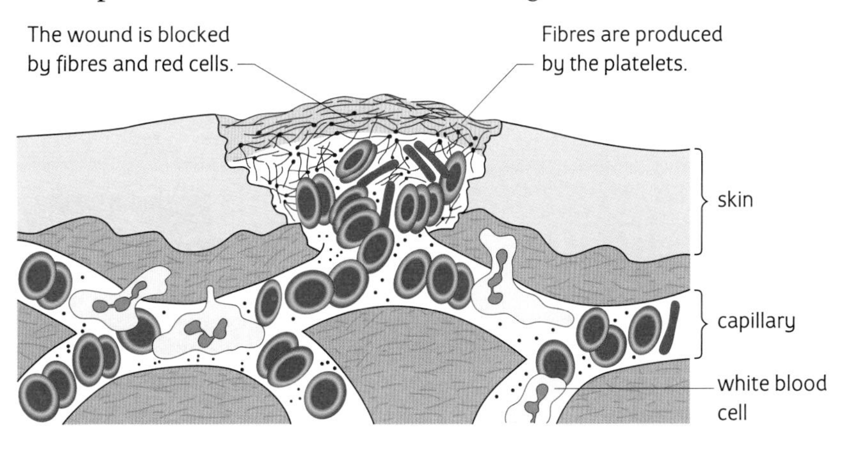

Questions

5 Give two differences in shape between red cells and white cells.

6 What are antibodies?

7 What is the difference between an antibody and an antitoxin?

8 Describe what platelets do in making a blood clot.

1.16 What blood does

Objectives

This spread should help you to

- describe what blood does
- explain how things are exchanged between blood and tissues

Blood carries...

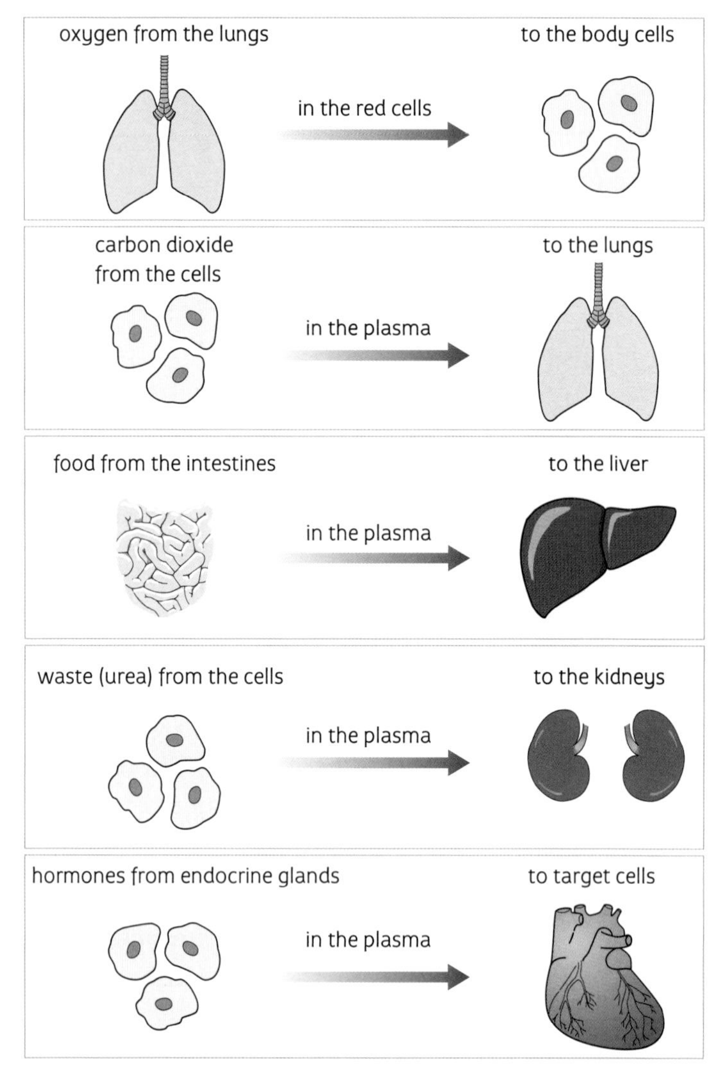

Blood protects us by...

...healing cuts ...and destroying microorganisms.

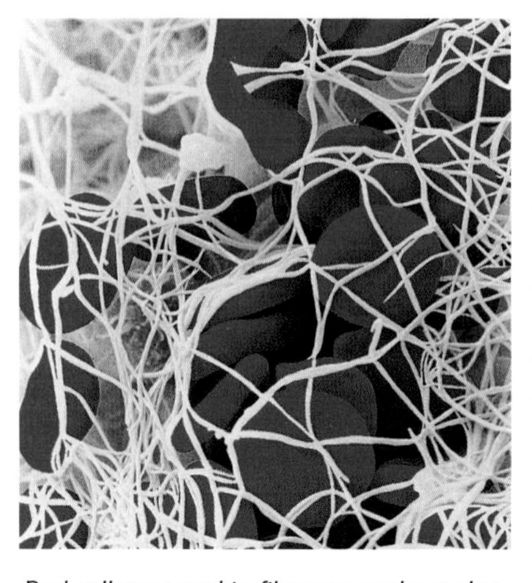

Red cells trapped in fibres to make a clot.

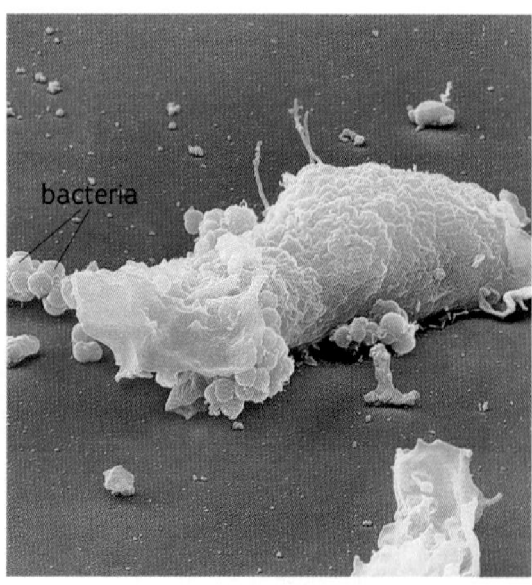

A white blood cell eating bacteria.

Questions

1. List four things carried by the blood.
2. Which things on your list are carried:
   - **a** by the red cells
   - **b** in the plasma?
3. **a** What is urea?
   - **b** Where is it made?
   - **c** Which organ removes it from the body?
4. Explain how blood protects us.

How things get in and out of the blood

The walls of the smallest blood vessels (capillaries) are very thin with tiny holes in them. A liquid called **tissue fluid** leaks out. Tissue fluid is mainly water. It makes a continuous link between the water in the blood and the water in the cells.

Tissue fluid is mainly water. This is why fresh meat always looks (and feels) wet.

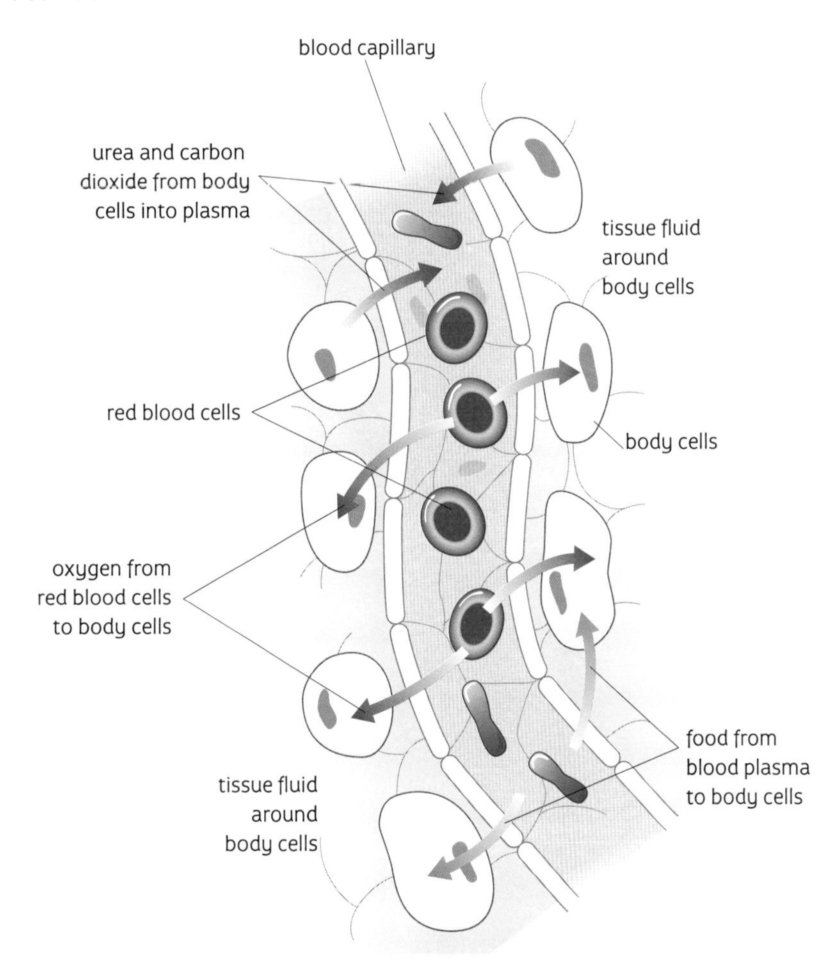

Tissue fluid carries oxygen and food, in solution, from the blood to the cells. Wastes such as carbon dioxide and urea travel in the opposite direction. They dissolve in the tissue fluid, which carries them into the blood.

Questions

5 What is tissue fluid?

6 Why does tissue fluid leak out of blood capillaries?

7 Explain why fresh meat is wet.

8 Describe how:

a oxygen gets from a red blood cell to a body cell

b carbon dioxide gets from a body cell to the blood plasma.

1.17 Microorganisms and disease

Objectives

This spread should help you to

- name the microorganisms that cause diseases
- describe how microorganisms cause diseases
- describe how diseases spread
- describe some ways of stopping the spread of diseases

Microorganisms are tiny living things that can only be seen with some sort of microscope. Diseases can result when microorganisms such as viruses and certain bacteria get into the body.

Bacteria

Bacteria are very small living cells. They can reproduce very quickly – about once every 20 minutes. Bacteria can make you feel ill in two ways:

1 They damage living tissue. Tuberculosis is a disease caused by bacteria that damage lung tissue.
2 They produce toxins (poisons). Food poisoning is caused by bacteria that produce toxins in the digestive system.

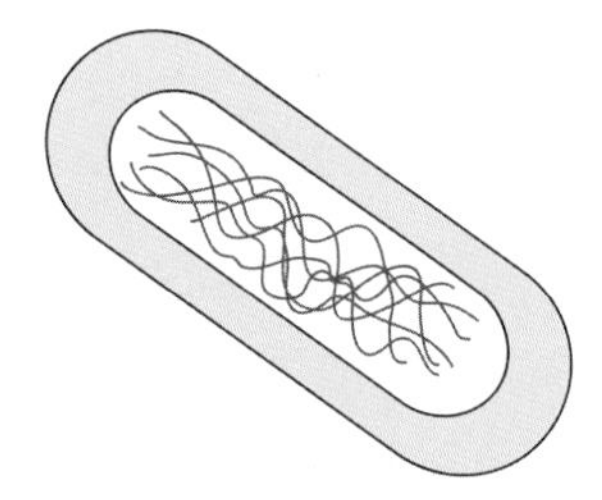

Bacteria are cells with no nucleus.

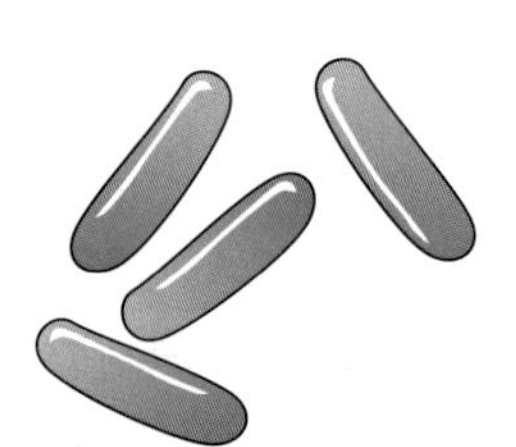

rod shaped bacteria

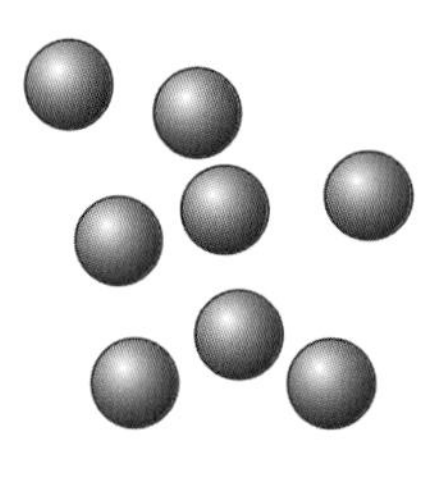

round bacteria

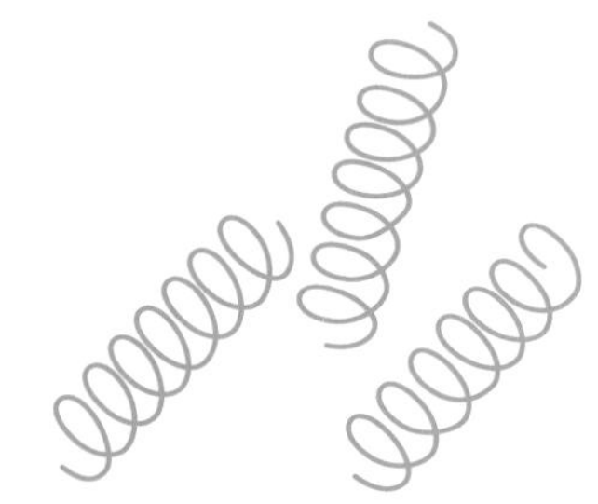

spiral shaped bacteria

Viruses

Viruses are not cells. They are about 100 times smaller than bacteria so they can't be seen under an ordinary microscope. Viruses can only reproduce and they need other living cells to do it in. They make you feel ill by damaging living tissue. The common cold is caused by a virus. If you breathe these viruses in, they get into the cells in your nose and throat. The viruses reproduce quickly and infect other cells.

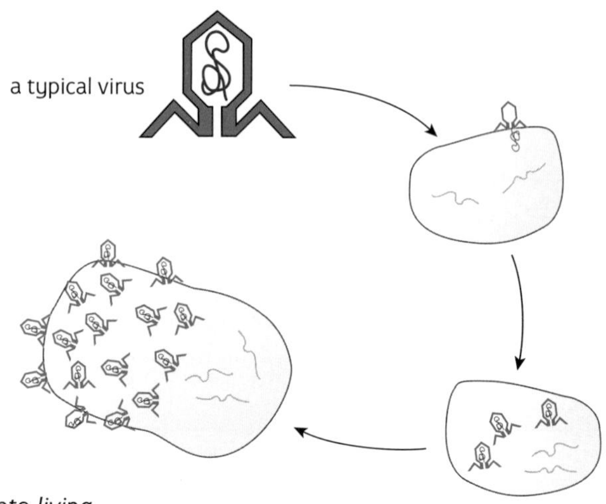

A virus gets into living cells and reproduces.

Questions

1 What are microorganisms?

2 How are bacteria different from other cells?

3 Explain how bacteria can make you feel ill.

4 Some people say viruses are not living things. Give one reason why:

a they could be right

b they could be wrong.

Coughs and sneezes spread diseases.

How diseases spread

Diseases are more likely to result if lots of microorganisms get into the body at once. There are a few ways this can happen:

- through the skin – touching an infected person (or things they have used) can spread viruses that cause chickenpox and measles
- through the digestive system – food and drink can be infected by coughs and sneezes, dirty hands, or unhygienic cooking methods. Food poisoning happens if you eat food and drink infected with bacteria.
- through the breathing system – colds and flu are caused by breathing in viruses from people who cough and sneeze into the air
- through the reproductive system – the bacteria that cause gonorrhoea and the virus that causes AIDS pass from one person to another during sexual intercourse.

How to stop diseases spreading

There are lots of things you can do to avoid catching a disease.

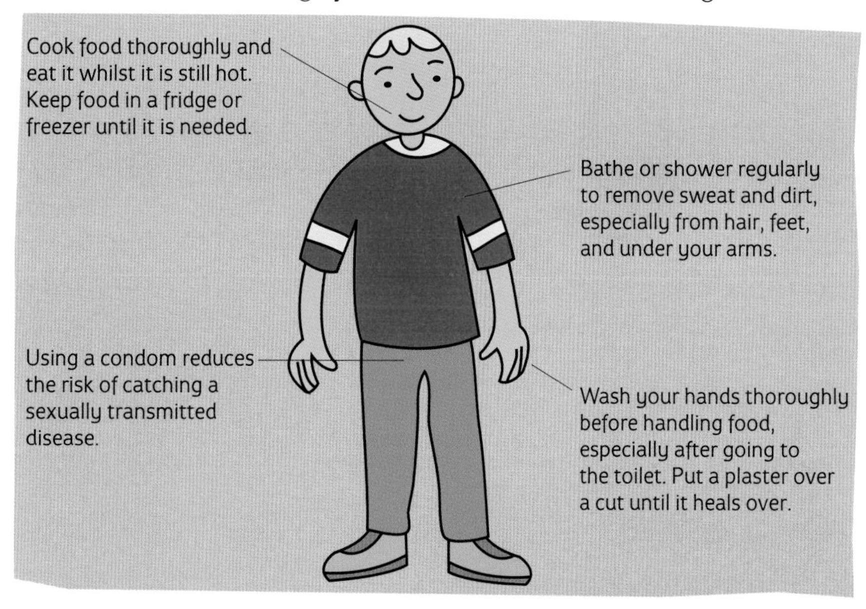

Questions

5 What causes you to have a disease?

6 List four ways that microorganisms can enter the body.

7 Why is it not a good idea to use a stranger's towel?

8 Why should you always cough or sneeze into a handkerchief?

9 Name two diseases caused by:
a bacteria **b** viruses.

10 Suggest why make-up should be washed off every night.

1.18 Defences against disease

Objectives

This spread should help you to

- describe how the body defends itself against disease
- describe how antibodies work
- explain the difference between natural and artificial immunity
- describe how a vaccine works

Lines of defence

Your body has lots of ways of defending itself against microorganisms.

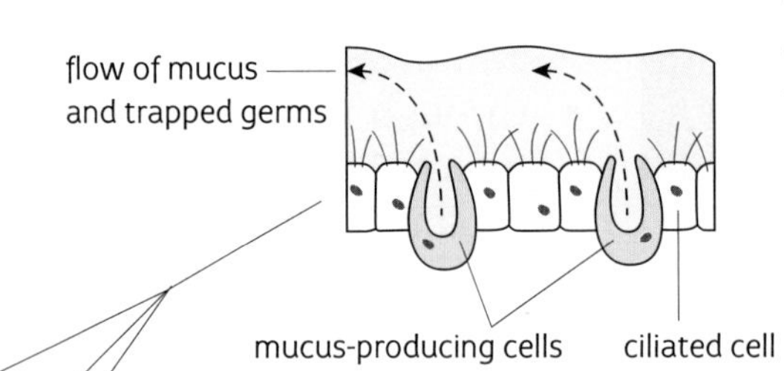

The nose and air passages into the lungs are lined with cells that produce a sticky **mucus**. Dirt and bacteria in the air get trapped in this mucus. Tiny hair-like **cilia** move the mucus up to the throat where it is swallowed.

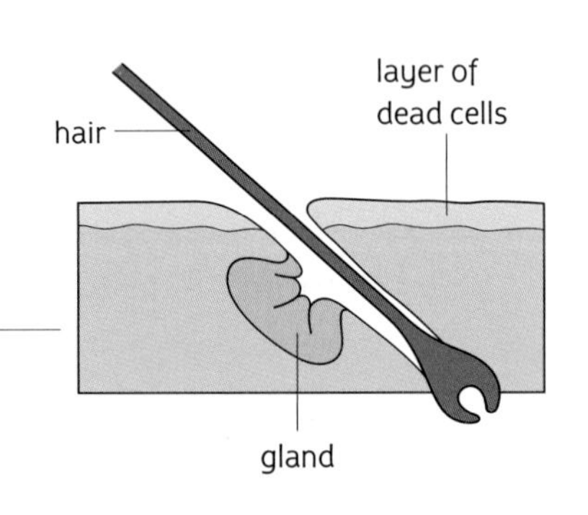

The skin is a very effective barrier. If it gets damaged, the blood clots quickly to seal the hole and keep microorganisms out.

The stomach produces hydrochloric acid which kills the microorganisms in dirty food or drink.

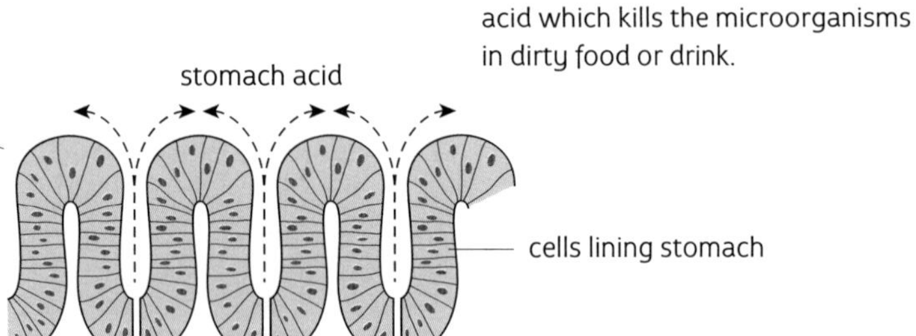

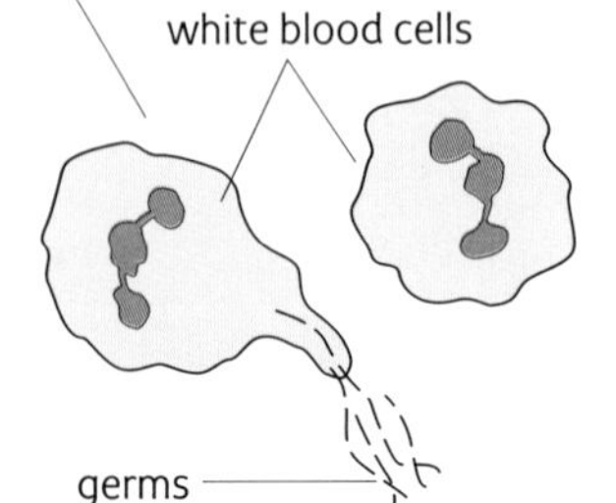

Microorganisms that get through these defences and enter the bloodstream are attacked by white blood cells. Some microorganisms are eaten and some are killed by antibodies. Poisonous toxins are dealt with by antitoxins. We call this our **immune system**. It gives us protection or natural immunity against disease.

Questions

1. How does the stomach stop microorganisms getting into the body?
2. Describe how the cilia and mucus keep our lungs clean.
3. Give three ways that white blood cells protect the body against disease.
4. What do we call the body's natural protection against disease?

More about antibodies

Antibodies are your body's chemical weapons against disease. They work like this.

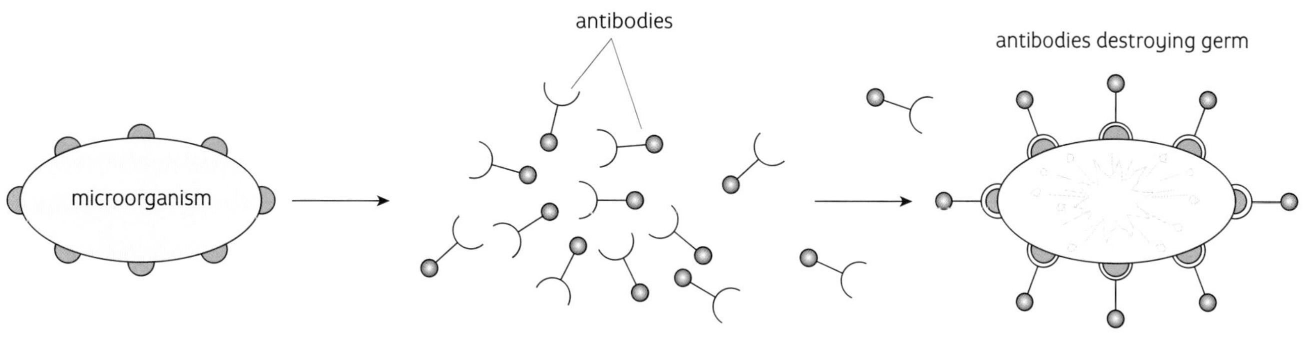

Each type of bacteria or virus has a different chemical on its surface, similar to the way all humans have different fingerprints.

When the body detects these chemicals, white blood cells produce antibodies.

Antibodies combine with the chemicals. This kills the microorganisms by making them burst.

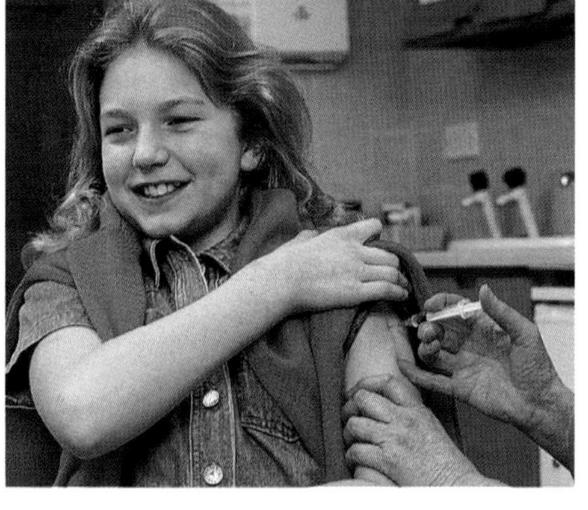

You have probably had quite a few vaccines injected into your body. This is how you are protected against diseases such as polio, tetanus, mumps, and measles.

Antibodies stay in your body to protect you just in case you get infected by the same bacteria or virus again. You have **immunity** against that disease.

Artificial immunity

Some diseases can make you very ill. You could die before your white cells can make enough antibodies to save you. Only **immunization** can protect you.

Dead or weakened bacteria or viruses are injected into the body. This is called a **vaccine**. You get a mild form of the disease, so mild you probably never notice it. However, the vaccine is enough to cause the white cells to make antibodies against the disease. If live bacteria or viruses of the same type get into your body, you are already protected from them.

Questions

5 What makes each type of microorganism different?

6 How do antibodies kill microorganisms?

7 Explain why you rarely catch the same disease more than once.

8 Explain why people are immunized.

9 What is a vaccine?

10 Explain how a vaccine works.

11 List four diseases that you can be vaccinated against.

12 A woman going to work in Africa had a vaccination against the serious disease cholera. She felt weak and ill for about 12 hours after the injection but then began to feel much better.

a Suggest why the injection made her ill.

b Why was it important for the woman to have the vaccine?

1.19 In and out of cells – diffusion

Objectives

This spread should help you to

- describe how particles move by diffusion
- describe how particles are moved by active transport

Diffusion – spreading out

When a farmer puts animal manure on his fields, it is not long before the smell spreads over the surrounding area. The closer you are to the manure, the stronger the smell. Fortunately the smell is not so strong as you move further away. After a time the smell seems to disappear. This is an example of **diffusion**.

Molecules are always moving about. The smell molecules move (diffuse) from where there are a lot of them to where there are not so many of them. In other words, the molecules move from an area of high concentration to an area of low concentration. We call this a **concentration gradient**.

Eventually the smell molecules are so diluted by the molecules of the air that the smell is hardly noticeable.

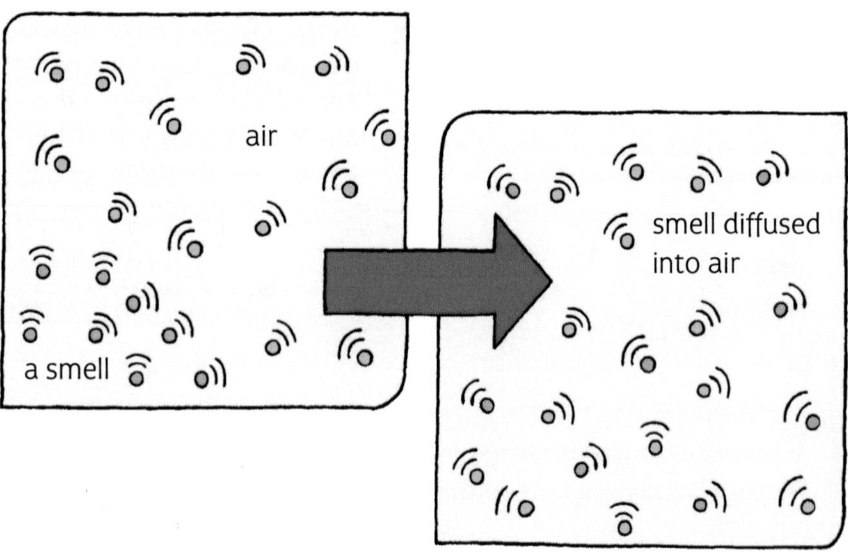

Diffusion also takes place in liquids. You can show this by adding a drop of ink to a bowl of water. Notice how the colour of the ink gets weaker as it diffuses through the water. Eventually it might seem to disappear altogether, just like the smell of the manure.

Questions

1. Molecules move down a concentration gradient. Explain what this means.
2. What is diffusion?
3. Sharks are attracted by the smell of blood in the water. They can smell the blood of a wounded animal several miles away. Explain this, using your knowledge of diffusion.

Diffusion in and out of cells

Cells in the human body need a regular supply of food and oxygen for respiration. There is a lot of food and oxygen in the blood and not so much in the cells. So the food and oxygen pass into the cells by diffusion.

During respiration, carbon dioxide is produced. This will poison the cell if it is not taken away. Since the blood contains less carbon dioxide than the cell, the carbon dioxide diffuses out of the cell into the blood.

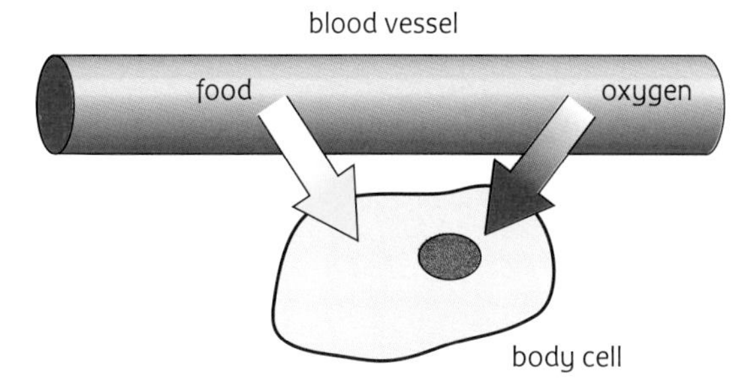

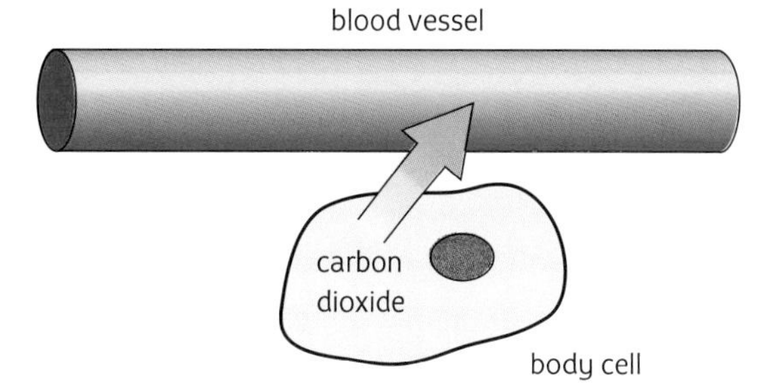

Active transport – moving against a concentration gradient

Sometimes substances are taken into cells when there is already more inside the cell than outside. For example, plants need minerals such as nitrates from the soil to make protein and grow. There is more nitrate inside plant root cells than there is in the surrounding soil. Diffusion won't work because the concentration gradient is the wrong way round. So the plant cells use a process called **active transport**. Active transport needs energy to make it work, just as you need energy to push your bike up a gradient.

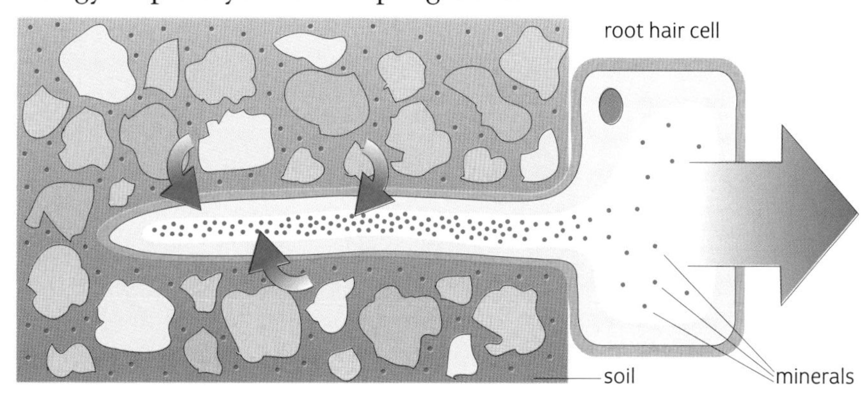

4 Give two examples of diffusion in human body cells.

5 Why can't plants take in nitrate by diffusion?

6 Why does active transport need energy?

1.20 William Harvey

Early ideas

An old Roman physician called Galen believed that blood was made in the liver from food. It then went through the heart once before being used up by the body tissues. According to Galen, blood leaked through tiny holes from the left to the right side of the heart. The right side was just another way of getting blood to the lungs and other tissues. Simple as that!

Doctors still believed in Galen's theory right up to the 1600s.

Who was William Harvey?

William Harvey was an English doctor who lived in the early part of the seventeenth century. He was interested in the way that the heart moved blood around the body. His views were very controversial and lost Harvey many patients. But this was a major step for science and changed the way people thought about blood circulation.

William Harvey (1578–1657) was the man who showed us how blood flows around the body. Despite a lot of opposition he was able to change the way that doctors thought about the heart and blood circulation. Harvey was doctor to King Charles I of England and was recognized as a medical leader in his day.

William Harvey investigating blood circulation in a dog. He also used humans!

Harvey demonstrating his discovery to a group of doctors in London in the early seventeenth century.

Harvey's work

Harvey completely smashed Galen's idea that blood only goes through the heart once. By taking his pulse, Harvey estimated the how much blood passed out of the heart. He calculated that the heart pumps out three times our body weight of blood every hour. Imagine eating food fast enough to make that much blood!

Harvey examined blood vessels in humans and other animals. He showed that veins had valves in them, which only let blood move in one direction – back to the heart. This is how he did it.

Harvey didn't know about oxygen, but he saw that the blood flowed from one side of the heart to the other through the lungs, where it was recycled.

But Harvey wasn't able to show how blood passed from arteries to the veins. He guessed that there were some tiny blood vessels that made the connection, but he couldn't see them. It wasn't until four years after Harvey died that the newly invented microscope showed the tiny blood vessels that we call capillaries.

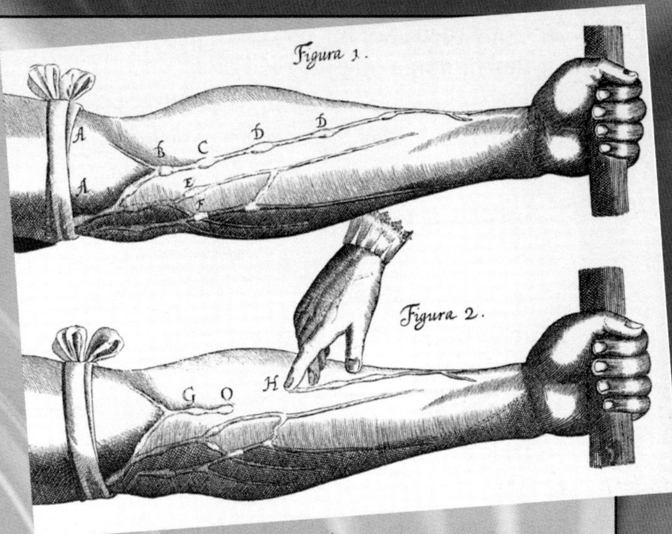

Finding valves in blood vessels.

Only when the microscope was invented were doctors able to prove Harvey's theory about the existence of capillaries.

Talking points

1 Why is William Harvey a famous scientist?

2 Doctors used to think that blood went in both directions along blood vessels. How did Harvey show this wasn't true?

3 Why couldn't Harvey prove that capillaries exist?

Practice questions

1 a Copy the list of life processes and write the correct description alongside each one.

Life processes: *movement, respiration, sensitivity, feeding, excretion, reproduction, growth*

Descriptions:

producing more of the same kind

changing position

increasing in size

releasing energy from food

responding to something

getting rid of waste

taking in food or raw materials

b A car moves, takes in fuel, releases energy from fuel, and gets rid of waste through the exhaust pipe. Is the car a living thing? Explain your answer.

2 The diagram shows a nerve cell from an animal.

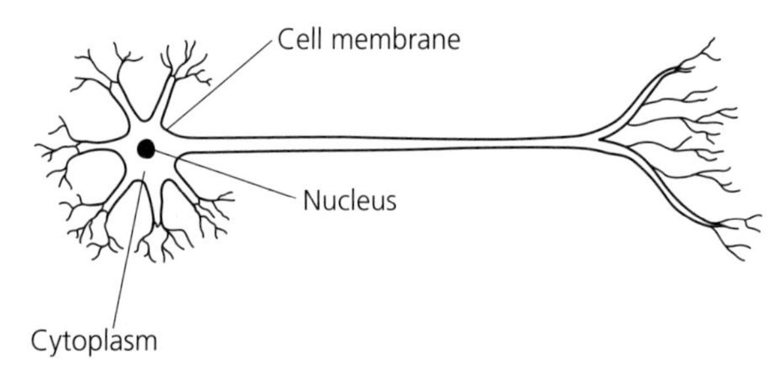

a What is the job of:

i the nucleus

ii the cell membrane?

b What happens in the cytoplasm of a cell?

c i What is the job of a nerve cell?

ii Describe how a nerve cell is suited to its job.

3 a Copy the list of organs and alongside each one write its job.

Organs: *stomach, kidney, lung, brain, eye*

Jobs:

gets information about the surroundings

exchanges oxygen and carbon dioxide

controls what the body does

gets rid of urea

digests food

b The diagrams show four organ systems in the human body.

A

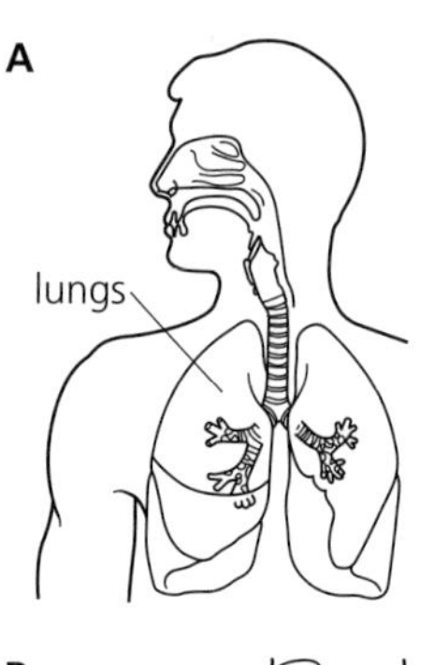

C

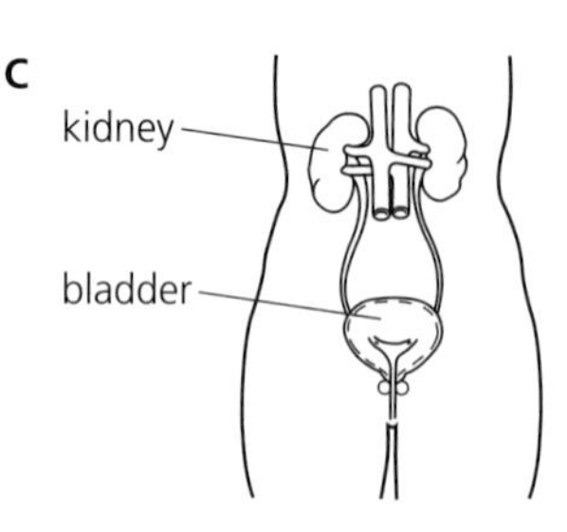

B

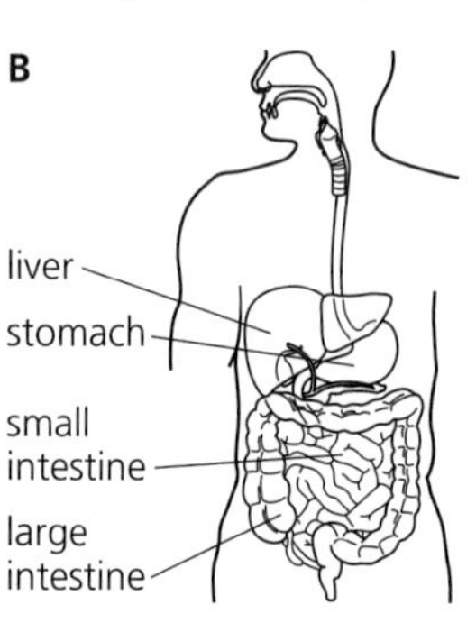

D

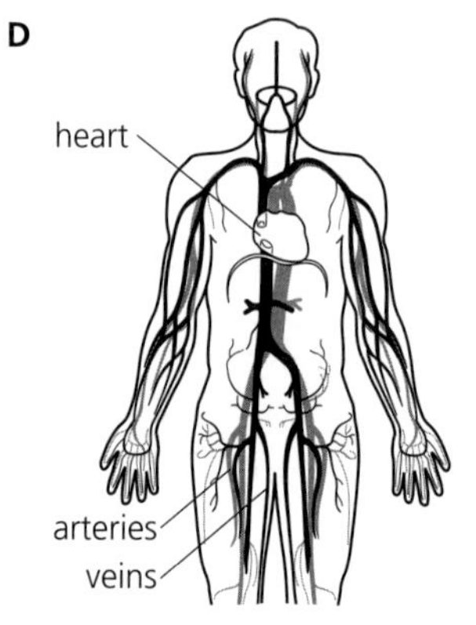

What is the name of:

i organ system A

ii organ system B

iii organ system C

iv organ system D?

4 The table was on a breakfast cereal packet.

NUTRITION INFORMATION per 100 g	
ENERGY	1403 kJ
	331 kcal
PROTEIN	10.2 g
CARBOHYDRATE	67.1 g
of which sugars	17.2 g
starch	49.9 g
FAT	2.4 g
of which saturates	0.5 g
SODIUM	0.6 g
FIBRE	14.1 g
VITAMINS	
NIACIN	18 mg
VITAMIN B_6	2 mg
RIBOFLAVIN (B_2)	1.6 mg
THIAMIN (B_1)	1.4 mg
FOLIC ACID	400 μg
VITAMIN D	5 μg
VITAMIN B_{12}	1 μg
IRON	14 mg

a i How many vitamins are there in the cereal?

ii Why do we need vitamins?

b **i** Why do we need dietary fibre (roughage) in our food?

ii Explain why this cereal is advertised as 'high fibre'.

c A girl eats 50 g of this cereal.

i How much energy does this give her?

ii How much protein does this give her?

5 The diagram shows the human digestive system.

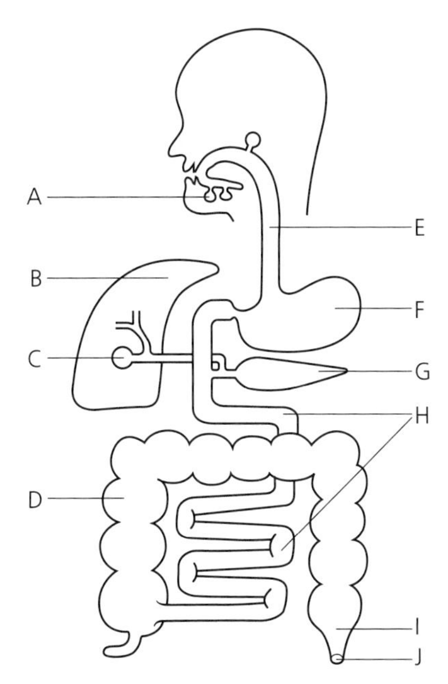

a List the letters A–J and write the correct label beside each one from the words below.

anus, gall bladder, gullet, large intestine, liver, mouth, pancreas, rectum, small intestine, stomach

b Which of the parts listed above does each of these jobs? Write the letter.

i absorbs water

ii produces bile

iii contains hydrochloric acid

iv absorbs digested food

v produces enzymes

vi chews food

vii stores faeces

6 Copy these sentences and fill the gaps using these words:

amino acid, amylase, fatty acid, glucose, glycerol, lipase, protease

a A starch molecule is made from lots of _____ molecules.

b A protein molecule is made from lots of _____ _____ molecules.

c A fat molecule is made from _____ _____ and _____ molecules.

d ____ is an enzyme that breaks down starch.

e _____ is an enzyme that breaks down fat.

f _____ is an enzyme that breaks down protein.

7 Glucose is a very small sugar molecule which can pass through a partially permeable membrane. A mixture of starch and glucose solution is put into a bag made from partially permeable membrane. The bag is put into a beaker of water.

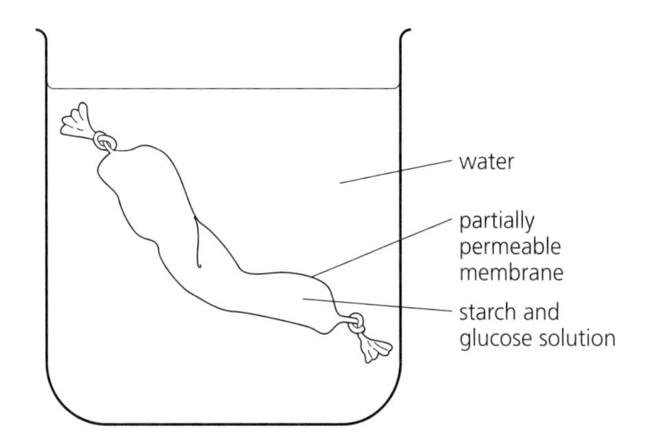

After an hour the water around the bag was tested for starch and for glucose. The results are shown in the table.

Water around bag	Starch	Glucose
at start	no	no
after one hour	no	yes

a Describe how you would do:

i a starch test **ii** a test for glucose.

b Which molecules can get through the partially permeable membrane? Explain why.

c Suggest why the starch molecules can't get through the partially permeable membrane.

d Which part of the digestive system works like the partially permeable membrane?

e Which part of the body does the water represent?

Practice questions

8 The walls of the small intestine are lined with thousands of tiny villi.

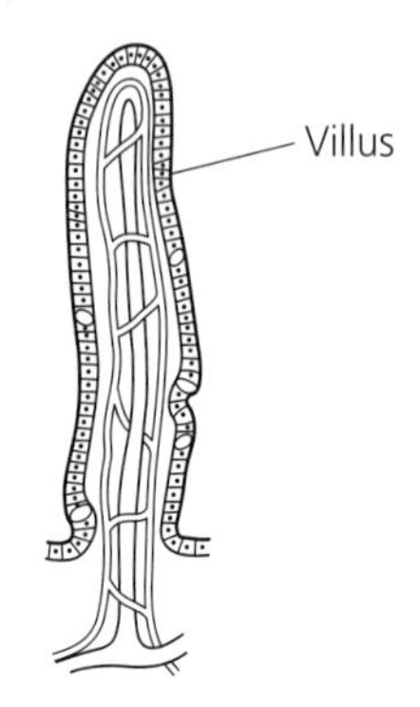

a What is the job of these villi?

b How do the following help the villi do a good job?

i a wall one cell thick

ii a large surface area

iii a good blood supply

9 The diagram shows the human breathing system.

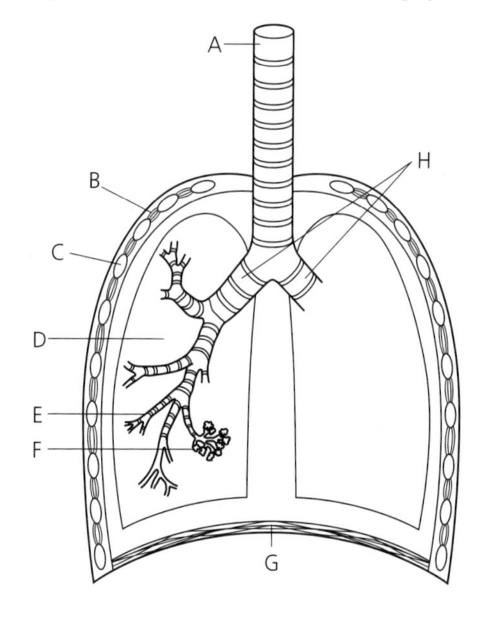

a List the letters A–H and write the correct label beside each one from the words below.

alveoli, bronchi, bronchiole, diaphragm, lung, muscle, rib, trachea

b Gas exchange happens in the alveoli.

i Name the gases that are exchanged in the alveoli.

ii What are the gases exchanged between?

c How do these help the alveoli do a good job?

i thin walls

ii moist lining

iii surrounded by blood vessels

10 Write these two headings:

Breathing in **Breathing out**

Write these statements under the correct heading and in the right order.

air pushed out of lungs

diaphragm relaxes

rib muscles contract

air sucked into lungs

chest gets bigger

rib muscles relax

diaphragm contracts

chest gets smaller

11 The graphs show the breathing rate of a runner before and after a race. The height of the curves tells you how much air is breathed in (volume of air) in one breath.

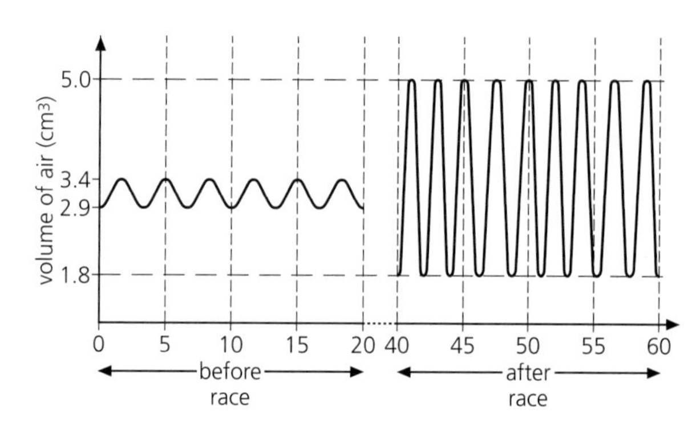

a How many breaths did the runner take in 20 seconds:

i before the race **ii** after the race?

b How much air was breathed in (volume of air) in one breath:

i before the race **ii** after the race?

c Explain why there is so much difference between the amounts of air breathed in before and after the race.

d Work out how much air goes in and out of the runner's lungs in 20 seconds after running the race.

12 Copy this passage and fill the gaps using these words:

carbon dioxide, cramp, emergency, energy, glucose, lactic acid, muscles, oxygen, oxygen debt, water

During respiration, _____ is broken down to release _____. If there is plenty of _____ around, the respiration is called aerobic respiration. In aerobic respiration, _____ _____ and _____ are given off as waste. Anaerobic respiration is not as efficient but is useful in an _____. The waste product of anaerobic respiration in animals is _____ _____. This causes painful _____ in the _____. The animal stops moving and breathes deeply until an _____ _____ has been repaid.

13 The diagram shows the inside of a human heart.

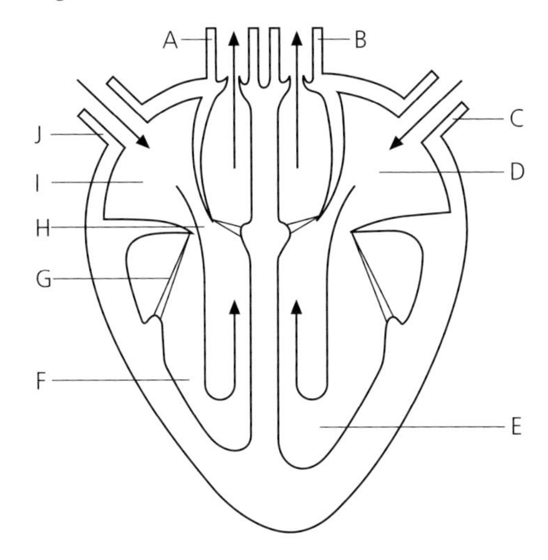

a List the letters A–J and write the correct label beside each one from the words below.

artery taking blood to lungs
artery taking blood to body
vein bringing blood from lungs
vein bringing blood from body
left ventricle
right ventricle
left atrium
right atrium
valve
valve strings

b Which part of the heart pumps blood to:

i the body **ii** the lungs?

c i Which parts of the heart carry blood full of oxygen?

ii Which parts of the heart carry blood full of carbon dioxide?

14 a Copy this list of words and write the correct description alongside each one.

Word list: *antibodies, antitoxin, haemoglobin, plasma, platelets, red cells, white cells*

Descriptions:

help the blood to clot
chemicals that kill microorganisms
joins with oxygen
have a big nucleus
a straw-coloured liquid
break down poisons
made in the bone marrow

b Name two things that are dissolved in the plasma.

c Antibodies and antitoxins are made by white blood cells. Describe one other way that white cells defend the body against microorganisms.

15 a i What are microorganisms?

ii Name two microorganisms.

b Explain why each of the following is a good way of stopping the spread of disease.

i covering your mouth when you cough or sneeze
ii cooking food thoroughly
iii washing your hands before eating
iv showering or bathing regularly
v never using a towel borrowed from a stranger

c Give three ways in which microorganisms can get into the body.

d Copy this passage and use these words to fill the gaps.

antibodies, antitoxins, artificial immunity, natural immunity, vaccine, white cells

The body has its own _____ _____ against disease. This is given by _____ _____, _____, and _____. _____ can be injected into the body to give protection against serious diseases. This is called _____ _____.

e Jack had chickenpox in 2000. His younger sister Jill caught chickenpox in 2001 but Jack didn't catch it again. Explain why. (Use some of the words in the first part of this question.)

absorption the movement of digested food through the walls of the digestive system into the blood

active transport using energy to move substances against a concentration gradient

aerobic respiration getting energy out of food with oxygen

alveoli air sacs in the lungs where gas exchange happens

amino acids the molecules that proteins are made of; proteins are broken down into amino acids during digestion

anaerobic respiration getting energy out of food without oxygen

antitoxins chemicals made in the blood that destroy poisons made by microorganisms

bacteria very small living cells; some bacteria can get into other living things and cause disease

blood liquid plasma with red cells, white cells, and platelets floating in it

bronchi tubes connecting the trachea with each lung

bronchioles small tubes carrying air through the lungs

capillaries the smallest blood vessels in the body; they have very thin walls

cell membrane a thin skin surrounding a cell; it controls what goes in and out of the cell

circulatory system the heart and blood vessels together

cytoplasm a jelly-like substance in a cell, where chemical reactions happen

diaphragm a sheet of muscle at the bottom of the ribcage

diffusion the natural movement of molecules from an area where there is a lot of them to an area where there is less of them

digestion the breaking down of food into smaller, soluble pieces

digestive system the organ system where digestion happens

enzymes substances that break down (digest) food into tiny, soluble pieces

fats solid and liquid (oil) foods that give energy and insulate the body

fatty acid one of the two chemicals produced when fats are digested

gall bladder place in the liver where bile is stored

gas exchange swapping oxygen with carbon dioxide in the lungs

gastric juice a mixture of hydrochloric acid and protease enzymes made in the stomach wall

glucose the simplest sugar produced by the digestion of starch and sugars

glycerol one of the two chemicals produced when fats are digested

haemoglobin a red chemical in red blood cells that joins with oxygen and carries it around the body

heart a muscular pump that sends blood around the body

hydrochloric acid part of gastric juice; it kills bacteria in food

immune system white blood cells and antibodies which destroy microorganisms that get into the body

immunity the body's protection against disease

immunization a vaccine injected into the body

lactic acid a substance produced by anaerobic respiration in muscles

large intestine part of the digestive system where water is absorbed, leaving semi-solid faeces

life processes things that all plants and animals do

liver the largest body organ; chemical reactions happen here; food is stored and heat produced

micro-organism a very tiny living thing that can only be seen by using a microscope

multicellular an animal or plant made of lots of cells

nucleus the control centre of a cell

organ a group of different tissues working together

organ system a number of organs working together to do one big job

organism something that is alive and able to survive on its own

oxygen debt caused when energy is 'borrowed' from glucose during anaerobic respiration in muscles

oxy-haemoglobin a chemical made when oxygen joins up with haemoglobin

peristalsis muscular contractions that push food along the digestive system

platelets bits in the blood that help it to clot

protein body-building foods

recovery time how long it takes for the heartbeat to get back to normal after exercise

red cells disc-shaped cells in the blood that carry oxygen around the body

roughage food that can't be digested (usually plant cell walls); roughage or fibre is important for bowel movements

saliva a slippery liquid made in the salivary glands and released into the mouth; it contains amylase

small intestine part of the digestive system where digestion is finished and digested food is absorbed into the blood

trachea the windpipe, which carries air from the mouth and nose to the lungs

vaccine dead bacteria or viruses that cause the body to make antibodies

valve strings strings that hold valve flaps in position in the heart

villi finger-like projections on the wall of the small intestine

viruses microorganisms that live inside the cells of other living things and cause disease

vitamins important substances in the diet, needed for chemical reactions in the body

white cells colourless cells in the blood that protect the body against disease

Maintenance of life

Like animals, green plants are made of cells, but plant cells are not quite the same as animal cells. Green plants are very special living things. They have one big advantage over animals – they can make their own food by photosynthesis. Plants get the raw materials for photosynthesis from the air and the soil. Leaves take in carbon dioxide and roots take in water and minerals. Inside plants, materials are transported to where they are needed.

Plants need to be able to respond to light, water, and gravity so that their roots and shoots can grow in the right direction to get the things the plant needs.

Humans have a variety of sense organs and can respond to changes in their surroundings in lots of different ways. Information gets around the body in two ways – by nerves and by hormones. The nervous system is very fast, helping us to respond quickly. The hormone system is much slower. The two systems interact to keep the body aware of what is going on around it and inside it.

To survive, our bodies must be kept at the right temperature and contain the right amounts of water and sugar. Our bodies have automatic systems which control these things.

Module 2

2.01 Plant cells

Objectives

This spread should help you to

- describe the structure of a plant cell
- link the shape of a plant cell with its job
- describe how plant cells and animal cells are the same and how they are different
- know how microscopic things are measured

Some simple organisms are made of only one cell. These are called **unicellular**. Plants and animals are **multicellular**. This means they are made up of lots of cells.

Plant and animal cells carry out similar jobs. They take in food, release energy, get rid of waste, grow, and reproduce. However, plant and animal cells are not the same.

All plant cells have these parts.

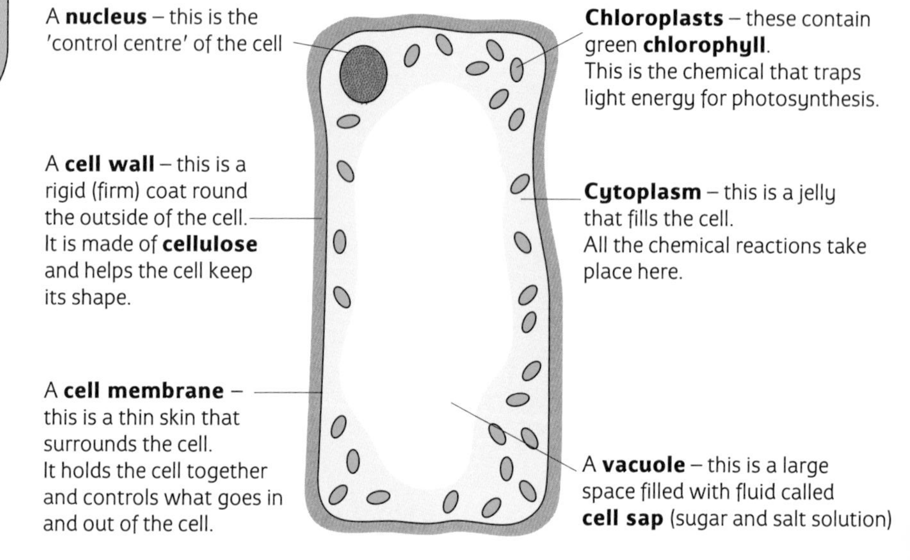

There are many different kinds of plant cell. Each one is suited to its job.

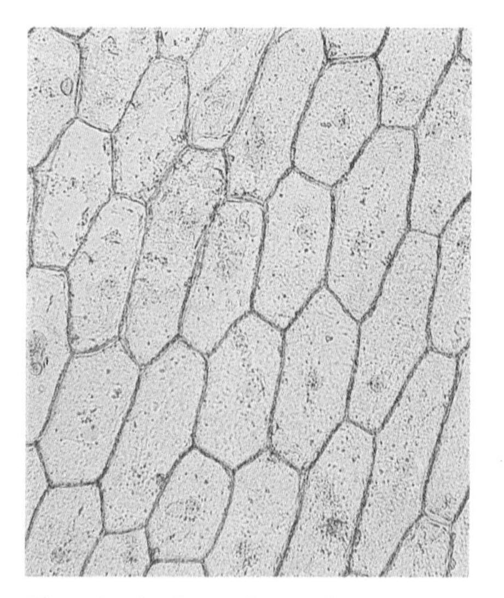

These leaf cells are firmer than animal cells. This is because of their cell walls.

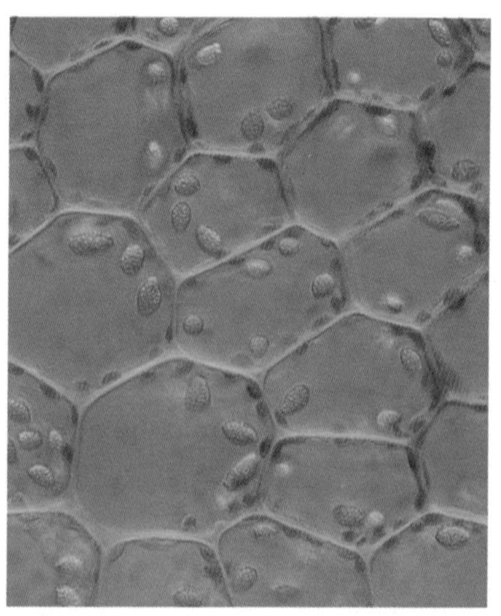

Leaf cells also have lots of chloroplasts inside them for photosynthesis.

Root hair cells increase the surface area of roots to help them take up water from the soil.

If you compare plant cells with animal cells, you will see that in some ways they are the same and in other ways they are different.

How animal and plant cells are the same

- Both have a nucleus.
- Both have a cell membrane.
- Both have cytoplasm.

How animal and plant cells are different

- Plant cells have cellulose cell walls. Animal cells don't.
- Plant cells have chloroplasts. Animal cells don't.
- Plant cells always have vacuoles. Animal cells sometimes have a vacuole.
- Plant cells have few different shapes. Animal cells have lots of different shapes.

How big are cells?

Cells are microscopic things. This means that you need a microscope to see them. The sizes of microscopic things are measure in **micrometres**, written **μm** for short. One μm is $\frac{1}{1000}$ of a millimetre.

This diagram shows the sizes of some microscopic things in μm.

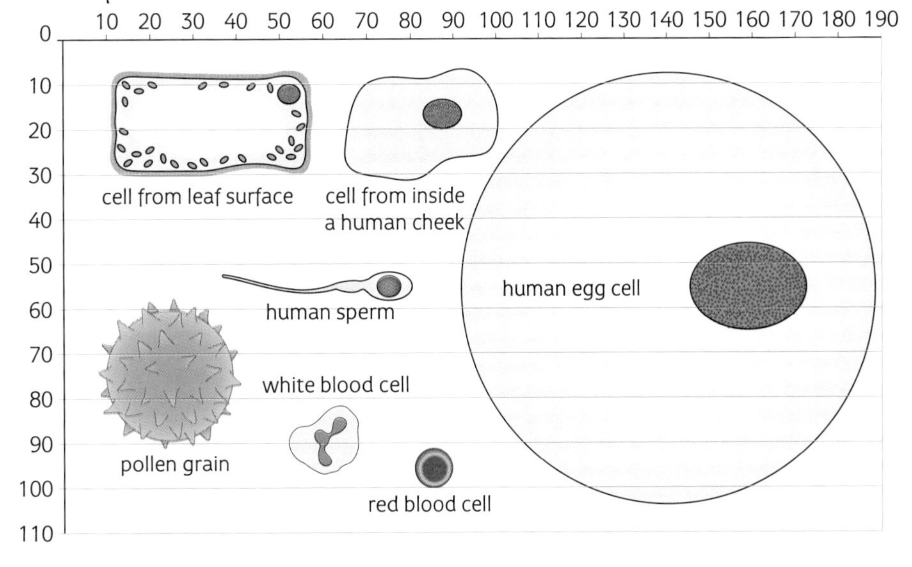

Questions

1 What is a unicellular organism?

2 **a** What are chloroplasts? **b** What do they do?

3 Explain the difference between a cell membrane and a cell wall.

4 Explain how root hair cells help a plant.

5 List four things found in both animal and plant cells.

6 Why are some things described as microscopic?

7 **a** What is a micrometre?

b What is the symbol for a micrometre?

8 What is the diameter of:

a a cell from inside a human cheek

b a pollen grain **c** a human egg cell?

9 How long is:

a a cell from a leaf surface **b** a human sperm?

2.02 Plant tissues and organs

Objectives

This spread should help you to

- describe how a plant is built

Plant tissues...

Plants are multicellular, and are made up of different types of cell. Each type of cell does a different job. Cells that do the same job group together to make a **tissue**.

Cells that contain chloroplasts, for example, make photosynthetic tissue. Its job is to carry out photosynthesis.

Leaves have lots of photosynthetic tissue.

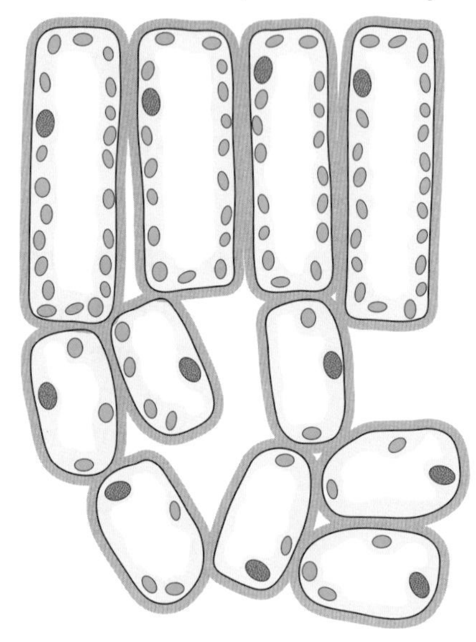

Other examples of plant tissues include...

...protective tissue which covers the surface of roots, stems, and leaves. It is a thin, transparent skin that protects the plant from damage or infection.

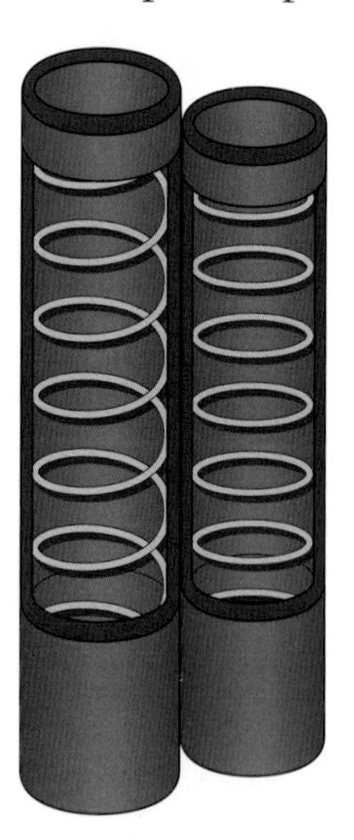

*... **xylem** tissue which carries water and minerals from the roots to the leaves. The spiral patterns are caused by thickening which gives strength.*

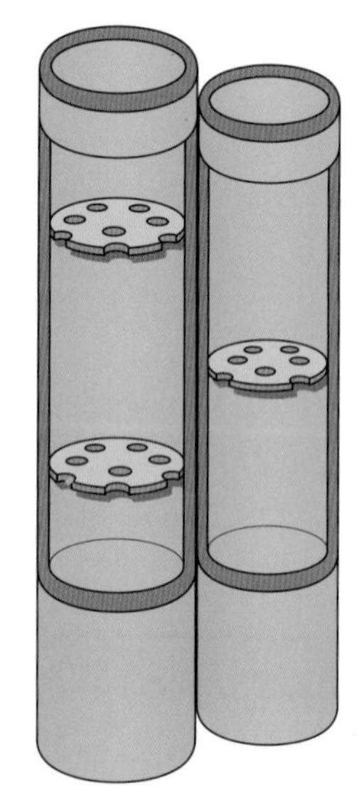

*... **phloem** tissue which carries food from the leaves to other parts of the plant. There are holes in the ends of each cell to let things pass from one cell to the next.*

...make up organs...

Different tissues combine to make an **organ**. A **leaf** is a plant organ. Its job is to make food.

A leaf is made from these tissues.

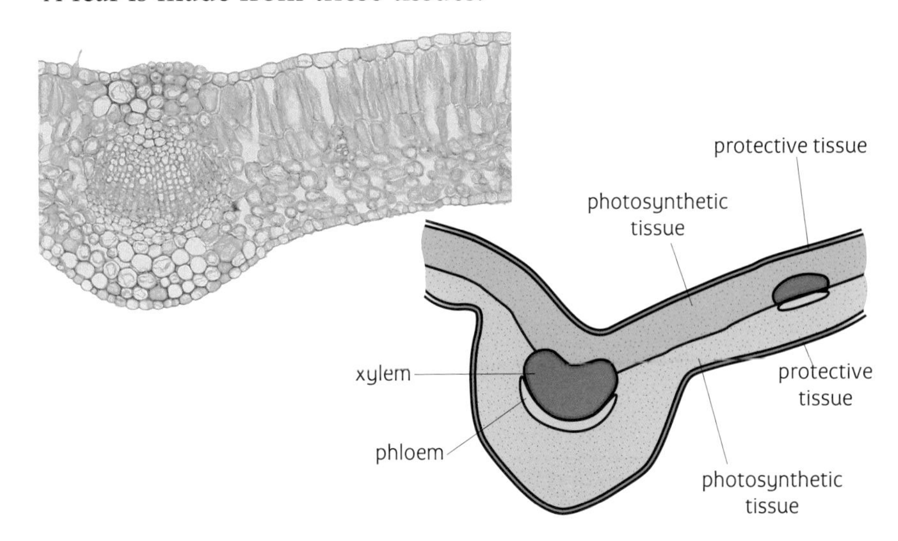

Other plant organs include...

*... **roots**. Their job is to hold the plant in the ground. They also take in water and minerals from the soil.*

*... **stem**. The stem keeps the plant upright and holds leaves so they can get sunlight.*

*... male and female sex organs. These are in the **flowers**. Flowers are usually brightly coloured and smell nice to attract insects.*

...which make up an organism

Organs together make a plant, which is an **organism**. An organism is something that is able to survive on its own.

Questions

1. What do we call a group of cells that do the same job?
2. Name four kinds of plant tissue.
3. How are xylem tissue and phloem tissue:
   a the same **b** different?
4. What is an organ?
5. Name three plant organs and say what job each one does.
6. What is an organism?
7. Explain why a bunch of flowers isn't a bunch of organisms.

2.03 Roots and stems

Objectives

This spread should help you to

- describe the structure of a root and a stem
- describe the job of each part of a root and a stem

Transport in plants

Roots and stems are plant organs which contain the **transport system** of a plant. They have thin tubes inside them which carry liquids up and down the plant. There are two kinds of tube – xylem and phloem. Xylem tubes carry water and minerals from the soil up to the parts of the plant above the ground. They have thick, strong walls to help support the plant. Phloem tubes carry sugar solution from the leaves to every other part of the plant. Xylem and phloem tubes are bundled together in **vascular bundles**.

Inside a root

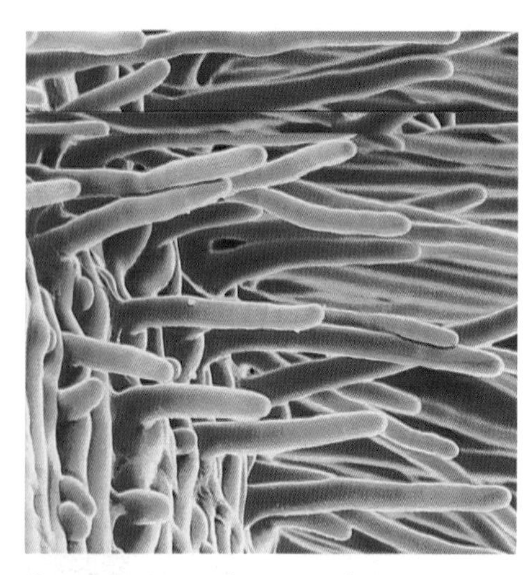

Root hairs give a bigger surface area to absorb water.

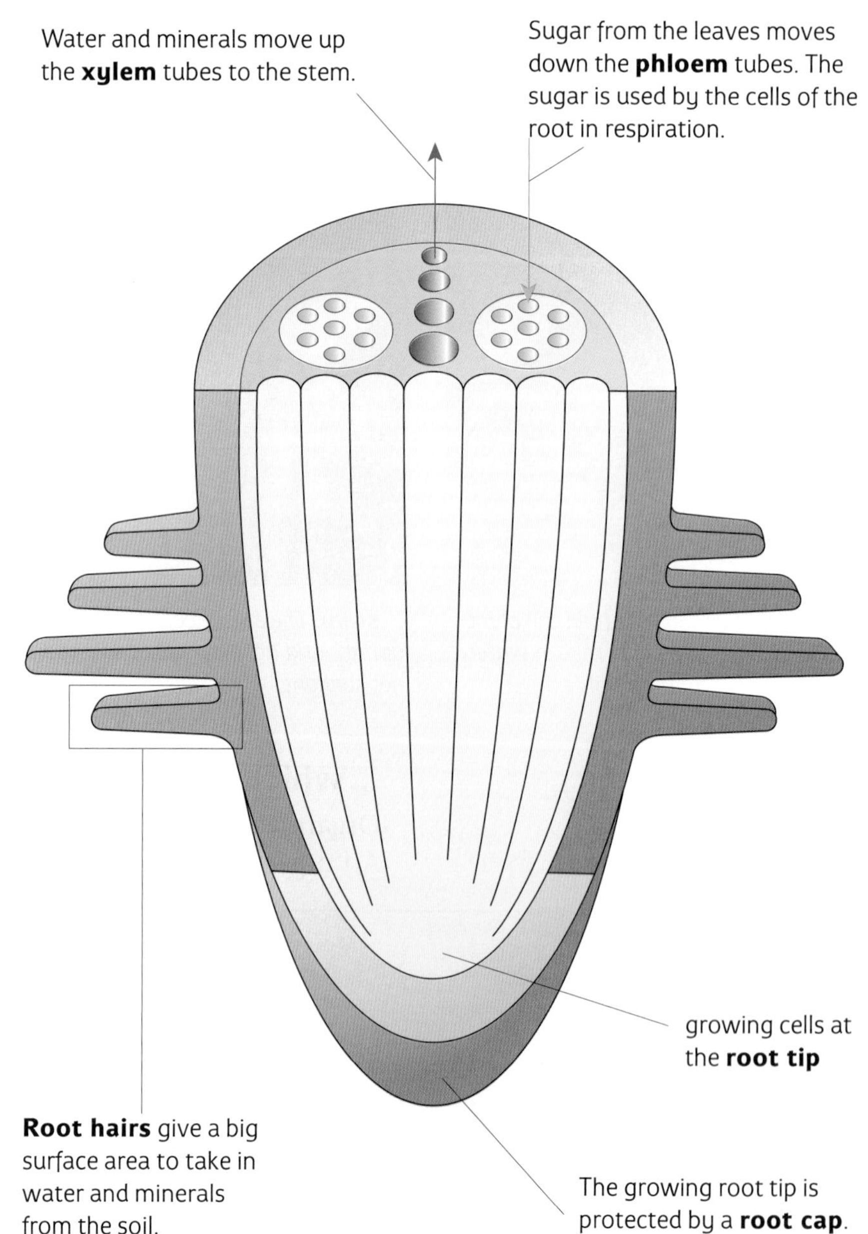

Questions

1. What is a transport system?
2. **a** Which tubes in a plant carry water and minerals?
   b Where do they carry them from?
   c Where do they carry them to?

Inside a stem

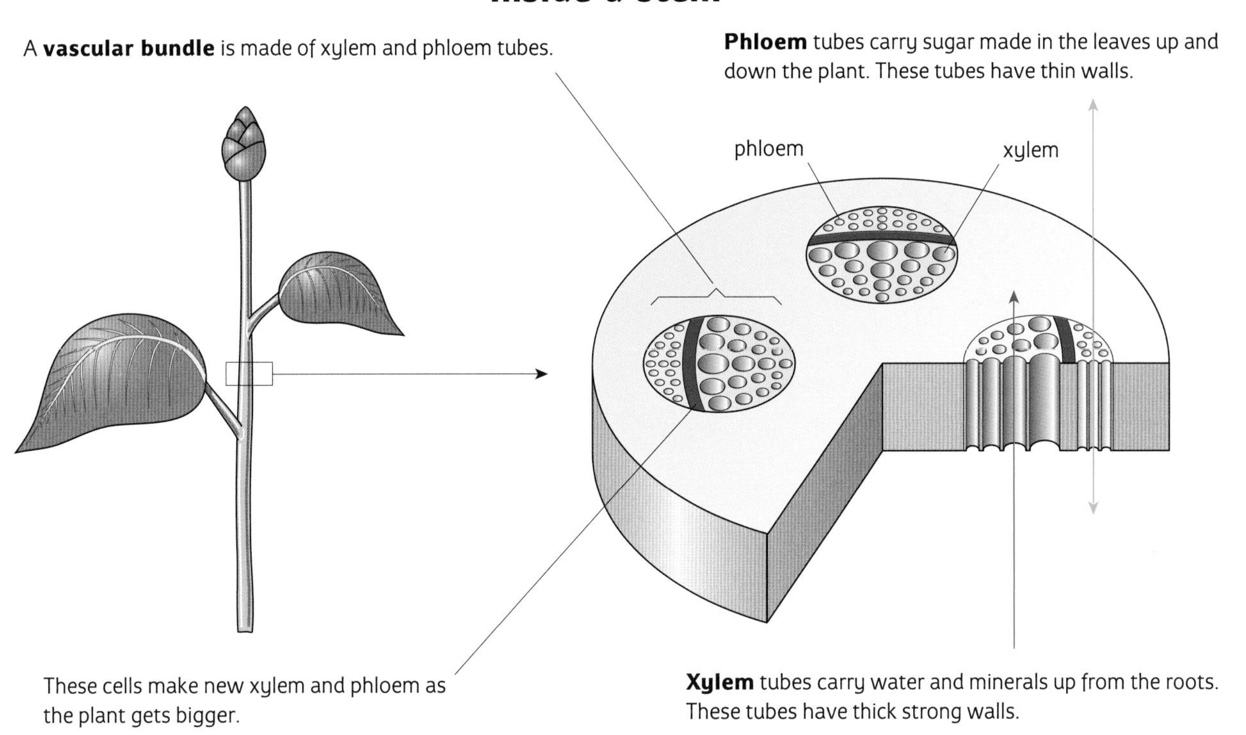

Trees have lots of xylem – the trunk is made of it. Xylem is wood!

Questions

3 **a** Which tubes in a plant carry sugar?
  b Where do they carry it from?
  c Where do they carry it to?

4 What is a vascular bundle?

5 Explain how xylem tubes help support a plant.

6 How do root hairs help a root do its job?

7 What happens at the root tip?

8 What is the job of the root cap?

2.04 Leaves

Objectives

This spread should help you to

- describe the structure of a leaf
- describe how a leaf is suited to its job

Leaves are the 'food factories' of plants. They are where photosynthesis happens.

Leaves are well suited to their job

- Leaves are broad and flat to absorb lots of light.
- They are thin so gases can get to the cells easily.
- They have holes called **stomata** underneath to let gases in and out.
- They have lots of **veins** to carry water to the photosynthesizing cells, and carry sugar away.

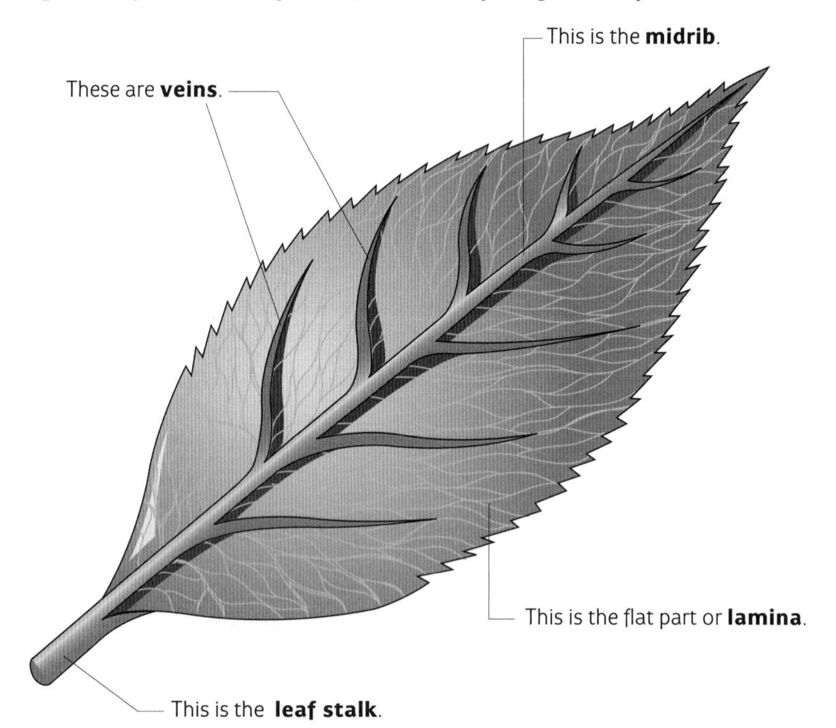

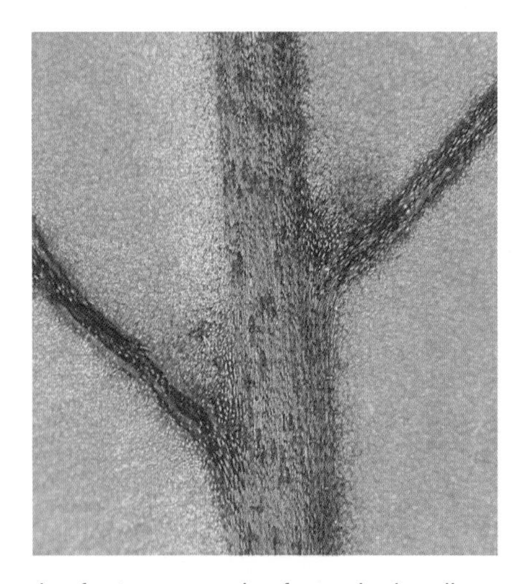

Leaf veins are made of vascular bundles. Xylem tubes carry water and minerals into the leaf from the stem. Phloem tubes carry sugar to all parts of the plant.

Rhubarb leaves are very broad to absorb lots of sunlight.

Grass leaves are narrow but very thin so gases can get to every cell.

Inside a leaf

Holly leaves have a thick, waxy cuticle.

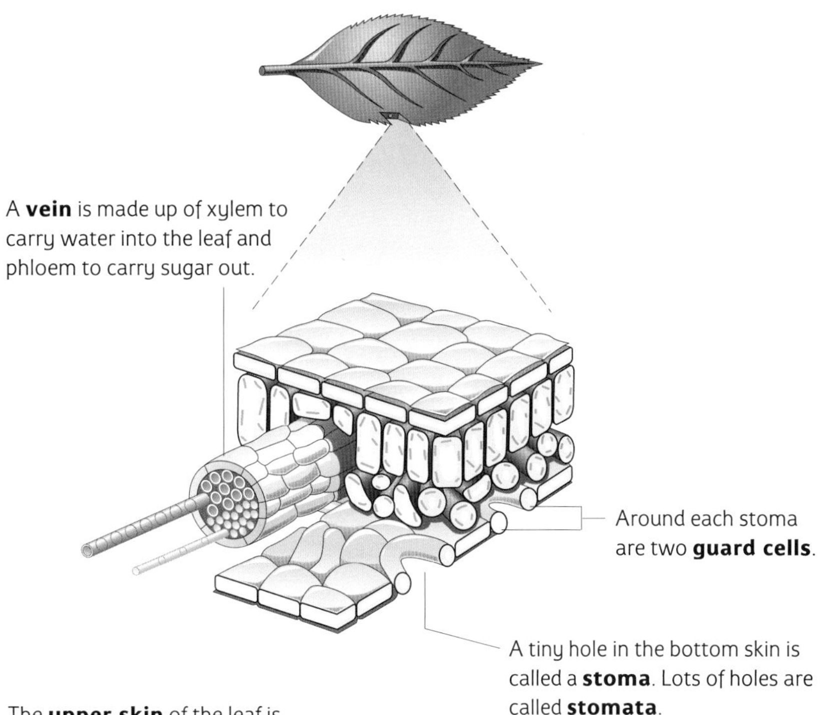

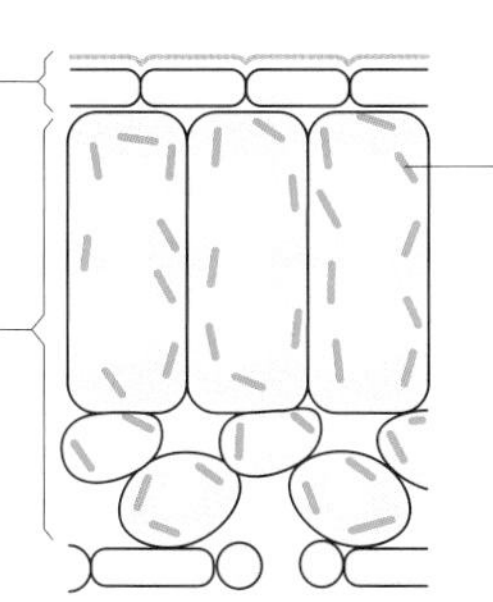

Questions

1 What is the job of leaves?

2 Why are leaves able to:

- **a** absorb lots of light
- **b** get gases to the cells easily
- **c** get gases in and out
- **d** get water in and sugar out?

3 **a** What are leaf veins?

- **b** What are they for?

4 Explain why narrow leaves can still do their job well.

5 What makes a leaf waterproof?

6 Explain why the upper skin of a leaf is transparent.

7 Why do the cells in the middle of a leaf have lots of chloroplasts?

8 **a** What is a stoma?

- **b** What surrounds each stoma?

2.05 Photosynthesis

Objectives

This spread should help you to

- describe what happens in photosynthesis
- describe how to test a leaf for starch

Making food – photosynthesis

Plants make food by **photosynthesis**. They use light energy to make food from carbon dioxide and water.

sunlight energy

Carbon dioxide gets into the leaf through the stomata by diffusion. Water is carried to the leaf in the xylem tubes from the stem.

oxygen

carbon dioxide

Carbon dioxide and water go into the chloroplasts where photosynthesis happens. In the chloroplasts light energy is absorbed by chlorophyll.

glucose converted to **starch** for storage

water

The light energy makes the carbon dioxide and water join together to make glucose and oxygen. Glucose is usually changed to starch for storage, then later used for food. Oxygen diffuses out into the air through the stomata. It is this oxygen that makes human life on Earth possible.

We can represent what happens during photosynthesis by this word equation:

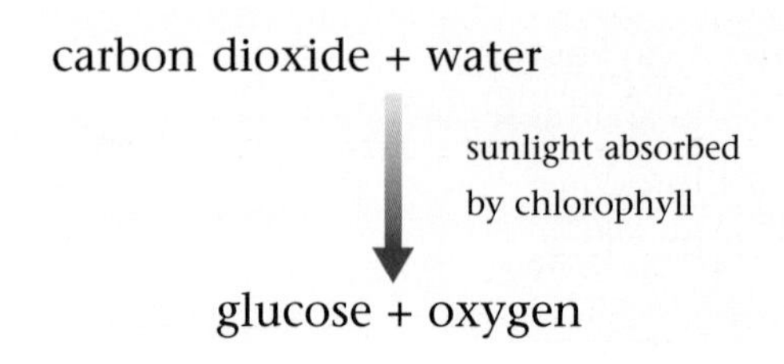

Questions

1 **a** Where does the energy for photosynthesis come from?

b What green chemical absorbs this energy?

2 **a** What two simple chemicals are needed for photosynthesis?

b How do these chemicals get into the leaf?

3 What large molecule is made during photosynthesis?

4 What happens to the oxygen produced during photosynthesis?

Testing a leaf for starch

An easy way of finding out whether a plant is photosynthesizing is to see if its leaves have made any starch or not.

Remember to wear safety glasses.

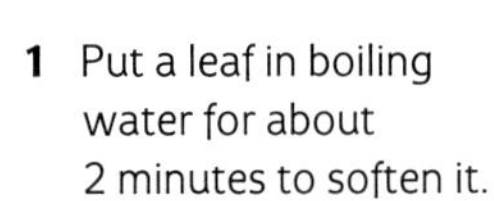

1 Put a leaf in boiling water for about 2 minutes to soften it.

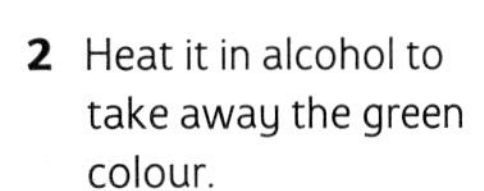

2 Heat it in alcohol to take away the green colour.

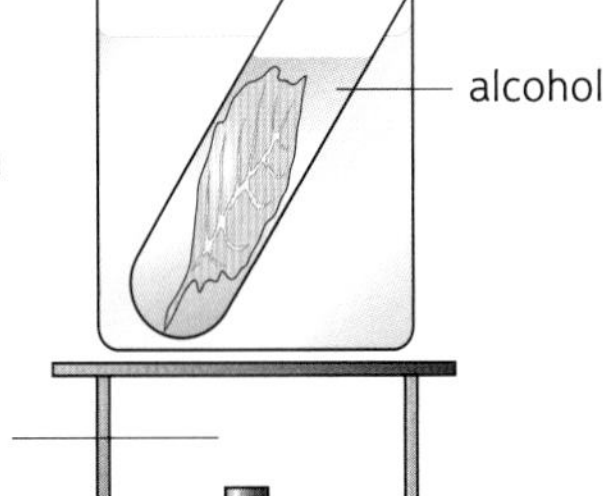

The flame must be put out. Alcohol burns!

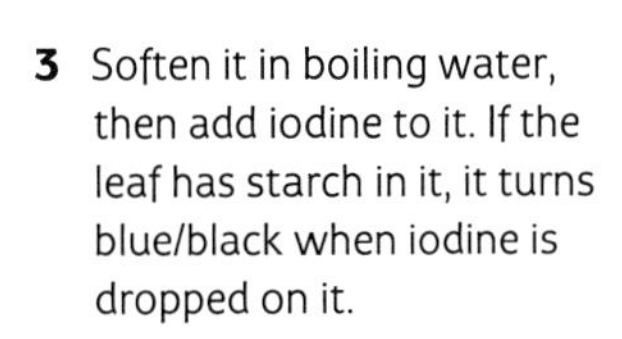

3 Soften it in boiling water, then add iodine to it. If the leaf has starch in it, it turns blue/black when iodine is dropped on it.

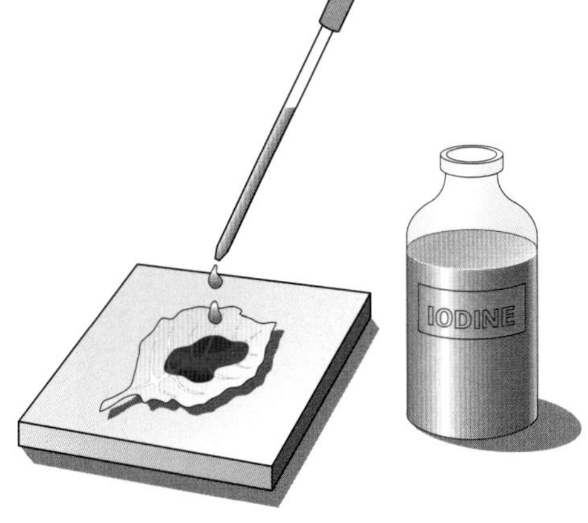

Starch reacts with **iodine**. Iodine turns from brown to blue/black when added to starch. If a leaf contains starch, it will turn blue/black when iodine is dropped on it. This shows that the leaf has been photosynthesizing. If a leaf hasn't been photosynthesizing, it won't turn blue/black.

Questions

5 Explain how the starch test can show if a plant has been photosynthesizing.

6 **a** What colour is iodine?

b What colour does iodine go when mixed with starch?

7 During the starch test:

a Why is the leaf boiled in water for a while?

b What is the job of the alcohol?

c Why is the leaf dipped in the water again before adding the iodine?

8 **a** Describe one important safety precaution that must be taken when doing the starch test.

b Explain why this is important.

2.06 The rate of photosynthesis

Objectives

This spread should help you to

- describe three things that can affect the rate of photosynthesis
- explain what limiting factors are

Photosynthesis doesn't happen at the same rate all the time. Three things, or factors, that can change the rate of photosynthesis in a plant are described below.

1 How much light there is: Chlorophyll uses light energy for photosynthesis. The more light energy there is, the faster photosynthesis will be. The colour of light is important in photosynthesis. Plants use red and blue light most. Green light isn't used at all, it is reflected – that's why leaves are green!

2 How much carbon dioxide there is: Carbon dioxide and water are the raw materials for photosynthesis. Without them, a plant won't be able to make any food. There is usually lots of water available. But there isn't much carbon dioxide in the air (about 0.03%). More carbon dioxide means photosynthesis can speed up and a plant will make more food.

These plants are given red light.

Carbon dioxide can be added to the air in greenhouses.

Plants make lots of food on warm, sunny days.

3 The temperature: Photosynthesis is a chemical reaction. It is controlled by enzymes. Most enzymes work best when it is warm. This is why the rate of photosynthesis is higher on warm days than cold days. There is a problem though. If it gets too hot the enzymes are destroyed or **denatured**. When this happens photosynthesis stops.

Limiting factors

If one of the factors isn't right for some reason, photosynthesis will slow down. The factor that isn't right is called the **limiting factor**.

For a plant to have the best rate of photosynthesis it needs:

- enough carbon dioxide
- enough light
- a perfect temperature.

Questions

1. Name three things that affect the rate of photosynthesis.
2. Explain why leaves are green.
3. Describe the best conditions for photosynthesis.
4. Explain why photosynthesis will stop if a plant gets too hot.

Light is a limiting factor in a wood. Ivy grows up tree trunks to get to the light.

If the amount of light goes up, the rate of photosynthesis will also go up, but only up to a certain point. Beyond this point, the amount of light makes no difference. This is because there isn't enough carbon dioxide or the temperature is too low. In this case carbon dioxide and temperature are limiting factors.

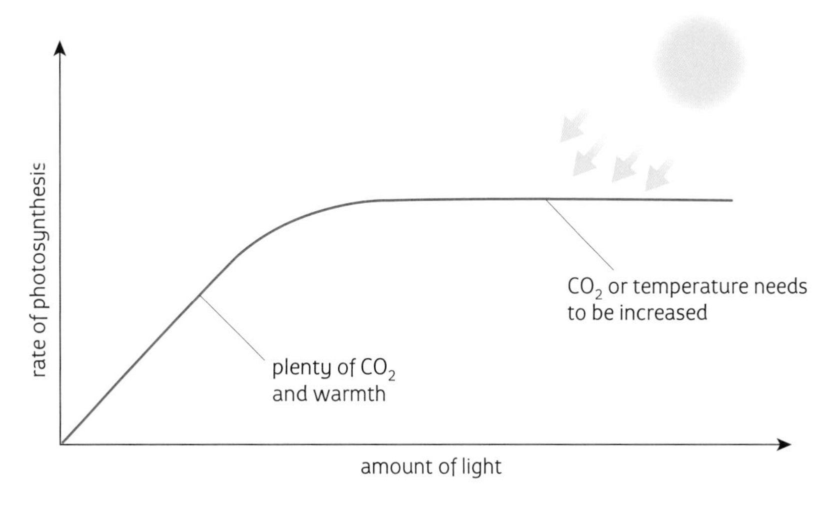

These bluebells flower early before the leaves grow above them and cut out the light.

If it is warm and there is plenty of carbon dioxide, photosynthesis will only go as fast as the amount of light will allow. Light is the limiting factor now.

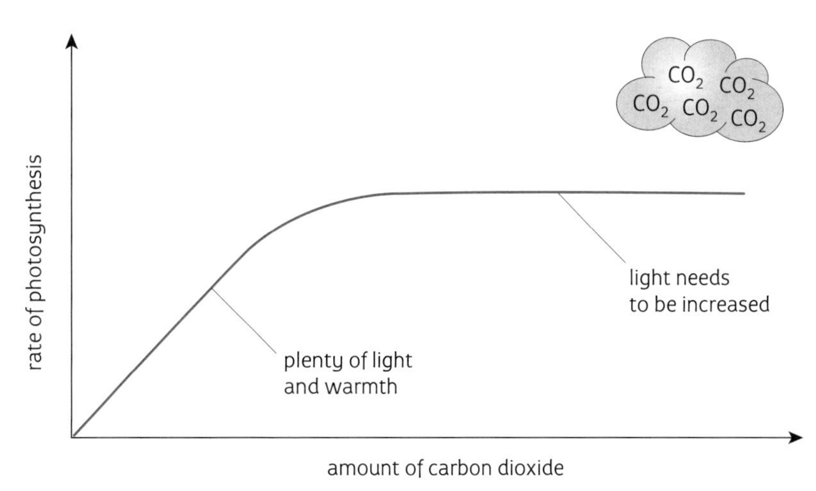

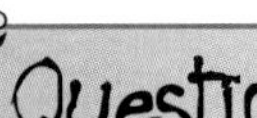

Questions

5 Explain what is meant by 'limiting factor'.

6 A plant has plenty of carbon dioxide and light but it is a cold day. What is the limiting factor in this example?

7 Light is often a limiting factor for woodland plants. Describe one way that a woodland plant might cope with this problem.

8 Most plant enzymes are destroyed at about 45°C. Draw a graph showing how the rate of photosynthesis is affected by temperature. (Put temperature along the bottom of the graph and rate of photosynthesis up the side.)

2.07 More about photosynthesis

Objectives

This spread should help you to
- describe what happens to the glucose made by photosynthesis
- describe the link between photosynthesis and respiration

What happens to the glucose?

Glucose is the 'food' made in photosynthesis. A plant can do lots of things with glucose.

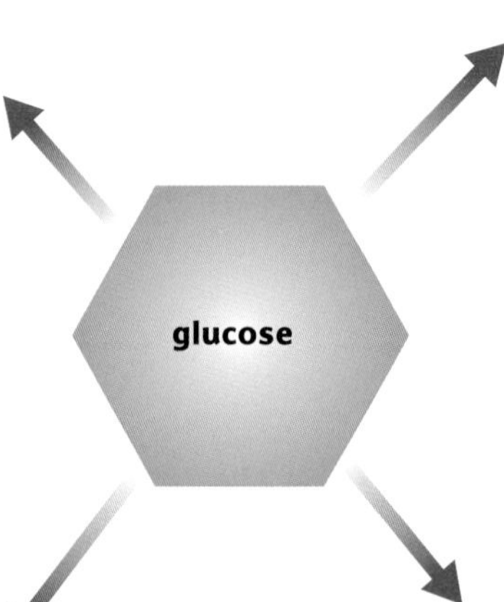

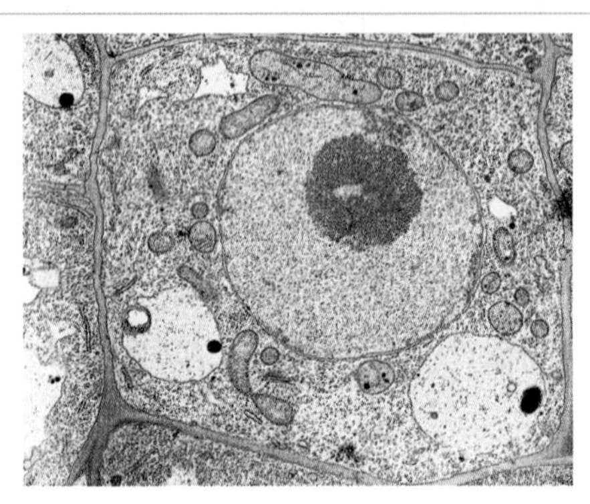

Some of it will be used straight away to produce energy. Respiration happens inside every living plant cell all the time – even at night.

Some glucose will be stored. First the glucose makes a strong sugar solution with water. It is then transported to where it will be stored. Finally the sugar is changed into starch or oil and stored in roots, stems, seeds, and fruits. When it is needed, it is changed back into sugar.

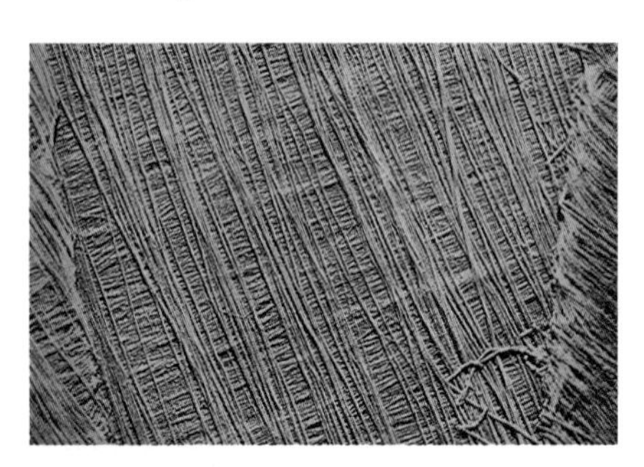

Some glucose is changed into cellulose to make new cell walls. Cellulose gives a strong outer coat to plant cells. It helps a plant support itself.

Some is joined up with minerals from the soil. Nitrogen is joined up with glucose to make protein. Plants need protein for growth. To make chlorophyll, magnesium is joined up with glucose. These minerals come from the soil and are taken in with water by the roots.

Photosynthesis and respiration

Remember the word equations for photosynthesis and respiration?

Photosynthesis:

carbon dioxide + water ⟶ glucose + oxygen (ENERGY needed)

Respiration:

glucose + oxygen ⟶ carbon dioxide + water (ENERGY released)

The sea is full of blue–green algae. These tiny plants help keep the balance of gases in the air.

Notice that the respiration equation is the opposite of the photosynthesis equation. Between them, respiration and photosynthesis keep the levels of oxygen and carbon dioxide in the air steady.

All living things respire so respiration goes on all the time, day and night. However, because plants need light for photosynthesis, photosynthesis only happens in daylight. So, to keep the balance, there must be more photosynthesis than respiration during the day. At night there is more respiration than photosynthesis. The graph shows how this works.

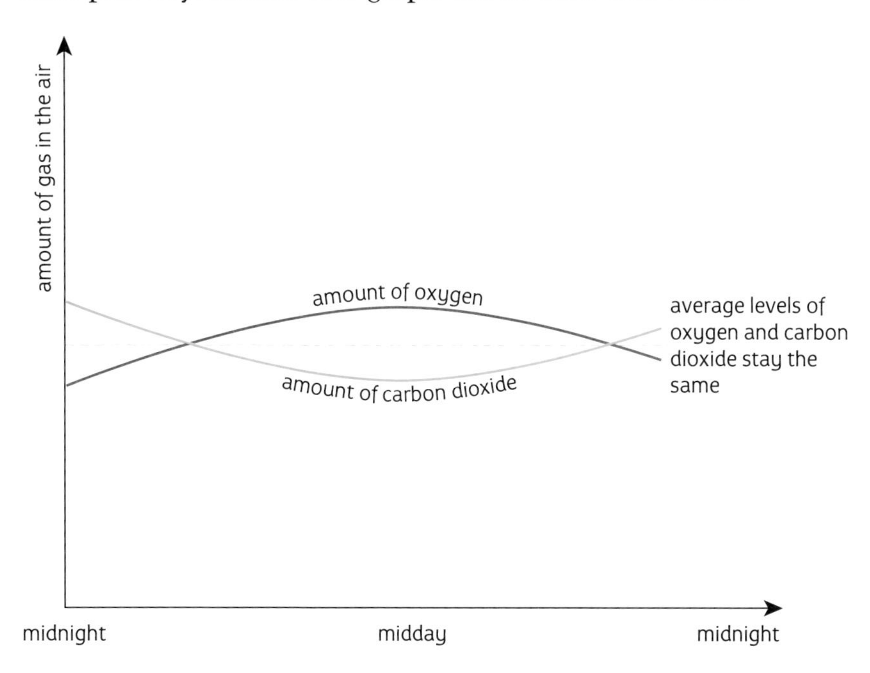

Questions

1 **a** Where does respiration happen?
  b When does respiration happen?

2 Explain how a plant stores glucose.

3 What does a plant do with cellulose?

4 Explain how a plant makes:
  a protein **b** chlorophyll.

5 Why does a plant need:
  a protein **b** chlorophyll?

6 Write the word equation for:
  a photosynthesis **b** respiration.

7 Explain why photosynthesis and respiration are not exactly opposite.

8 An aquarium contains fish and green plants.
  a What gas is given out when the fish respire?
  b What gas is taken in when the green plants photosynthesize?
  c What gas is given out when the green plants photosynthesize?
  d What gas is taken in when the fish respire?
  e The gases in the water are in balance during the day. Explain what happens at night.

9 Oil pollution in the sea kills green algae. Explain how oil pollution can affect the balance of gases in the air.

2.08 Transporting food

Objectives

This spread should help you to

- describe how sugar is transported in a plant
- describe what phloem tubes are like

Making and moving

Sugar is made in the leaves by photosynthesis. It mixes with water to make a strong sugar solution. Then it is transported in the phloem tubes out of the leaves and into the stem. From there it may go up to growing shoots, or down to the roots. The speed and direction of the sugar solution depends on how much is needed by the parts of the plant and when. In spring, when new shoots are growing, most of the sugar solution goes up to growing shoots. In the autumn, a lot of sugar is stored so it goes up to fruits and seeds or down to the roots.

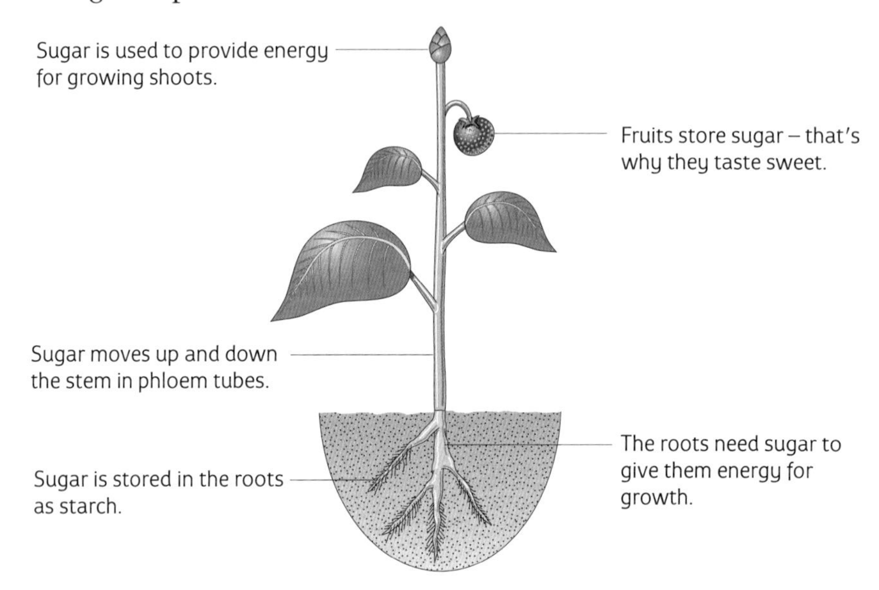

Did you know?

Phloem tubes are alive.
Xylem is dead.

Phloem tubes

Phloem tubes are made from living cells joined end to end. There are holes in the ends of each cell so sugar solution can get through. Sugar solution can go both up and down a phloem tube at the same time.

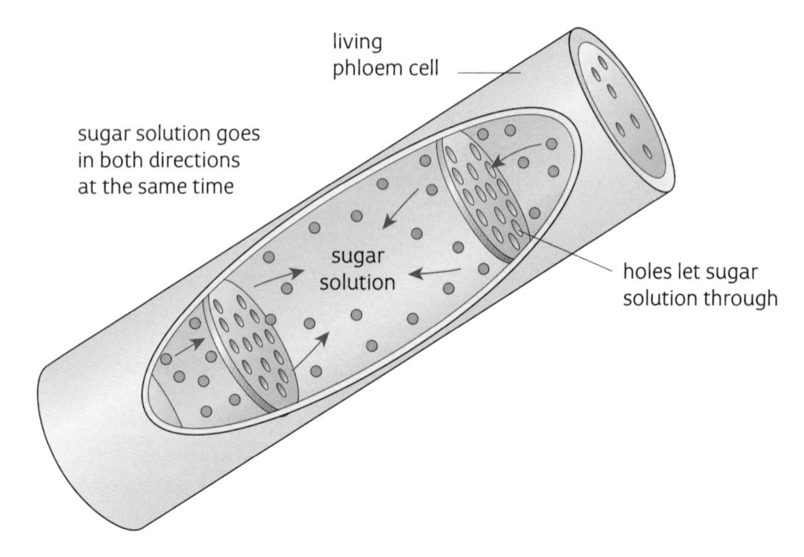

Questions

1. Why do growing shoots need sugar?
2. Why do fruits taste sweet?
3. What are the tubes called that transport sugar?

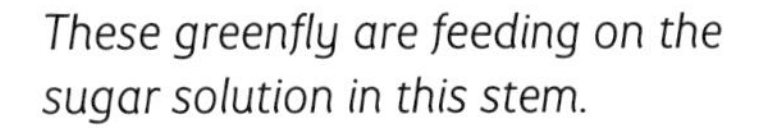

These greenfly are feeding on the sugar solution in this stem.

Phloem tubes are underneath the bark of a tree. Sticky 'sap' is sugar solution.

How sugar flows through a plant

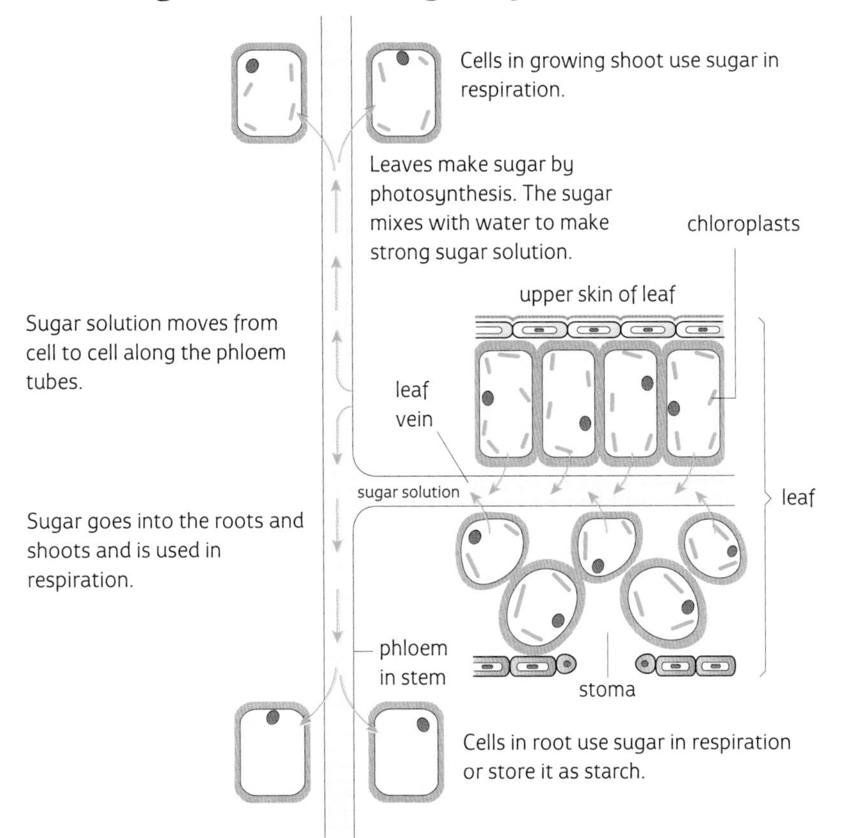

Questions

4 Why are there holes at the ends of phloem cells?

5 Explain why the speed and direction of sugar solution moving in the phloem changes.

6 Suggest and explain what happens to the movement of sugar in winter when a tree has no leaves.

7 In spring, sugar solution rises quickly up the phloem in plant stems. Give a reason for this.

2.09 Minerals from the soil

Objectives

This spread should help you to

- list some mineral elements used by plants
- describe what use plants make of some mineral elements
- describe how plants get mineral elements from the soil

Plants need minerals

Plants need more than carbon dioxide and water to survive. They need nitrogen and other **mineral elements** for normal, healthy growth.

Some of the most important mineral elements are shown in the table below.

Mineral element	What it is used for
nitrogen	making leaves
potassium	making flowers and fruit
phosphorus	making roots
magnesium	making chlorophyll for photosynthesis

If a plant doesn't get enough of these minerals, it will look like this.

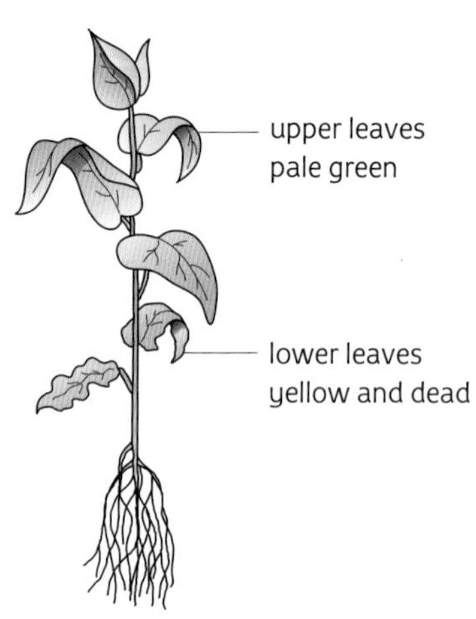

no nitrogen – yellow/green leaves, weak stem

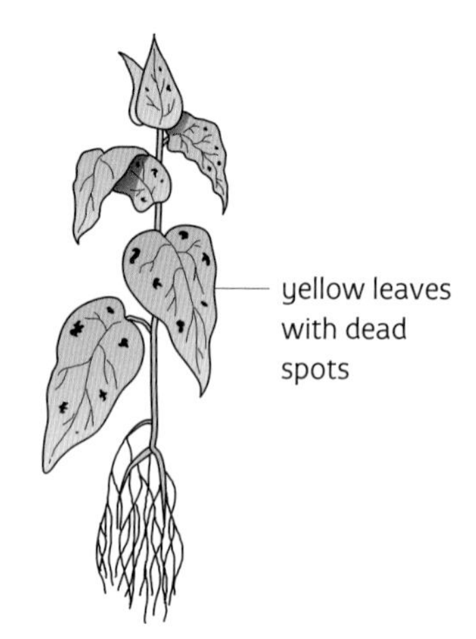

no potassium – poor flower and fruit growth

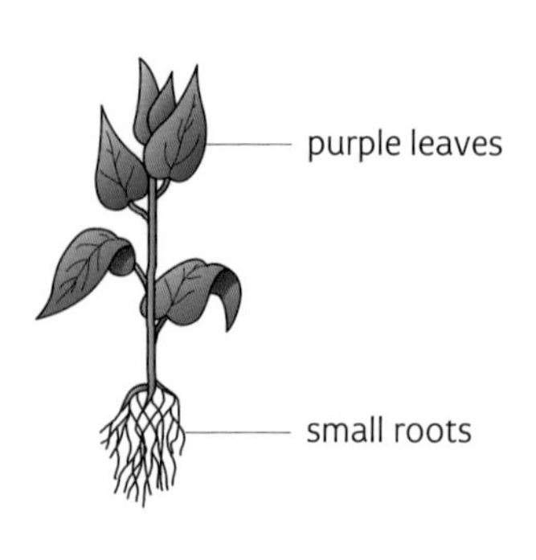

no phosphorus – poor root growth

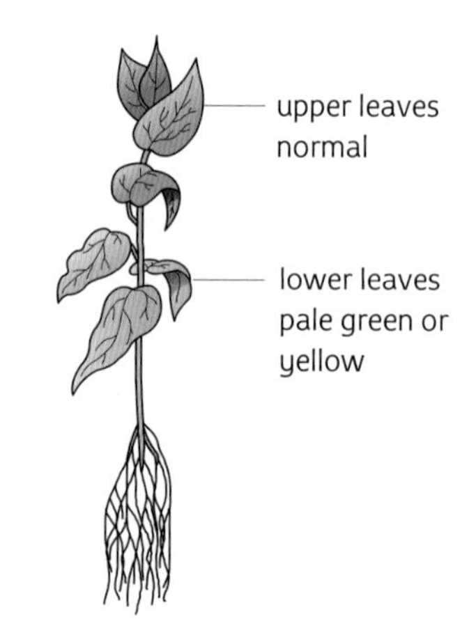

no magnesium – leaves turn yellow from bottom upwards

Growing the same crop in the same field every year removes the minerals needed by that plant.

Rainwater dissolves minerals from rocks.

Fertilizers contain mineral elements to improve plant growth.

Where do minerals come from?

Some minerals come from rocks. Rainwater is slightly acidic. This slowly dissolves some rocks, and minerals are washed into the soil. Other minerals come from the faeces of animals and the decay of dead plants and animals. Farmers add artificial **fertilizers** to the soil. These contain all the minerals that crop plants need for healthy growth.

How minerals get into plants

Mineral elements from the soil get into the roots of plants in solution. Some dissolve in the water in the soil and then move into the roots by diffusion. Most minerals however get into roots by active transport. Active transport uses energy to move minerals into the root cells. It means that a plant can get minerals into its root cells even when there is already more of the mineral inside the cells than outside.

Active transport works against diffusion. That is why energy is needed to move the minerals against a concentration gradient.

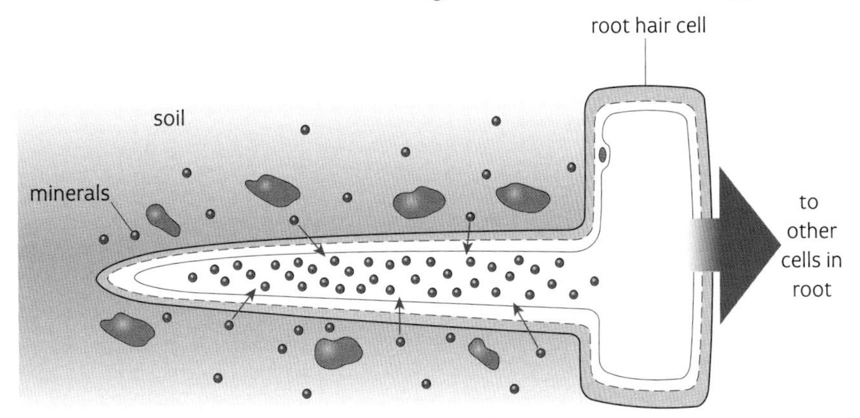

Questions

1 Why do plants need mineral elements?

2 Explain why a plant needs:
a nitrogen **b** potassium
c phosphorus **d** magnesium.

3 Describe the appearance of a plant that hasn't had enough:
a nitrogen **b** potassium
c phosphorus **d** magnesium.

4 Explain how minerals get into the soil from:
a rocks **b** dead animals and plants.

5 Describe two ways that minerals can get into roots.

6 What is the advantage of active transport?

7 Explain why active transport needs energy.

2.10 In and out of cells – osmosis

Objectives

This spread should help you to

- describe how and why osmosis happens
- describe how water enters a cell by osmosis

Moving water molecules

Osmosis is a special kind of diffusion. While diffusion can describe the movement of any kind of molecule, osmosis is only about the movement of water molecules.

Osmosis is the movement of water molecules from a region of high water concentration to a region of low water concentration across a **partially permeable membrane**. A partially permeable membrane has very tiny holes in it. Water molecules will fit through because they are very small. Bigger molecules, such as sugar molecules, are too big to get through the holes.

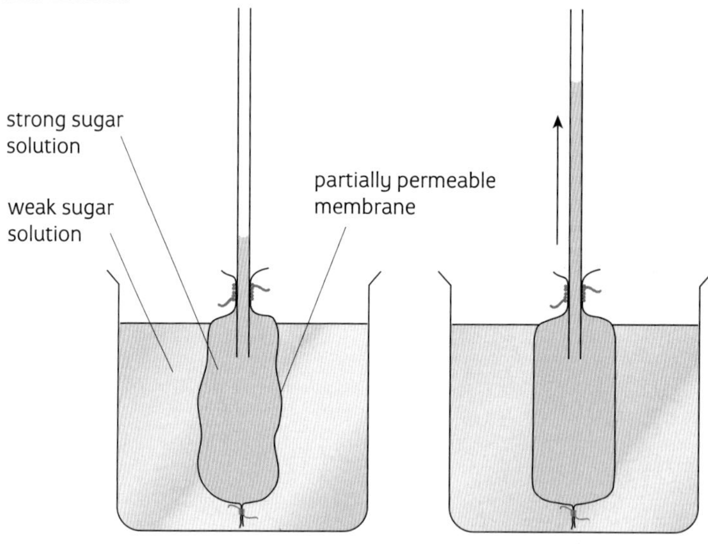

A bag made from partially permeable membrane is filled with strong sugar solution. The bag is tied to a piece of glass tubing and then put in a beaker of a weak sugar solution.

Soon the liquid begins to rise up the tube.

Water can pass both ways across the membrane. But because there are more water molecules in the weak sugar solution than in the strong sugar solution, water molecules will move mainly across the membrane into the strong sugar solution. This goes on until both sides have the same concentration.

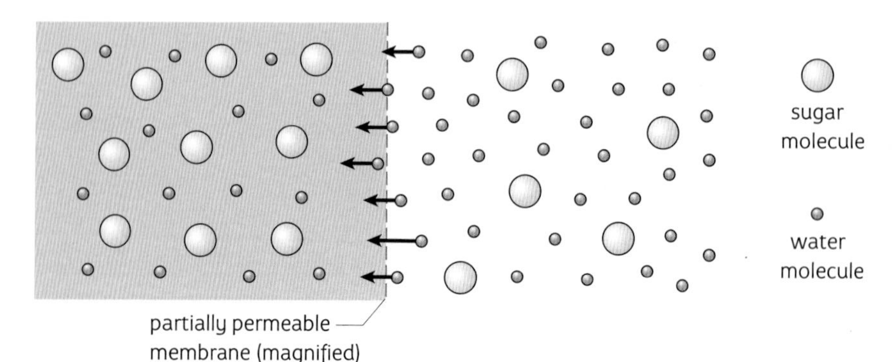

Questions

1. Why is osmosis a special kind of diffusion?
2. What is osmosis?
3. What is a partially permeable membrane?
4. Explain why the liquid rises up the tubing in the experiment shown on this page.

Osmosis in living cells

Living cells contain a solution of sugars and salts. The concentration of water inside cells is therefore fairly low. Cell membranes are partially permeable. When cells are put into water, water will enter the cells by osmosis.

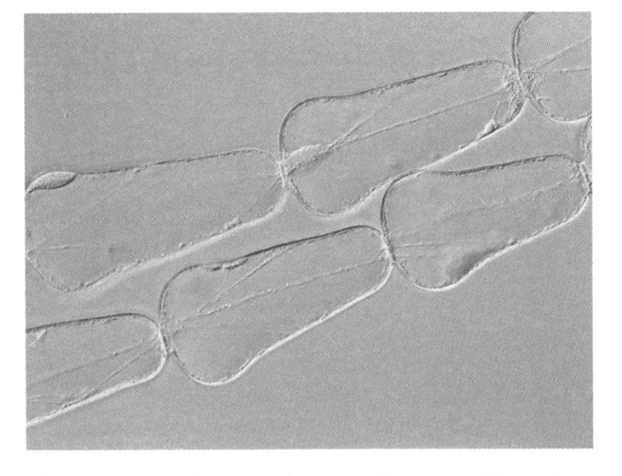

*Osmosis makes a plant cell expand so the contents push out against the cell wall. This makes the cell **turgid**, which is very good for giving the plant support.*

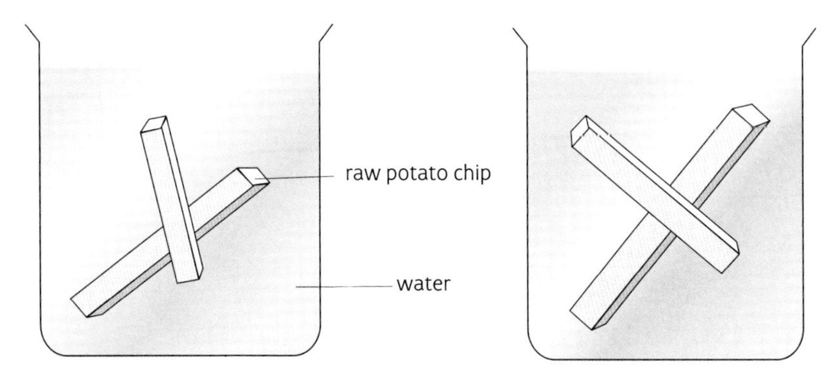

These freshly cut potato chips get longer if they are put into pure water. This is because water moves into each cell by osmosis and makes it bigger.

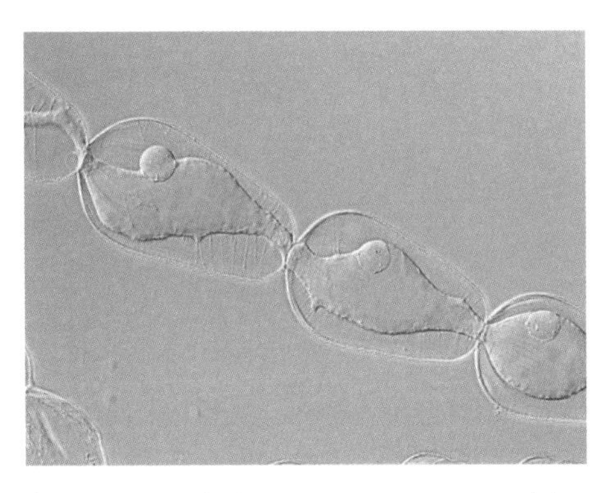

In a strong solution water moves out of the cells by osmosis. The cytoplasm shrinks and the cells can't support the plant.

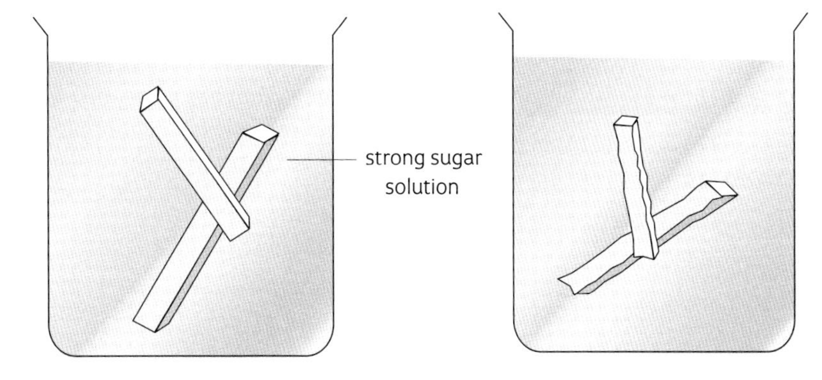

However, if the chips are put into very strong sugar solution, the opposite happens. This time there is more water inside the cells than outside, so water moves out by osmosis. The chips shrink.

Questions

5 Why is there a fairly low water concentration inside a plant cell?

6 Describe what happens to a plant cell when it is surrounded by pure water.

7 Describe what happens to a plant cell when it is surrounded by strong sugar solution.

8 Animal cells burst if they are put in water. Suggest a reason for this. (Hint: Look at the difference in the structure of plant and animal cells.)

2.11 Transpiration

Objectives

This spread should help you to

- explain what transpiration is
- describe the transpiration stream
- describe the things that affect the rate of transpiration

Transpiration is the process by which a plant loses water vapour from its leaves into the surrounding air. The water evaporates from cells inside the leaf. Water vapour then passes out of the leaf mainly through the tiny holes called stomata. Stomata are mainly found on the lower side of leaves.

Controlling transpiration

Each stoma is surrounded by two sausage-shaped guard cells. Guard cells control the size of the hole.

*Losing water faster than it can be taken in causes **wilting**.*

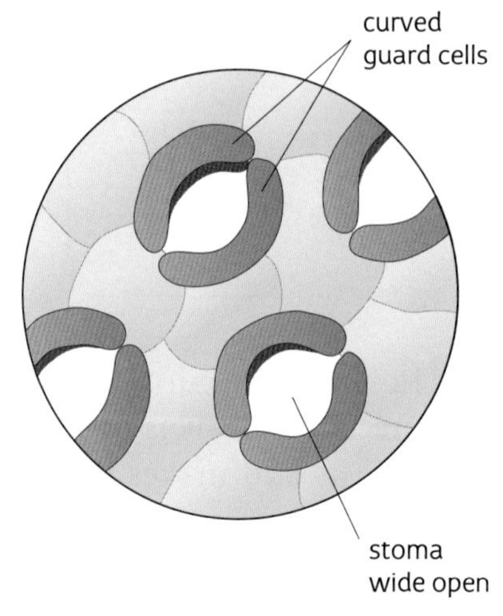

When a plant has plenty of water the pressure inside the guard cells rises and the stoma will open. Open stomata allow more transpiration.

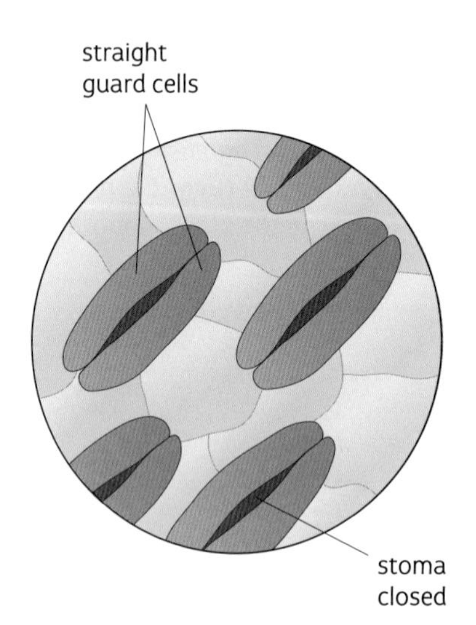

However, when guard cells lose water the pressure inside falls and the stoma will close. Closed stomata slow down transpiration.

What affects the speed of transpiration?

The speed, or rate, of transpiration can be affected by a number of things.

1 Temperature – on warm days more water will evaporate from the leaf cells, so transpiration will speed up.
2 Wind – on windy days water vapour will be blown away as it comes out through the stomata, so transpiration will speed up.
3 Humidity – on dry days the air can hold more water vapour, so transpiration will speed up.
4 Time of day – stomata are only open during the day in most plants so transpiration will be fastest then.

Questions

1 What is transpiration?

2 **a** What are stomata?

b Describe how guard cells change the shape of stomata.

3 List four things that affect the speed of transpiration.

4 Explain why a pot plant will soon start to wilt when it is put on a sunny window sill.

The transpiration stream

As a plant loses water by transpiration, more water is 'pulled' up the xylem tubes to replace it – just as a drink is pulled up a drinking straw when you suck on it. At the same time water enters the xylem from the roots. Roots are covered in root hairs – tiny extensions of the outer cells of the root. Root hairs increase the surface area of the roots so that more water can be taken in. This flow of water in the xylem from roots to leaves is called the **transpiration stream**.

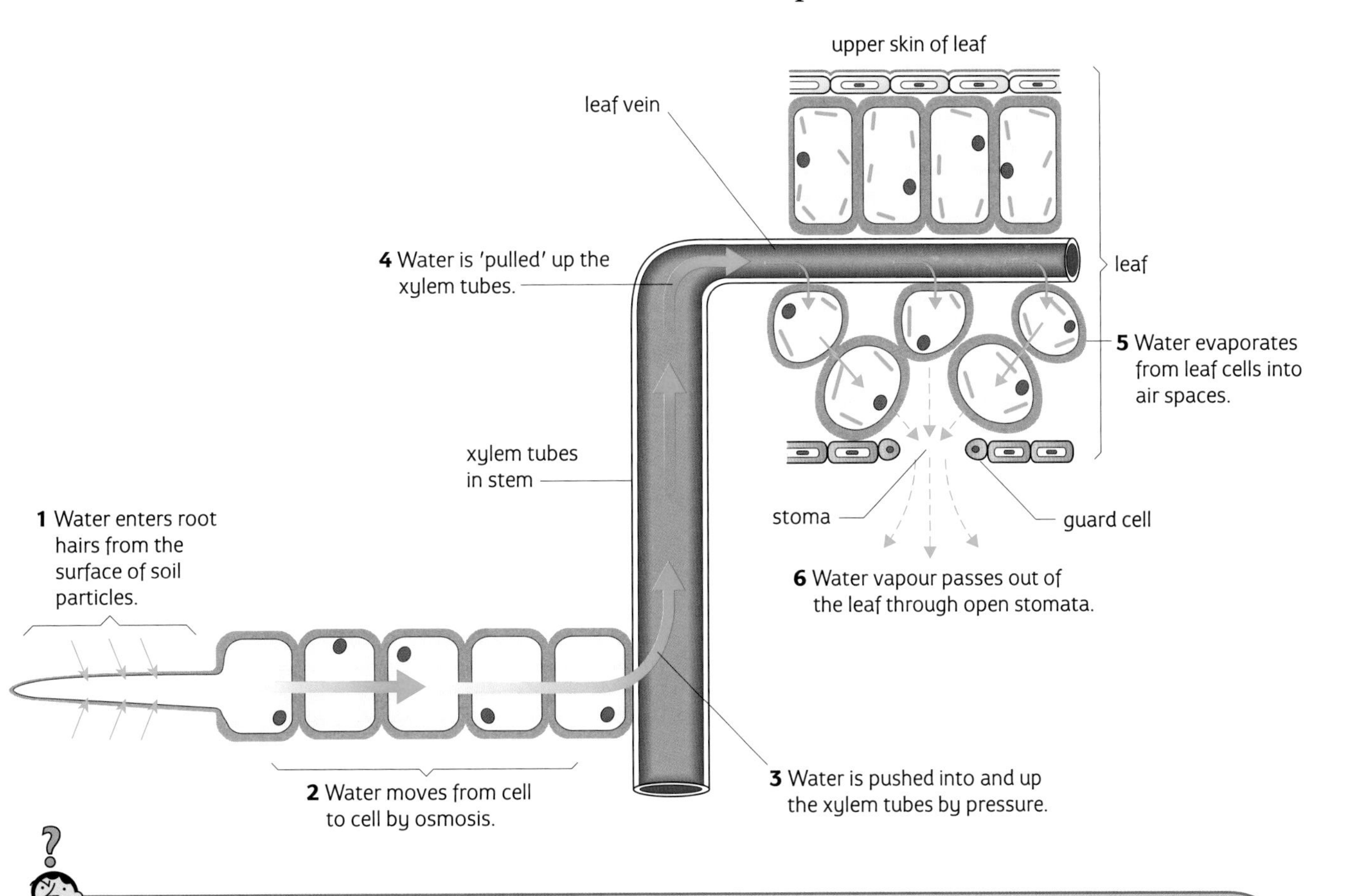

Questions

5 The table opposite shows the transpiration rate of a plant in different weather conditions.

What sort of day will cause:

a the fastest

b the slowest rate of transpiration?

6 Describe the path taken by water as it passes from the soil, through a plant, and into the air.

7 Carol tells her mother that plants lose water in the same way as clothes dry on a washing line. Is Carol right? Explain your answer.

Weather condition	Transpiration rate (mm/min)
warm air	15
cold air	5
still air	8
moving air	20
dry air	10
moist air	4

2.12 Support in plants

Objectives

This spread should help you to

- describe the structure of a vascular bundle
- explain how veins support leaves
- explain how turgid cells support a plant

The leaves of a plant need to get as much light as they can for photosynthesis. Flowers must be held up in the air so they can be pollinated by the wind or by passing insects. Plants are supported by a framework of vascular bundles. Cells inside this framework are supported by the water in their cells.

Vascular bundles

Vascular bundles (sometimes called veins) are made of xylem and phloem tubes bundled together. Xylem tubes have thick, strong walls ideal for their job of supporting a plant. Veins run all through a plant from the ends of roots to the tips of leaves.

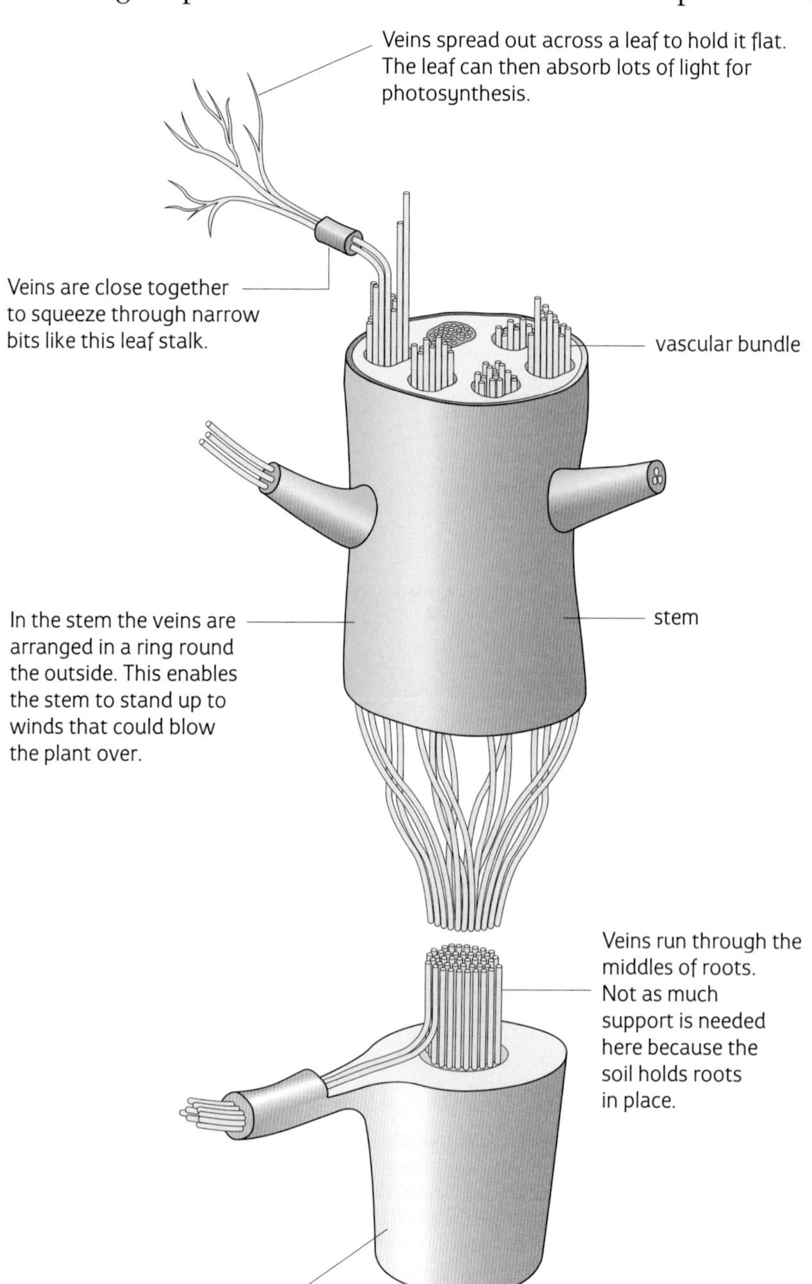

Questions

1 Why do:

a leaves **b** flowers

need support?

2 **a** What are vascular bundles made from?

b What makes vascular bundles ideal for supporting a plant?

Turgid cells and floppy cells

When a plant has plenty of water, the water will go into each of its cells by osmosis. This makes the cell expand so the contents push out against the strong cell wall. This makes the cell **turgid**. Turgid cells give lots of support, each one like a brick in a wall.

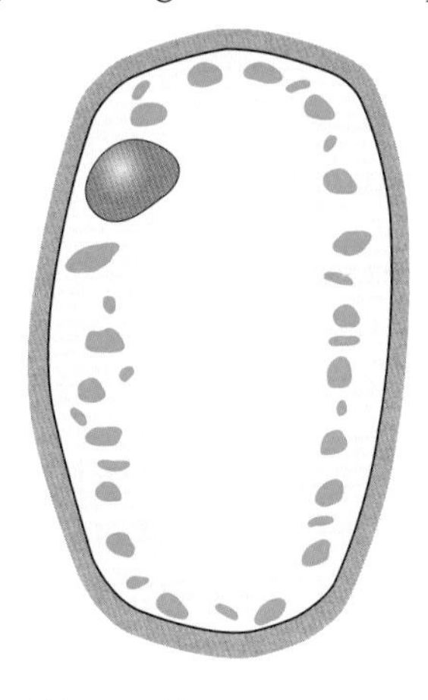

This cell is turgid.

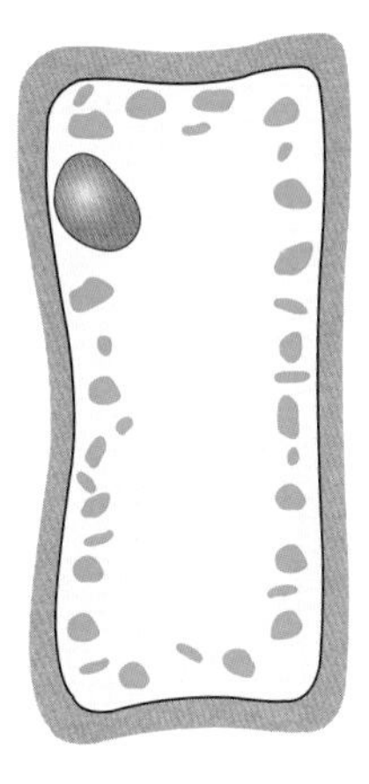

This cell has gone floppy.

If a plant doesn't get a good supply of water, water leaves the cells and they go floppy. Floppy cells give no support and the plant droops or wilts.

Turgid cells support the soft parts of plants like leaves and flowers.

A plant without water has floppy cells. Stems, leaves, and flowers wilt.

This leaf gets its support from the turgid cells around its 'skeleton' of veins.

Questions

3 Give one place in a plant where veins are:
a close together **b** spread out.

4 Explain why the veins in roots are in the middle but the veins in stems are round the outside.

5 How does water get into cells?

6 Explain how a cell gets turgid.

7 What happens when water leaves a plant cell?

8 What do we call a plant that has gone all 'droopy'?

9 Explain why some people call the network of veins in a plant a 'skeleton'.

2.13 Plant senses

Objectives

This spread should help you to

- describe how plants respond to light, gravity, and water
- explain how auxins cause plant responses

Growth movements – tropisms

Plants are sensitive to light, water, and the pull of gravity. They respond to these things by growing towards them or away from them. These growth movements are called **tropisms**.

Phototropism is the response to light. Plants need light for photosynthesis so they grow towards it. This makes sure that leaves get as much light as they can.

Geotropism is the response to gravity. Roots always grow down in response to gravity. This makes sure they get water and minerals for the plant. Shoots always grow up.

Hydrotropism is the response to water. Roots always grow towards water even if they have to grow upwards! The response to water is stronger than the response to gravity.

Phototropism: the cress seeds in the flowerpot on the left were grown with light coming from the left.

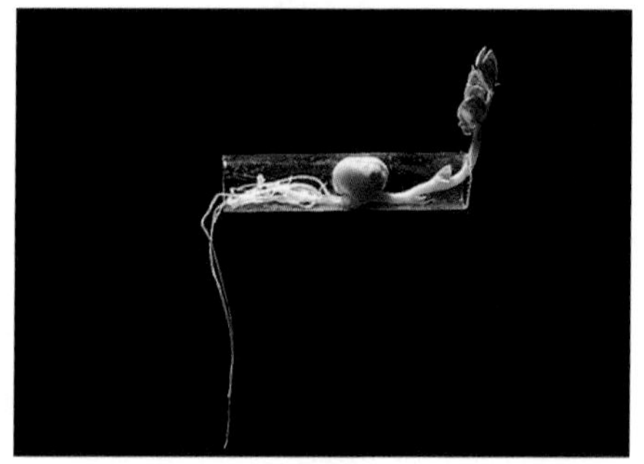

Geotropism: this pea seed was 'planted' on its side. The shoot grows up and the root grows down.

Hydrotropism: roots grow towards the water.

Auxins

Hormones called **auxins** are made in cells at the tips of roots and shoots. Auxins speed up growth in shoots but slow down growth in roots.

Auxins and phototropism

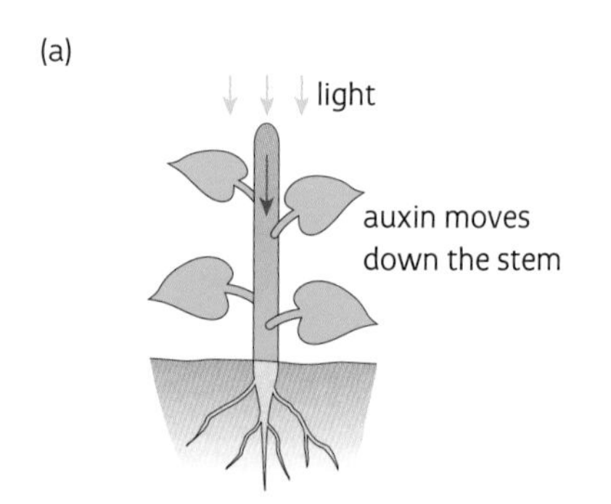

When a growing shoot gets light from above, the auxins spread out evenly and the shoot grows straight up.

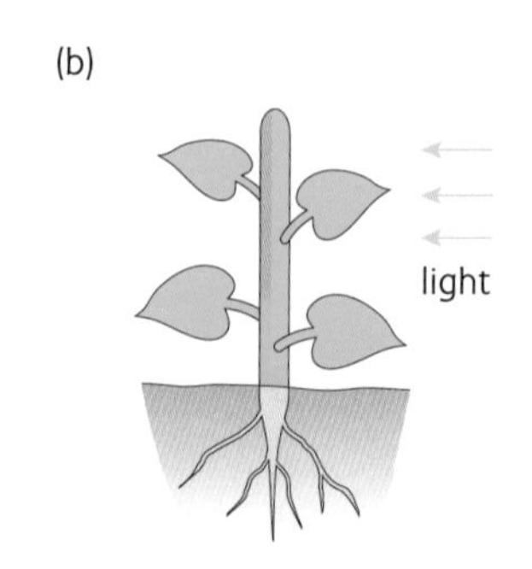

When light comes from one side, the auxins move over to the shaded side.

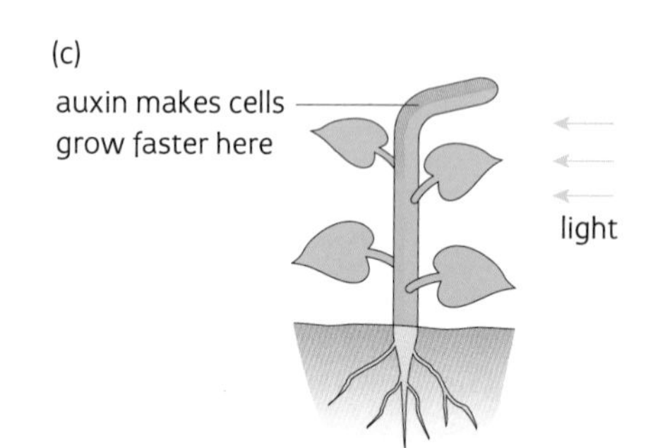

The shoot grows faster on the shaded side, making the shoot bend towards the light.

Auxins and geotropism

(a)

When a plant is on its side, the auxins sink to the lower side.

(b)

Auxins slow down growth on the lower side of the root. The root grows down.

Auxins speed up growth in the lower side of the stem, making it bend upwards.

Auxins and hydrotropism

(a)

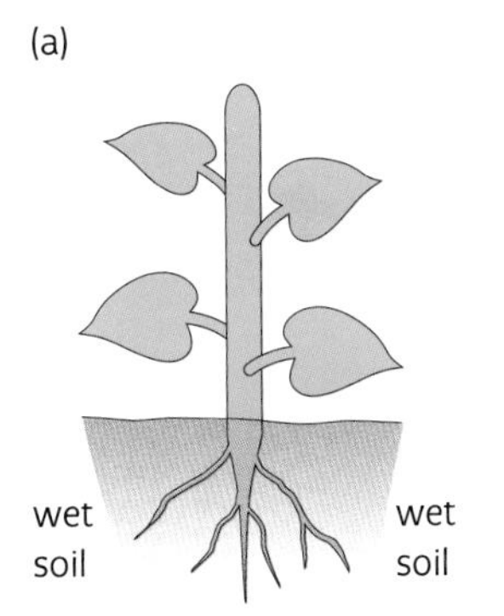

When roots are well watered, the auxins spread out evenly and the roots grow down.

(b)

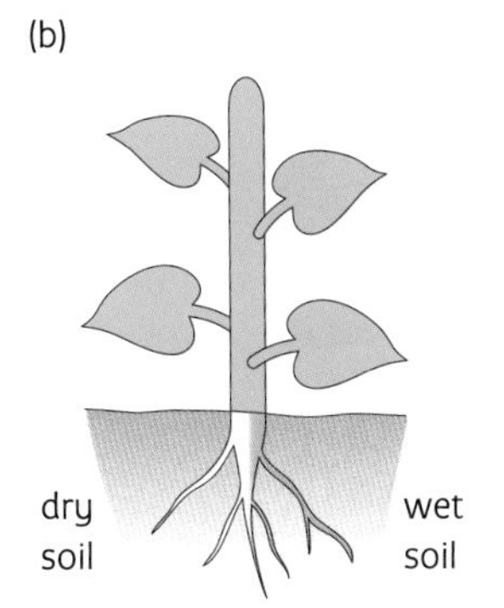

If there is more water on one side, auxins move across to the side nearest the water.

(c)

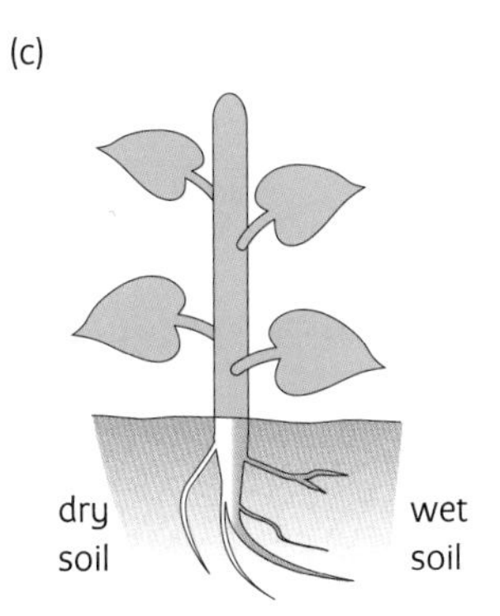

The root grows slower on the wetter side, making the root bend towards the water.

Questions

1 What are tropisms?

2 What is:
a phototropism **b** geotropism
c hydrotropism?

3 Explain why:
a phototropism **b** geotropism
is important to a plant.

4 Explain why the roots of a plant can sometimes grow upwards.

5 **a** What are auxins?
b Where are auxins made?

6 What do auxins do to:
a shoots **b** roots?

7 Explain how auxins make a plant bend towards the light.

8 Penny accidentally knocked a potted plant over onto its side. After a few days she noticed that the plant stem was growing upwards. Explain this as fully as you can.

2.14 Using plant hormones

Objectives

This spread should help you to

- describe how to take cuttings
- explain how selective weedkillers work
- describe how plant hormones can be used to control the ripening of fruit and in the production of seedless grapes

Plant hormones have lots of uses in the production of plants for sale and in the food growing business.

Growing cuttings

A **cutting** is part of a plant that has been cut off. Cuttings can be taken from stems, leaves, and even roots.

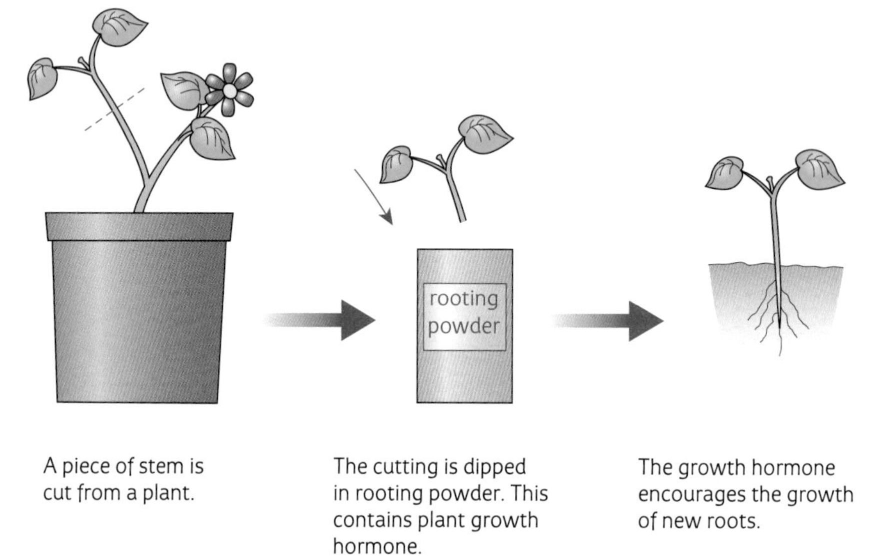

A piece of stem is cut from a plant.

The cutting is dipped in rooting powder. This contains plant growth hormone.

The growth hormone encourages the growth of new roots.

Taking cuttings enables growers to produce lots of clones of good plants quickly.

Killing weeds

Most weeds that grow in gardens and fields have broad leaves. Grasses like wheat and other cereal crops have narrow leaves. **Selective weedkillers** contain plant growth hormones at concentrations that only affect broad-leaved plants. The hormones upset the normal growth of the weeds and they soon die. The grass is unaffected.

Narrow-leaved plants are not affected by weedkiller.

Ripening of fruit

You have probably noticed that the fruit we buy in the shops is usually ripe and ready to eat. To make this possible the fruit is picked before it is ripe. Unripe fruit is usually hard and less likely to get damaged when being transported. The unripe fruit is sprayed with hormones during transport. These hormones ripen the fruit so it is perfect when it gets to the shops.

This fruit is ripe and ready to eat.

Most people prefer eating grapes without the 'pips' (seeds).

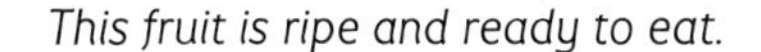

Did you know?
You should always wash fruit and vegetables before you eat them. This gets rid of any chemicals that might be left on their surface.

Seedless grapes

Sexual reproduction in plants produces fruits with seeds inside. If a flower doesn't get pollinated, the fruits and seeds don't grow. But if growth hormone is sprayed onto unpollinated flowers, the fruits will grow but the seeds won't. This is what happens to grape flowers before they can be pollinated by insects. Seedless grapes are the result.

Questions

1. Why are cuttings dipped into plant growth hormone?
2. What is the advantage of taking cuttings?
3. Name a plant with narrow leaves.
4. **a** What is a selective weedkiller?
   b Explain how selective weedkillers work.
5. Give an advantage of transporting fruit unripe.
6. When is unripe fruit sprayed with ripening hormone?
7. Explain how fruits can grow even though there has been no sexual reproduction in a plant.
8. Name two seedless fruits sold in shops.

2.15 The eye

Objectives

This spread should help you to

- describe the structure of the eye
- describe how eyes are protected from damage
- describe how we see things

Did you know?

There are about 120 000 000 cells in the retina. People who don't get enough vitamin A in their diet (there's lots in carrots) suffer from poor night vision.

Eyes are the sense organs of sight.

Parts of the eye

The **sclera** is a tough protective outer layer. It is the 'white' of the eye.

The **ciliary muscles** are a ring of muscles that change the shape of the lens during focusing.

The **lens** helps focus an image on the retina. It is clear and can change shape.

The **cornea** is the clear part of the sclera at the front of the eye. It is clear to let light through and helps make an image on the retina.

The **iris** controls how much light gets into the eye.

The front part of the eye is filled with watery liquid.

The **pupil** is the hole in the middle of the iris.

The **suspensory ligaments** hold the lens in place and connect it to the ciliary muscles.

The eye is filled with a jelly which helps the eye keep its shape.

The **retina** is a layer of light-sensitive cells which send messages to the brain.

The **optic nerve** carries messages to the brain from the cells on the retina.

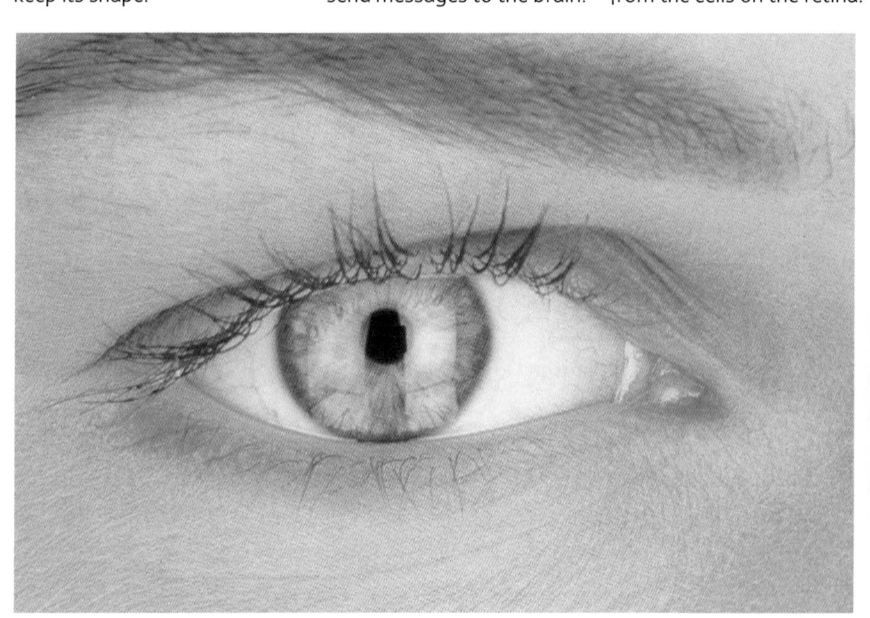

Can you spot the pupil, iris, sclera, and cornea?

Questions

1 What part of the eye:

a helps focus an image on the retina

b controls how much light gets in

c is the transparent part of the sclera

d changes the shape of the lens

e is a layer of light-sensitive cells

f carries messages to the brain?

Protecting the eyes

Our eyes are in **sockets** or holes in the skull. This helps protect them from damage. Muscles hold the eyes in place. They let you move your eyes up and down and from side to side.

How we see things

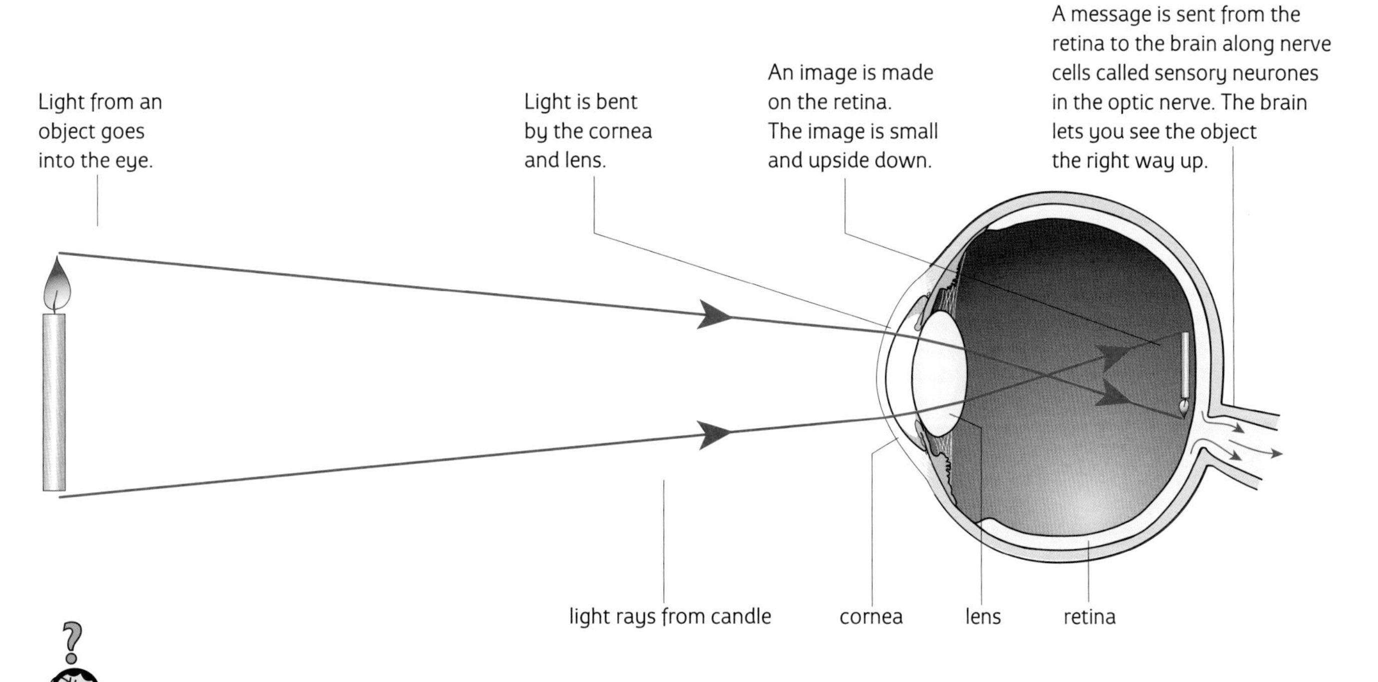

Questions

2 How are your eyes moved up and down?

3 How are your eyes protected from bright light?

4 Why do you make more tears when smoke gets in your eyes?

5 How do you know that your eyelashes are sensitive to touch?

2.16 Seeing things

Objectives

This spread should help you to

- describe how we see things a long way off and close up
- explain how we judge distance
- explain how we see in the light and the dark

Seeing clearly

Light that goes into the eye must be bent to **focus** it carefully onto the retina, otherwise we wouldn't see things clearly. The cornea does most of the bending as the light enters the eye. It is the lens that makes fine focusing adjustments. The shape of the lens is changed by the ciliary muscles.

Seeing things a long way off

To see an object a long way off, the lens has to be thin. A thin lens doesn't bend light very much. The ciliary muscles relax, making the muscle ring bigger. The suspensory ligaments are pulled tight and the lens gets thinner.

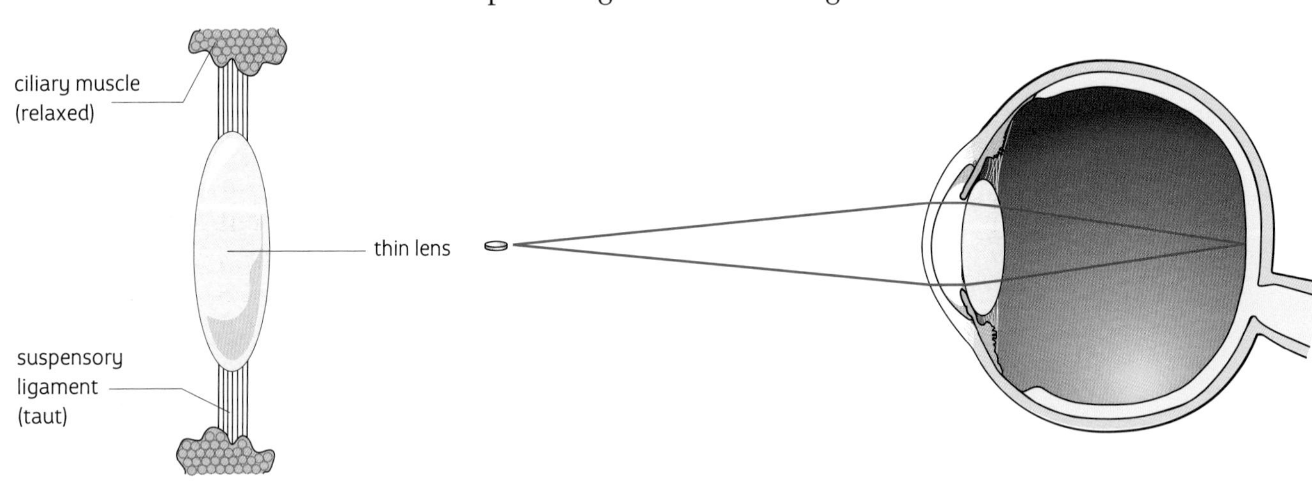

A thin lens doesn't bend light very much.

Seeing things up close

To see an object up close, the lens must be fat. A fat lens bends light a lot. The ciliary muscles contract, making the muscle ring smaller. The suspensory ligaments go slack and the lens gets fatter.

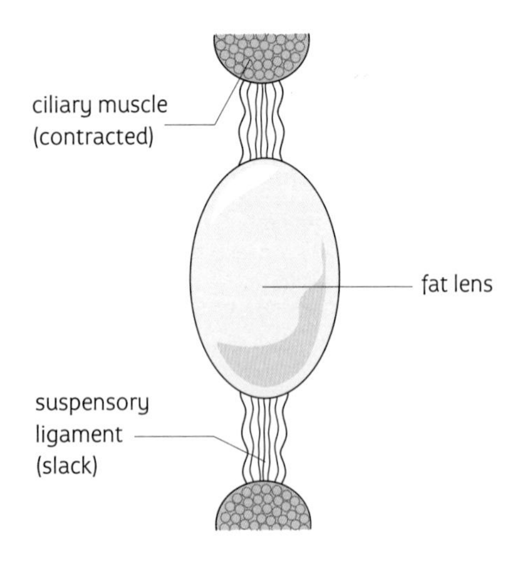

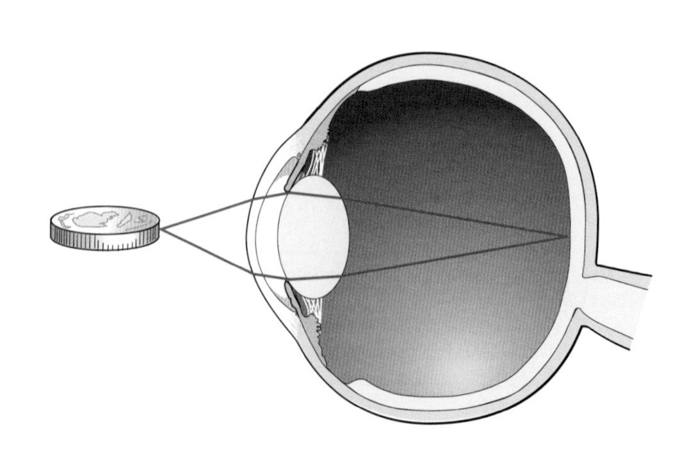

A fat lens bends light a lot.

Foxes look straight ahead.

Rabbits can see what's going on all around them.

Judging distance

We have two eyes a few centimetres apart. This helps us judge distance and find out easily where things are. The view from each eye is slightly different. The brain puts the two views together to give us a 3-D (three dimensional) view. This is called **stereoscopic vision**.

Animals with eyes that look straight ahead have good stereoscopic vision. They must judge distances accurately to catch prey.

Animals with eyes on the sides of their head can't judge distance very well, but they can see all round them. This is good for avoiding being caught and eaten.

Seeing in the light and in the dark

Light enters the eye through the pupil. If too much light gets in, the retina can be damaged. If there is not enough light, you can't see anything. The amount of light entering the eye is controlled by the iris. The muscles of the iris change the size of the pupil to allow more or less light to pass through.

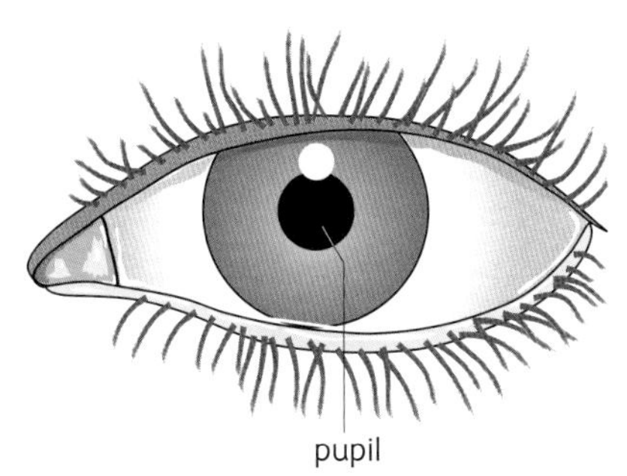

In bright light muscles pull the iris in to make the pupil smaller.

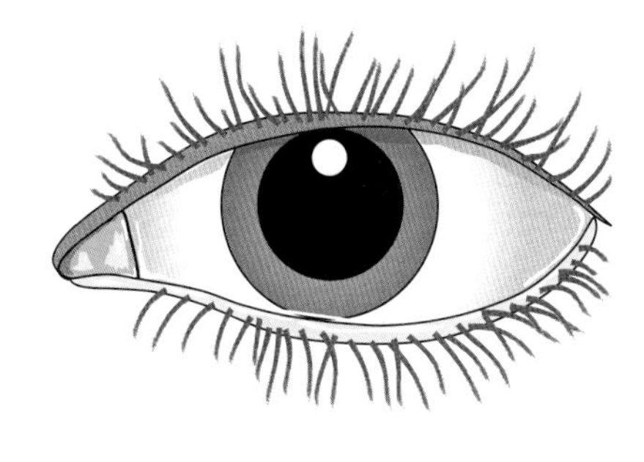

In dim light muscles pull the iris wider to make the pupil bigger.

Questions

1. **a** Which part of the eye is responsible for fine focusing?
   b Describe what happens in the eye when someone looks at a tree a long way off.
2. The diagram shows the range of vision of a rabbit. Draw a similar diagram showing the range of vision of a human.

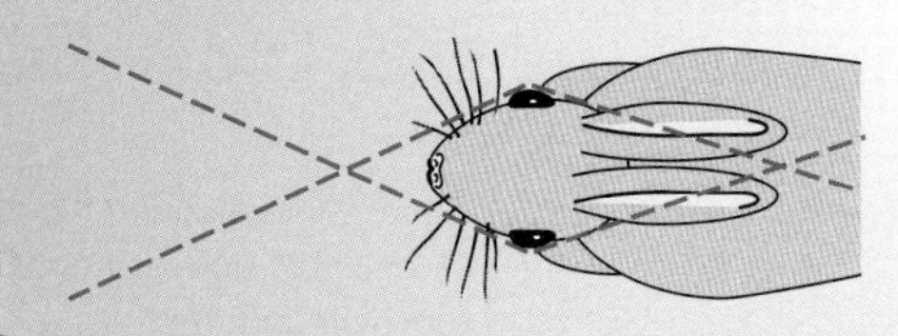

3. Barry is watching a cricket match on a bright sunny day. He goes into the tea room to buy a cup of tea. The tea room has no windows.
   a Describe the changes that take place to Barry's pupils as he moves from one place to the other.
   b Why are these changes necessary?

2.17 Other senses

Objectives

This spread should help you to

- describe the structure of the ear
- explain how we hear things
- explain how we keep our balance
- describe the skin as a sense organ
- explain how we taste things
- explain how we smell things

The eye is probably our most useful sense organ. However, humans have other sense organs as well.

Ears…and hearing…

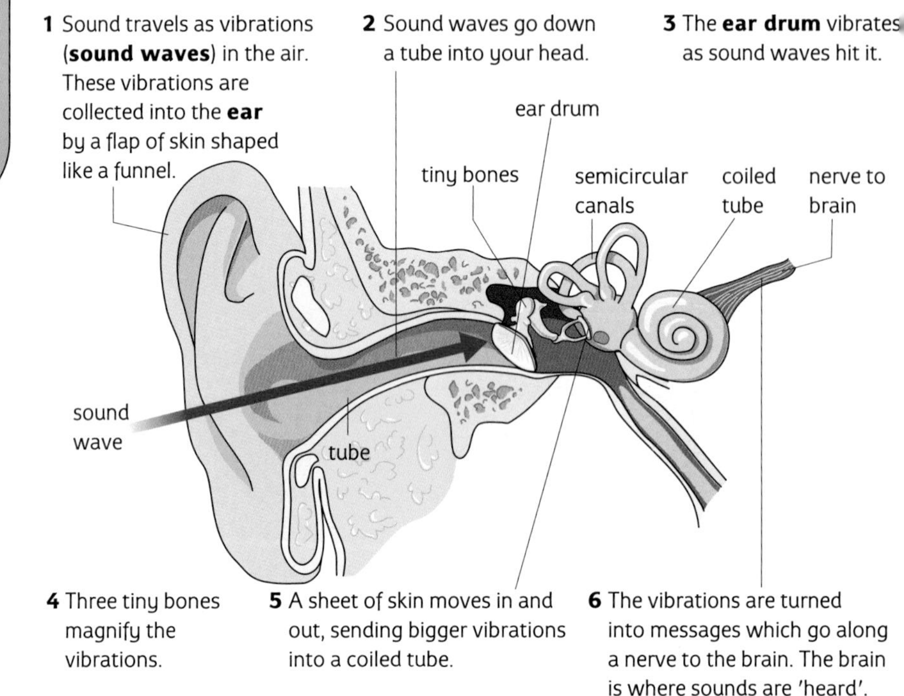

…and balance

Your eyes tell you if you are standing upright or not. **Semicircular canals** in the ear tell you if you are losing your balance. Liquid in the canals moves when you move. This moving liquid is detected by the brain. When you go off balance, the brain 'tells' your muscles to keep you upright.

Touch

The sense of touch is produced by **nerve endings** in the skin.

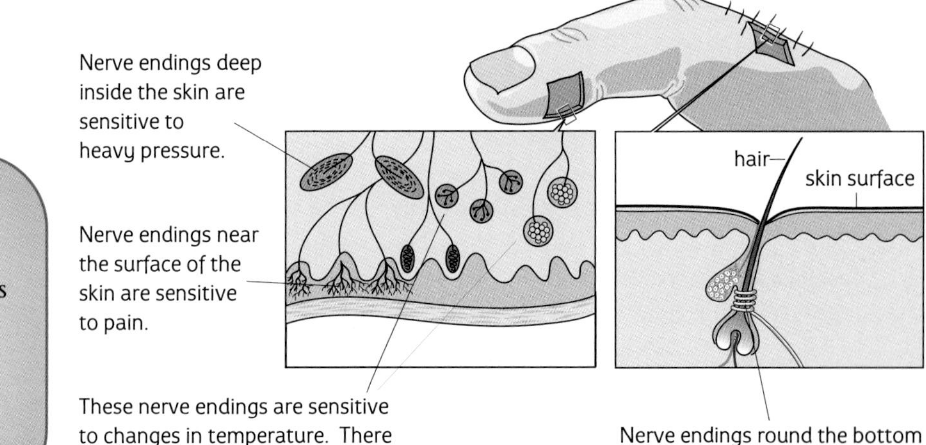

Questions

1. What are sound waves?
2. What do the three tiny bones in the ear do?
3. Where do we really 'hear' sounds?
4. Explain how you keep your balance.

Taste

Your tongue has **taste buds** in between the ridges. There are four kinds of taste bud. They taste bitter, salty, sour, and sweet things. The diagram shows where these taste buds are.

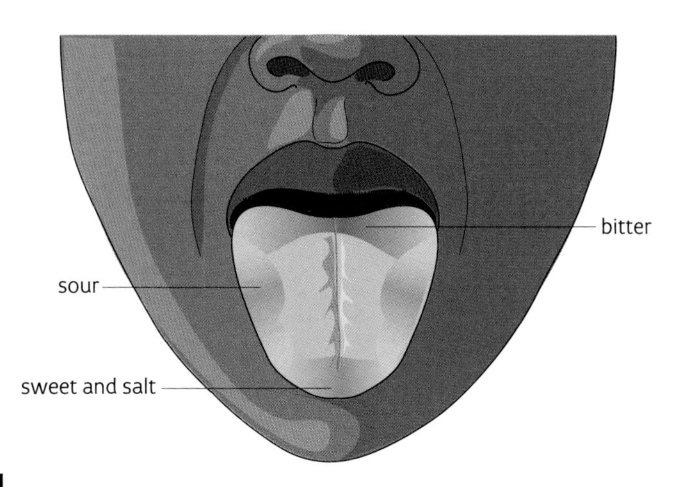

Did you know?

You've got about 9000 taste buds on your tongue and 20000000 smell cells in your nose.

Smell

Smells are chemicals in the air. These chemicals dissolve in the moisture lining your nose. Messages are sent to the brain which cause the sensation of smell. Flavours are smells – that's why you don't taste your food so well if a cold blocks your nose.

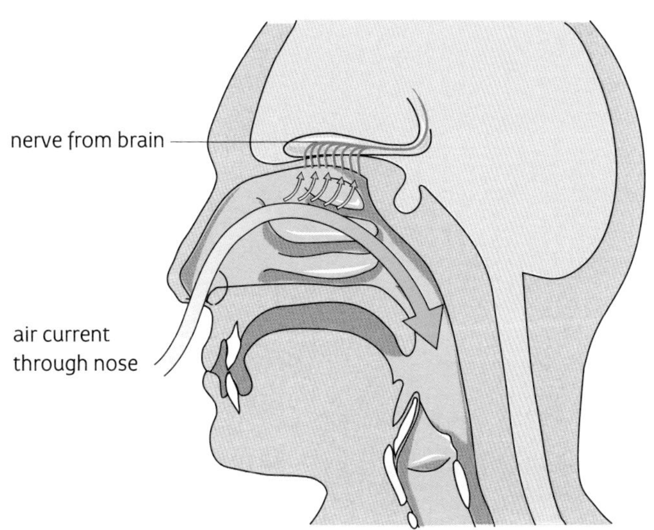

Questions

5 If someone gently strokes the hairs on the back of your neck, you feel a tingling sensation. Explain why.

6 Why is it an advantage to move food around your mouth before you swallow it?

7 Explain why you seem to lose your sense of taste when you have a cold.

8 Name the five senses and their sense organs.

2.18 The nervous system

Objectives

This spread should help you to

- describe the central nervous system
- explain the difference between a sensory nerve cell and a motor nerve cell
- describe how messages pass along nerve cells
- describe the structure of a nerve

Working together

All of the things that happen in the body must be **coordinated.** This means they must be made to work together, doing their jobs at the right time. Coordination is the job of the nervous system. The **brain** and **spinal cord** together are called the **central nervous system**. Millions of **nerves** branch off from the central nervous system to all parts of the body. The nervous system is made up of nerve cells or **neurones**.

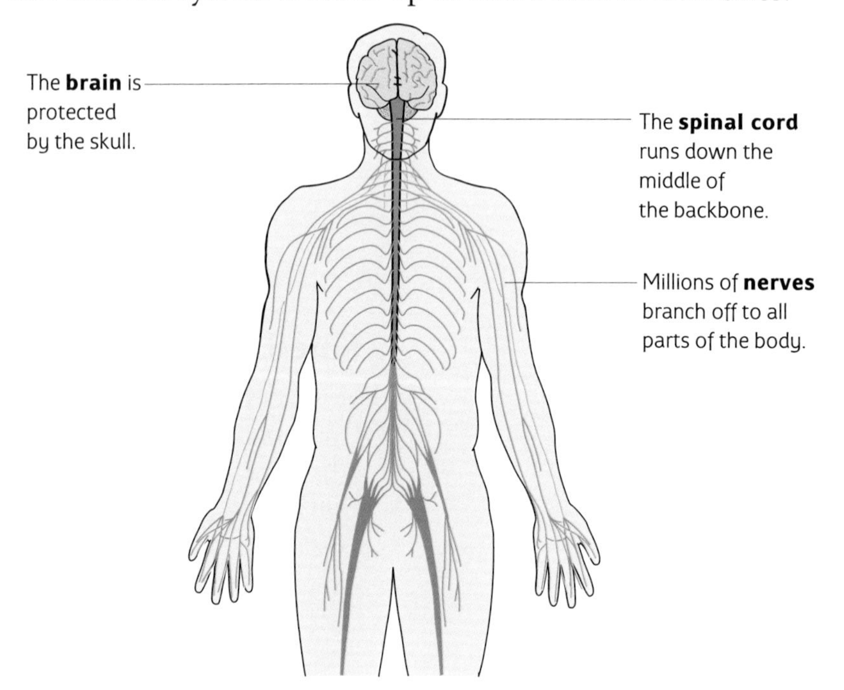

The human nervous system.

Nerve cells (neurones)

Nerve cells are long and thin – some are over a metre long. This means they don't take up much room and can spread all over the body. Messages travel along nerve cells as tiny electrical signals or **impulses**. Impulses can only go in one direction so there have to be two kinds of nerve cell.

Sensory nerve cells (or sensory neurones) carry impulses from sense receptors to the central nervous system.

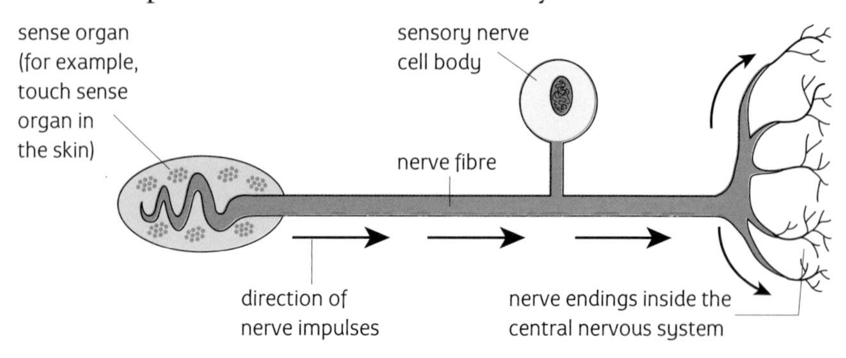

Questions

1. Why must the body be coordinated?
2. What is the nervous system made up of?
3. What is the central nervous system?
4. What is the scientific name for a nerve cell?
5. Why do you think the spinal cord runs down the middle of the backbone?

Motor nerve cells (or motor neurones) carry impulses from the central nervous system to muscles and glands.

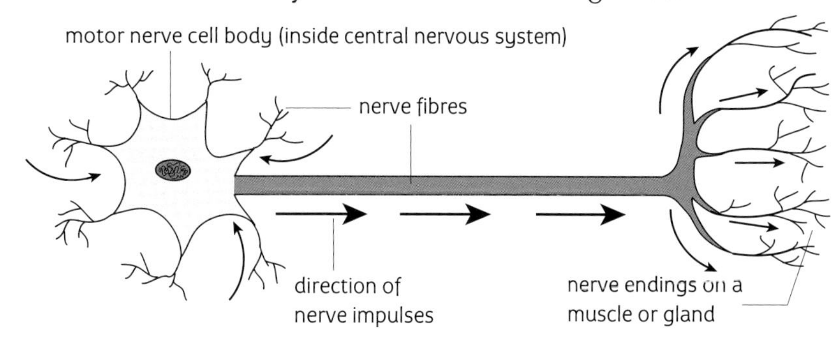

Nerves

A nerve is a bundle of nerve fibres. These nerve fibres can be all from sensory nerve cells, or all from motor nerve cells, or a mixture of both. The fibres are wrapped in fatty insulation. A nerve is a bit like an electric cable. The nerve fibres are the wires and the fatty insulation is the plastic around the cable.

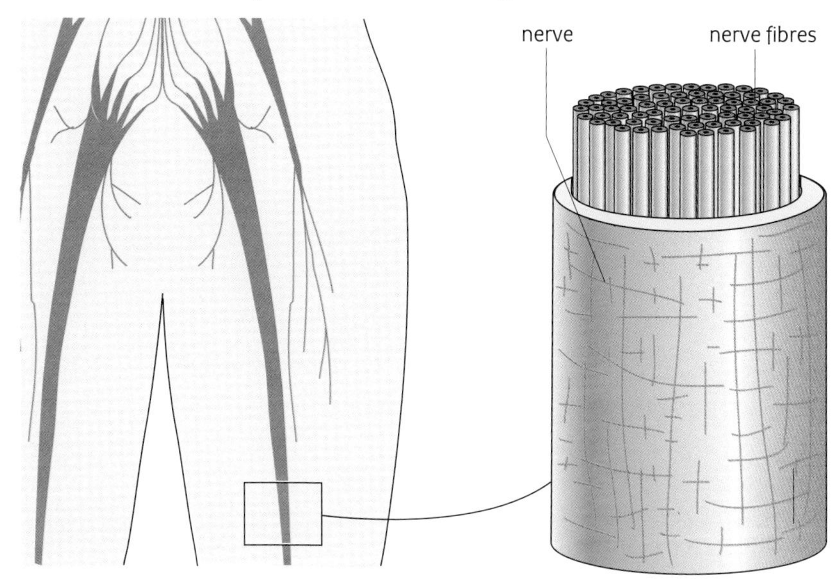

Questions

6 How are messages carried along nerve cells?

7 a Name the two kinds of nerve cell. Where do they carry messages to and from?

b Explain why there have to be two kinds of nerve cell.

8 How are nerve cells adapted to their job?

9 What is the difference between a nerve and a neurone?

10 Suggest why a nerve has fatty insulation around it.

2.19 Reflex actions

Objectives

This spread should help you to

- explain the difference between a voluntary action and a reflex action
- describe how a reflex action happens
- describe how synapses work
- give some examples of reflex actions

Voluntary and reflex actions

You need to think when you speak to a friend or write a letter. These are **voluntary actions**. Something you do without thinking is called a **reflex action**. Reflex actions are very fast and save your body from injury.

The diagram shows what happens if you accidentally touch a hot plate.

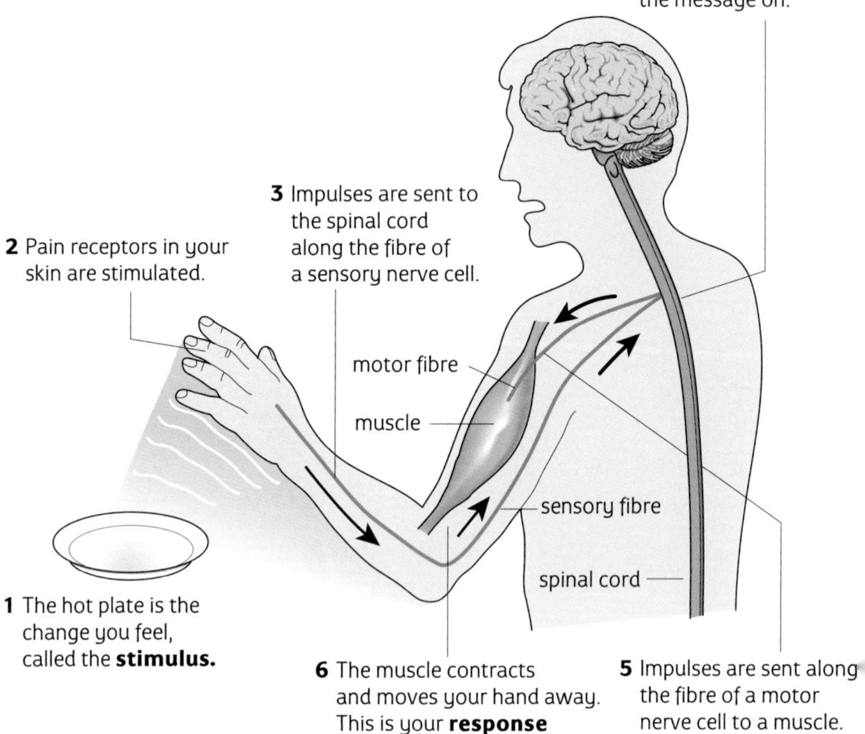

Passing the message between nerve cells

Impulses pass from one nerve cell to another by a special link called a **synapse**. The branching ends of one nerve cell lie very close to the cell body of another nerve cell. They do not actually touch. When an impulse moves along one nerve cell it releases a chemical that diffuses across the gap and stimulates the other nerve cell.

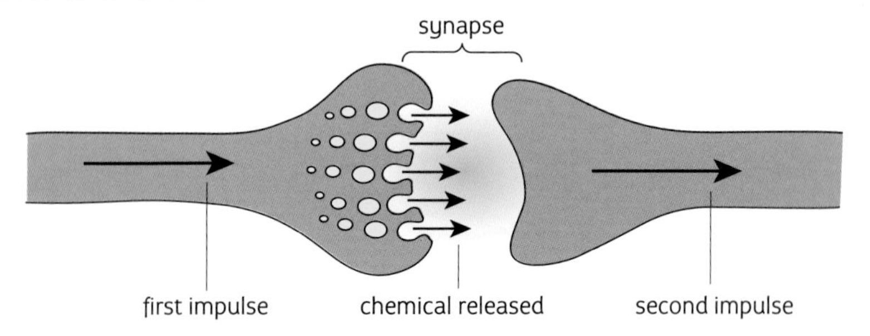

Questions

1. What is the difference between a voluntary action and a reflex action?
2. What is a stimulus? Give an example.
3. What is a response? Give an example.
4. Describe the path of a nerve impulse causing a reflex action when you sit on a drawing pin.

One nerve cell can have many synapses with other nerve cells. This means lots of connections can be made. This is why you can respond in lots of ways to one stimulus. As well as moving your hand away from a hot plate you might possibly yell out or even jump up and down in pain. All these actions need their own motor nerve cells to carry the message!

Reflex actions save the body from harm.

Did you know?

Babies are born with some reflexes. If you touch the palm of its hand, a baby will curl its fingers to grip you.

Examples of reflex actions

Stimulus	Reflex action
flash of bright light	pupil gets smaller
insect touches eyelid	eye blinks
food gets into the windpipe ('goes down the wrong way')	coughing
hand touches 'live' electric wire (electric shock)	arm moves away quickly
you are hungry and smell food	saliva in the mouth
body gets hot	sweating
body gets cold	shivering

Questions

5 **a** What is a synapse?
b Explain how a synapse works.

6 Explain why you can have many responses to one stimulus.

7 Which of the reflex actions in the table protect you from harm?

8 Explain why picking up a cup of tea and drinking it is not a reflex action.

2.20 Drugs and the nervous system

Objectives

This spread should help you to

- describe how drugs work
- explain the difference between useful and harmful drugs
- describe the effects that drugs have on the body
- explain what addiction is
- describe the effects of drinking alcohol

Drugs are substances that change the way the body works. They can act on the nervous system by altering the movement of nerve impulses along the fibres of nerve cells. Drugs also affect the way that synapses work.

Most drugs are useful but many are harmful and can easily kill if they are misused.

Useful drugs

Drugs such as aspirin and paracetamol are **painkillers**. They can be bought at a chemist's for headaches and other pains.

Tranquillizers and **sedatives** can only be obtained on prescription from a doctor. Tranquillizers calm people down. Sedatives include barbiturates and 'sleeping pills'. The body can easily become dependent on them.

Anaesthetics are used to numb parts of the body. Dentists use local anaesthetics to numb your mouth. A general anaesthetic is used to put you completely to sleep.

Some useful drugs.

Harmful drugs

Unfortunately some people use these drugs to give pleasant short-term effects. You can become dependent on or **addicted** to harmful drugs. Many drug addicts die from the drugs themselves or from accidents that happen whilst under their influence. The taking or possession of **controlled drugs** such as cannabis, cocaine, and heroin is illegal.

Stimulant drugs speed up nerve impulses. Some people take them to stay awake for a long time or to make them more 'alert'. Caffeine is a mild stimulant found in tea and coffee. It is harmless if taken sensibly. Amphetamines however are strong stimulants which cause serious depression when a person stops taking them.

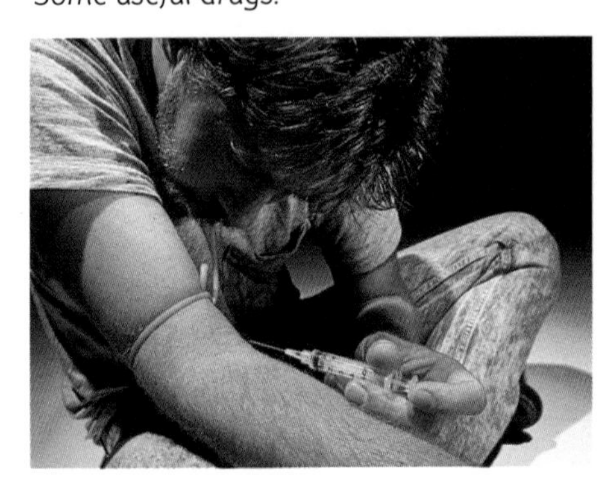

Using a harmful drug can kill you.

Depressant drugs slow down nerve impulses. They are taken to help people relax and reduce worry. Alcohol is a depressant, and so are barbiturates. If people take too much of a depressant they can't react as quickly as normal. They easily lose control and can't judge speed or distances. They definitely shouldn't drive. Heroin is a particularly nasty depressant drug. It can be used to give pain relief to terminally ill people, but it is very addictive.

Hallucinogens make people feel 'larger than life'. People talk about 'taking a trip' when using hallucinogenic drugs. Ecstasy and LSD are hallucinogenic. Addicts sometimes have 'bad trips' – they see horrifying things and get very scared. Hallucinogens are especially dangerous because of the long-term effects they have on the brain.

Questions

1. What are drugs?
2. Explain the difference between painkillers and tranquillizers.
3. What are anaesthetics used for?
4. What is a drug addict?
5. Describe how a drug addict may die.

There are two types of addiction. With **chemical addiction** the body gets used to the drug being there. When a user stops taking the drug there are unpleasant **withdrawal symptoms**. With **psychological addiction** a person feels the need to keep taking the drug.

More about alcohol

Large numbers of people drink alcohol at some time in their lives – but it can be dangerous.

It is legal for adults to drink alcohol, but it is a harmful drug. A small amount of alcohol makes a person happy and relaxed. Alcohol is a depressant drug that slows down the nervous system and affects judgement. People need to be careful how much they drink, otherwise they can become dependent – they become **alcoholics**. Alcoholics, like other drug addicts, risk serious damage to their health.

Alcohol is a poison. Drinking too much will damage the liver and the brain. **Hepatitis** (liver inflammation) and **cirrhosis** (scarring of liver tissue) are common in heavy drinkers. Too much alcohol in the blood causes blood pressure to rise. **High blood pressure** can lead to heart disease.

It is in everyone's interest either to avoid alcohol or to drink within sensible limits.

Did you know?

About 800 people die each year in Britain because of people driving after drinking alcohol.

These drinks may be different, but they all have the same amount of alcohol in them. This amount is called a ***unit*** *of alcohol. Doctors advise men not to drink more than 21 units in a week. Women shouldn't drink more than 14 units. People should avoid alchohol for at least two days each week.*

Questions

6 Name a drug that is a:

a stimulant **b** depressant **c** hallucinogen.

7 What are the two types of drug addiction?

8 What is an alcoholic?

9 Explain how alcohol can damage the body of an alcoholic.

10 What is the recommended 'safe' alcohol intake for:

a a man **b** a woman?

2.21 More about synapses

Bridging the gap

Nerve impulses pass from one nerve cell to another, or from a motor nerve cell to a muscle, across a tiny gap called a synapse. When an impulse passes along one nerve cell, it releases a chemical that diffuses across the gap. The chemical then stimulates the other nerve cell to pass the impulse on or the muscle to contract. The chemical is called a **neurotransmitter**. When the neurotransmitter has crossed the synapse it is broken down by an enzyme. If it weren't, then the nerve cell or muscle would continue to be stimulated.

Synapses allow lots of connections between different nerve cells. This is why we respond to a stimulus in lots of ways. Synapses can be upset by many things, causing problems in the nervous system.

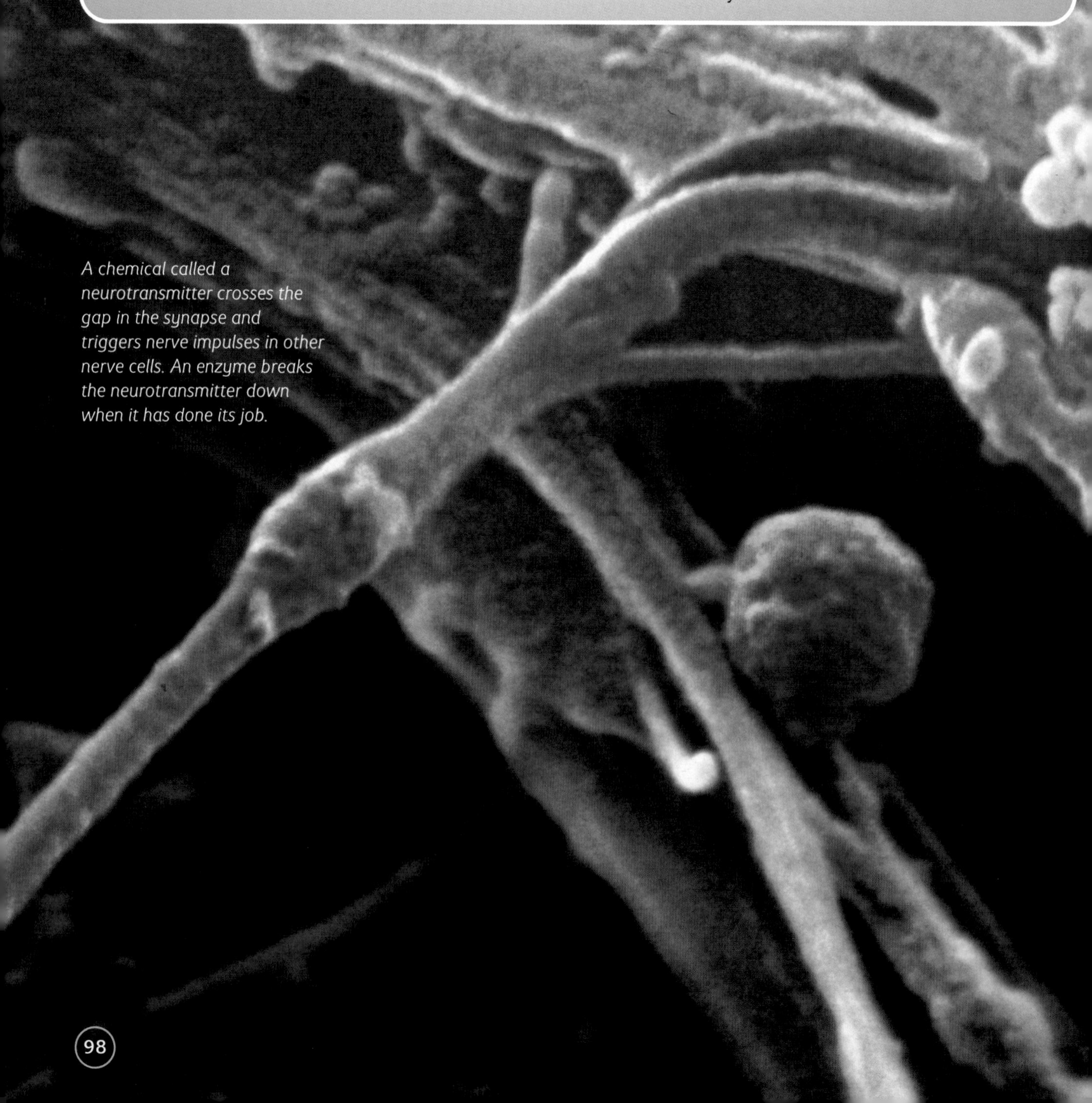

A chemical called a neurotransmitter crosses the gap in the synapse and triggers nerve impulses in other nerve cells. An enzyme breaks the neurotransmitter down when it has done its job.

Drugs

Nicotine in cigarette smoke helps the neurotransmitter stimulate the next nerve cell. Messages travel faster and responses are quicker. This is why nicotine is called a **stimulant** drug. In large doses, nicotine and other stimulant drugs can kill. Nicotine is used as an insecticide, killing insects by overstimulating the nervous system.

The nicotine in cigarette smoke speeds up the chemical message across a synapse.

*The alcohol in this drink slows down the chemical message across a synapse. This is why alcohol is called a **depressant** drug.*

Alzheimer's disease

Alzheimer's disease is a slow loss of memory which gets worse as time goes by. It tends to affect older people more than the young. The brain is full of synapses linking the millions of nerve cells. As some people get older, the cells that produce the neurotransmitter chemical don't work properly.

Animal poisons

Some poisonous animals such as the black widow spider make a poison called venom. This venom causes a sudden surge of neurotransmitter to be released, then no more. The victim suffers muscular spasms, and then is soon paralysed.

The venom of the black widow spider causes a sudden rush of neurotransmitter. This results in muscle spasms and death.

Chemical warfare

Nerve gases have been used since the First World War. Nerve gases work by stopping the enzyme from breaking down the neurotransmitter after it has done its job. This means that the nerve cells and muscles are continuously stimulated. Victims have convulsions, are paralysed, and then die.

Nerve gas stops the enzyme breaking down the neurotransmitter after it has done its job. Muscles are stimulated all the time, causing paralysis and death.

Plant poisons

Curare is a deadly poison. South American Indians put it on the tips of their arrows when they hunt for food. Only a small amount is needed to kill a monkey. If handled carelessly, it can quickly kill a human. Curare works by stopping the neurotransmitter getting into the next nerve cell. (Deadly nightshade is a plant found all over the UK. It works in the same way as curare.)

South American Indians use a plant poison called curare. Only a small amount is needed to kill a monkey.

Talking points

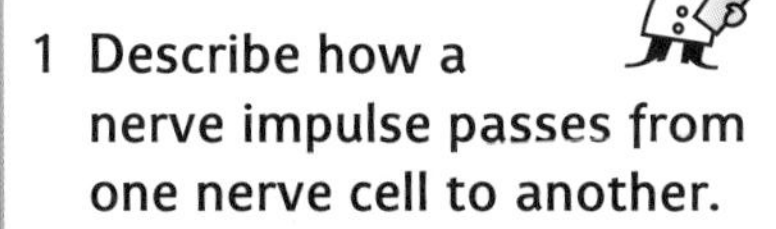

1. **Describe how a nerve impulse passes from one nerve cell to another.**
2. **a What is Alzheimer's disease?**
   b What causes it?
3. **How do you think deadly nightshade got its name? Explain your answer.**
4. **Explain how nerve gases work.**

2.22 Breathing can be dangerous

Objectives

This spread should help you to
- list the dangerous chemicals in cigarette smoke
- describe the diseases caused by smoking
- describe the effects of cigarette smoke on non-smokers and pregnant women
- describe the effects of sniffing solvents

Sometimes people deliberately breathe in harmful substances. Smoking tobacco and sniffing solvents can seriously damage health.

Smoking

Smoking kills about 100 000 people a year in Britain. Tobacco smoke contains lots of harmful chemicals.

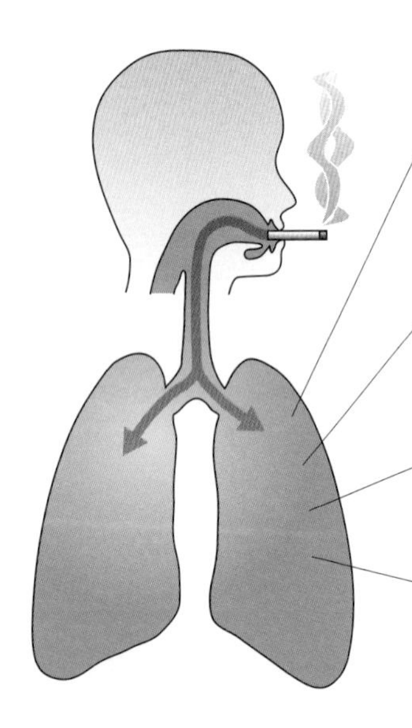

Nicotine is a poison which damages the heart, blood vessels, and nerves. It is an **addictive** drug. Smokers can't do without it and so find it hard to give up.

Tar sticks to the insides of the lungs when smoke cools. It affects the walls of the alveoli and causes lung cancer.

Carbon monoxide is a poisonous gas. It stops blood carrying oxygen around the body.

Other gases such as **ammonia** and **cyanide** irritate the insides of the air passages. This makes smokers cough.

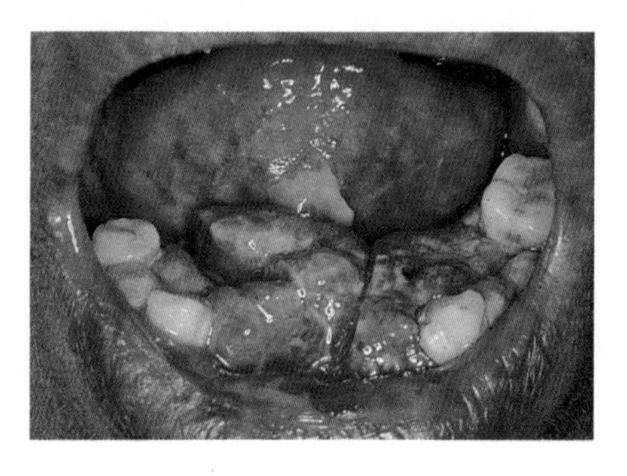

Most people who get lung or mouth **cancer** are smokers.

Diseases caused by smoking

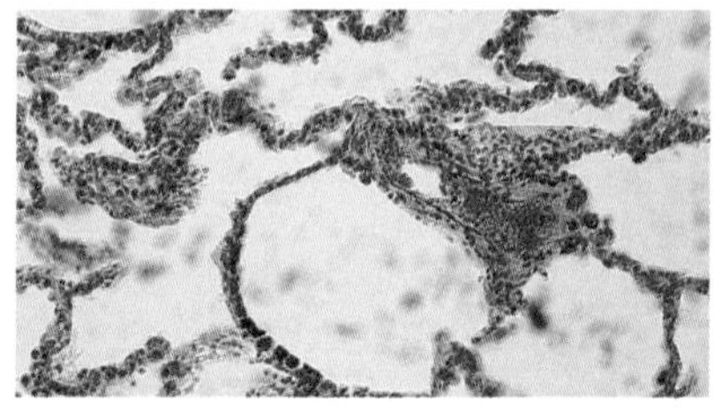

Normal alveoli under a microscope.

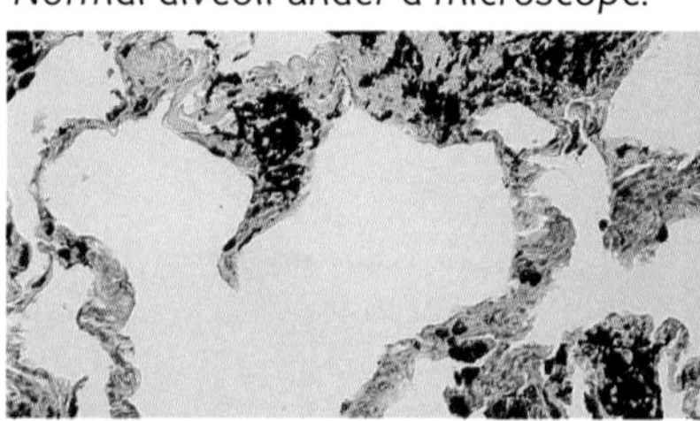

Emphysema.

Emphysema is caused by chemicals in tobacco smoke. Alveoli are destroyed, reducing the surface area of the lungs.

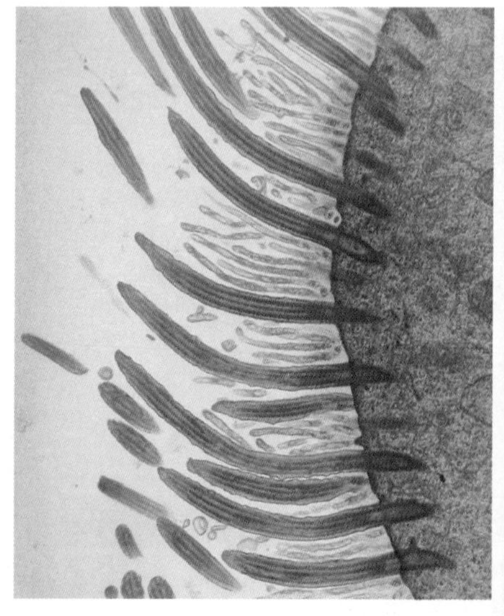

These tiny cilia which keep the lungs clean are killed by tobacco smoke. The lungs fill with sticky mucus and can become infected by bacteria. This causes **bronchitis**.

Smokers affect other people too

People who don't smoke find it unpleasant to be in a smoke-filled room. The smoke in the air stings the eyes and causes sore throats and headaches. Non-smokers who live or work with smokers risk getting all of the diseases associated with tobacco smoking.

Pregnant women who smoke have smaller babies. The carbon monoxide stops enough oxygen getting to the baby in the womb, upsetting its growth. Childbirth can also be more difficult.

Passive smoking harms the baby after birth as well.

Sniffing solvents is dangerous and can even kill you.

Solvents

Solvents are found in lots of household items such as glue, paint, and cleaning fluids. Some people breathe the vapour from the solvent to get 'high'. Solvents are extremely dangerous. They cause hallucinations and affect personality and behaviour. The lungs, brain, and liver can be seriously damaged. Most solvent sniffers are young people.

Questions

1. List five harmful things found in tobacco smoke.
2. Explain why a smoker can get bronchitis.
3. Name two other diseases caused by smoking.
4. Why is it bad for pregnant women to smoke?
5. Work out how much it costs to buy 20 cigarettes a day for a year.
6. List four things you would buy with this money instead of spending it on cigarettes.
7. Why do people sniff solvents?
8. Describe the effects of sniffing solvents.

2.23 Hormones

Objectives

This spread should help you to

- describe the hormone system
- name the hormones produced by each gland
- describe what each hormone does

What are hormones?

Hormones are chemicals made in special glands called **endocrine glands**. Like the nervous system, hormones coordinate the body, but in a different way. A nerve impulse travels from the brain or spinal cord to one particular muscle very quickly. Hormones travel in the bloodstream. This means hormones move more slowly and their effects are more general. Once released into the blood from a gland, a hormone travels round the body until it reaches a **target organ**. Target organs are those parts that respond to the particular hormone.

There are advantages in having a 'slow' communication system in the body. Hormones can be released into the bloodstream over a long period of time. This means that changes that take a long time, such as growing, can be carefully controlled. The liver takes hormones from the blood when they are no longer needed.

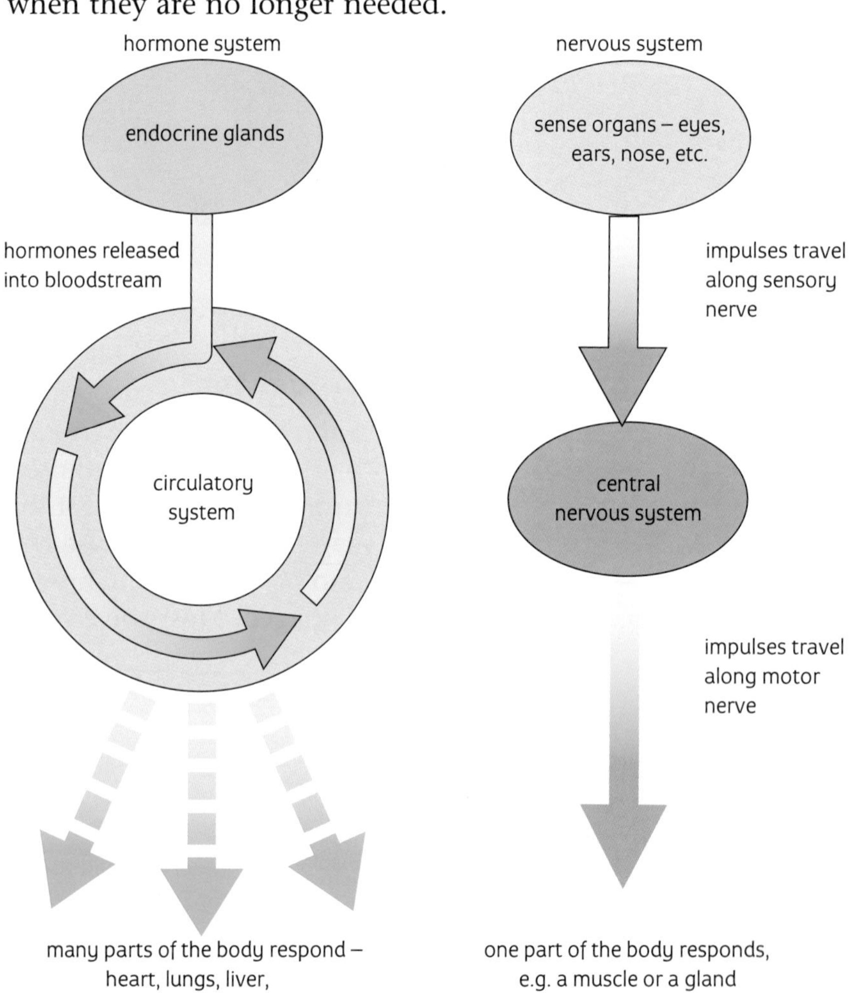

Comparing the hormone and nervous systems.

Questions

1. Where are hormones made?
2. Explain why hormones travel slowly round the body.
3. What is a target organ?
4. Why is it an advantage to have a slow communication system in the body?
5. What happens to hormones when they are no longer needed?

Glands and hormones

This diagram shows where the main endocrine glands are and tells you what each hormone does.

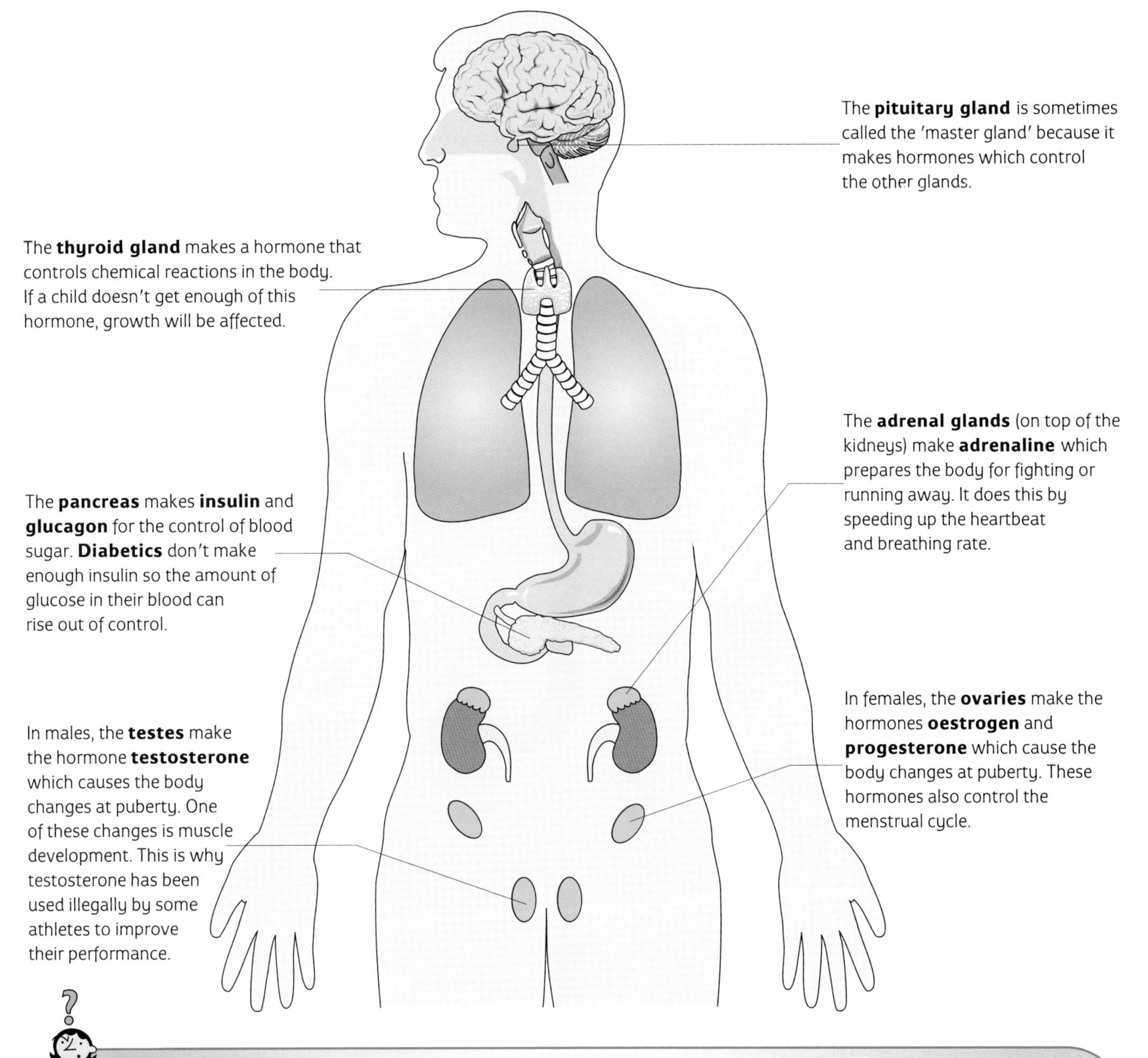

Questions

6 Where are the hormones made that cause these things to happen?

a controlling the amount of glucose in the blood

b controlling a woman's menstrual cycle

c giving a boy body hair at puberty

d controlling the growth of a child

7 Explain why the pituitary gland is sometimes called the 'master gland'.

8 Which hormone would be made if you were chased by a bull?

9 Explain why testosterone has been used as an illegal drug by some athletes.

2.24 Insulin and adrenaline

Objectives

This spread should help you to

- describe how the level of glucose in the blood is controlled
- explain how diabetics can control their condition
- describe the effects of adrenaline on the body

Insulin

Sometimes there is more glucose in the blood than we need. Sometimes there is not enough. Respiration uses up glucose, and the more exercise you do the more glucose is used up. But we aren't exercising all the time. So glucose must be stored ready for when it might be needed. Glucose is stored in the liver as **glycogen**.

To keep the level of glucose controlled there has to be a way of adding glucose to or taking glucose from the blood. This is where insulin comes in.

When the level of glucose gets too high...

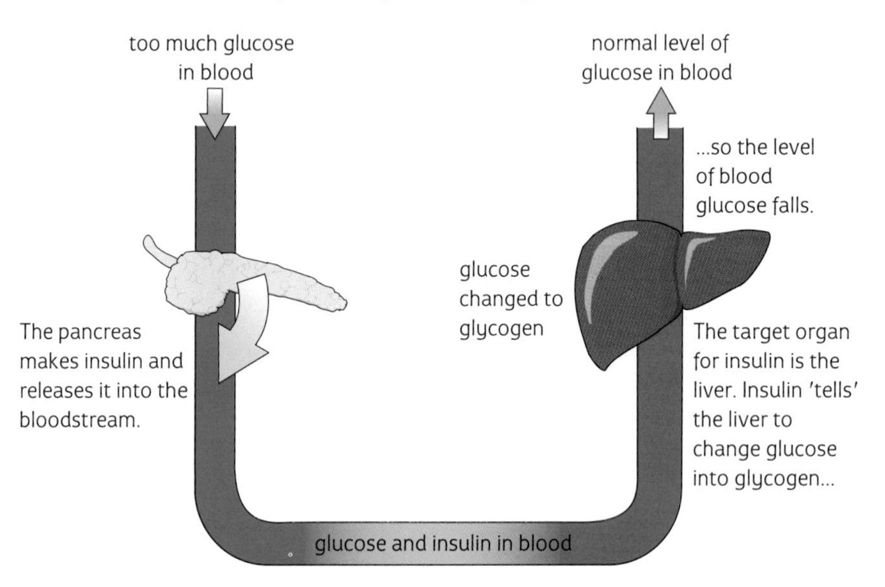

Did you know?

The pancreas has two jobs – making insulin and producing enzymes for digesting food.

When the level of glucose gets too low...

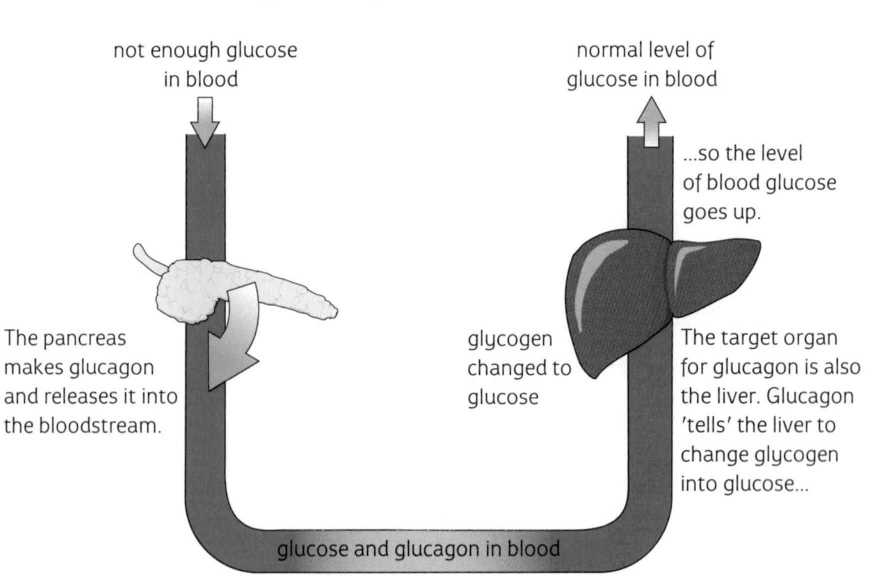

Questions

1 Explain why the level of glucose in the blood must be controlled.

2 Where in the body is glucose stored?

3 Name the hormone that:

a changes glucose into glycogen

b changes glycogen into glucose.

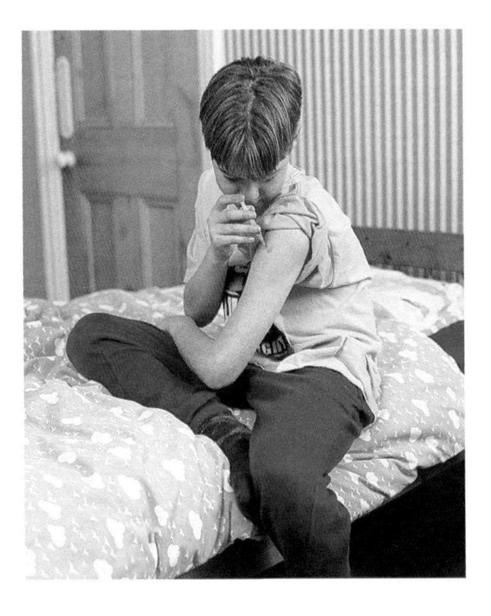

Diabetes

Diabetes is a disease caused when the pancreas doesn't produce enough insulin. If a person has too much glucose in their blood it can kill them. Someone who has diabetes is called a diabetic. Diabetics can get over their problem in one of two ways. They can avoid eating too many carbohydrates or they can inject themselves with insulin after a meal.

Adrenaline

Adrenaline is often called the 'fight or flight' hormone. This is because it prepares your body for action when you are in a dangerous, frightening, or exciting situation.

Adrenaline works by...

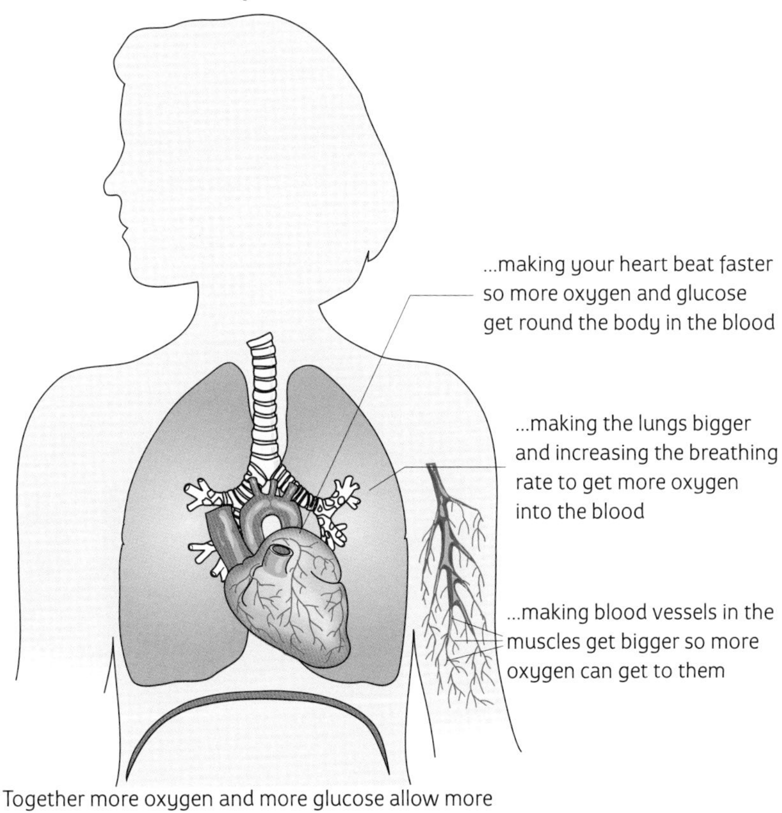

Together more oxygen and more glucose allow more respiration and therefore more energy ready for action.

Questions

4 What is a person with diabetes called?

5 Describe how a person with diabetes can control their disease.

6 Where is adrenaline made?

7 Explain why your heat beats faster when you are frightened.

8 Explain why getting more oxygen and glucose to muscle cells helps when you are running from danger.

9 Describe a situation when your body might produce lots of adrenaline.

2.25 The kidney and water control

Objectives

This spread should help you to

- describe the structure of a kidney
- describe how the kidney works
- describe how the amount of water in your body is controlled

The kidneys get rid of waste

You have two **kidneys**, just above your waist at the back. The kidneys are **excretory organs**. They get rid of poisons like **urea** and other unwanted substances from the blood. They also control the amount of water in your body.

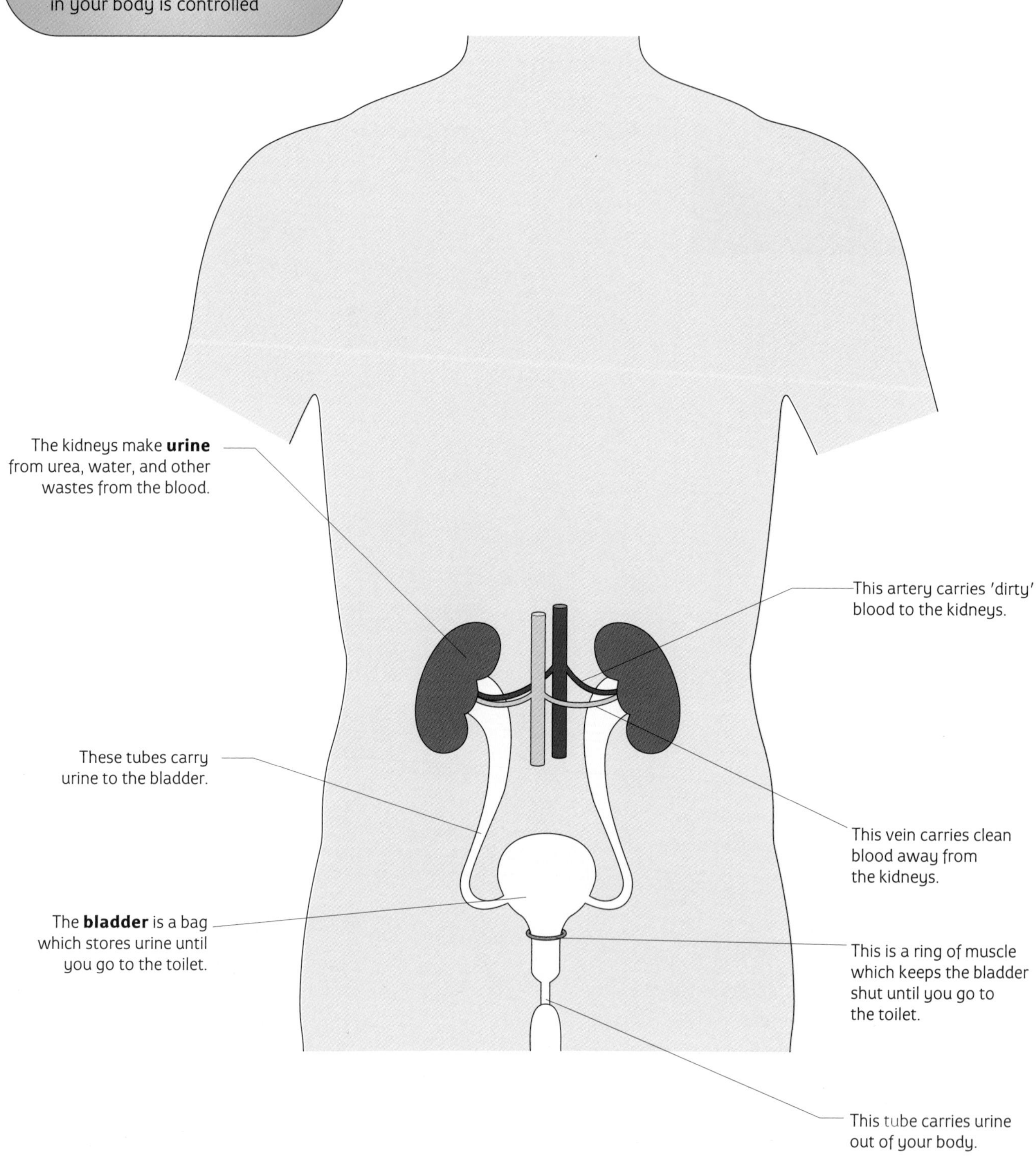

How the kidneys work

The kidneys are like filters that clean the blood. Filtering is done in millions of tiny tubes called **nephrons**.

The diagram shows a kidney cut in half. One nephron is enlarged.

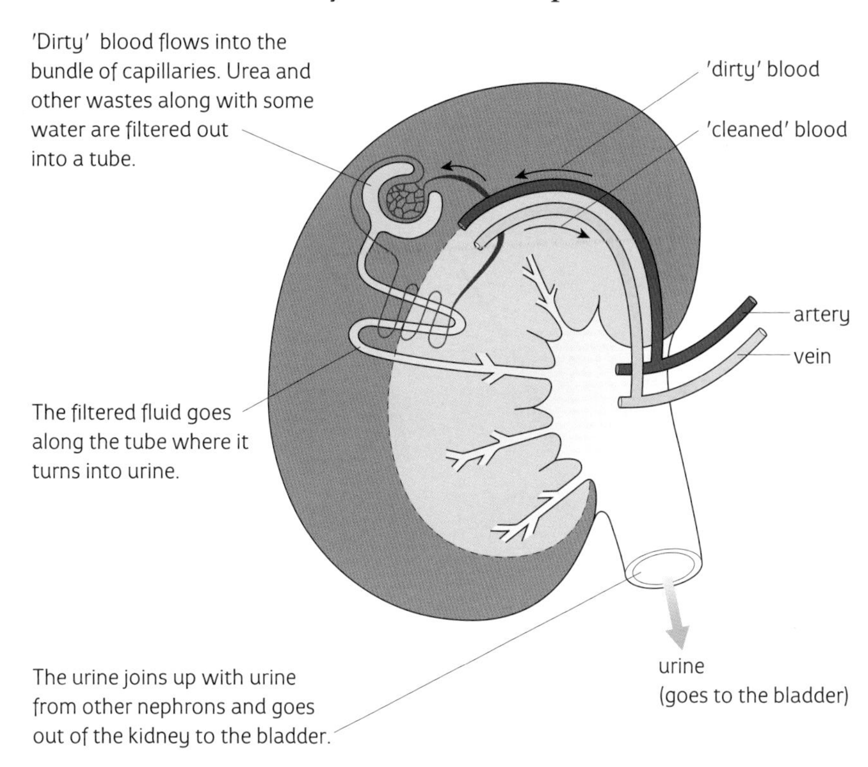

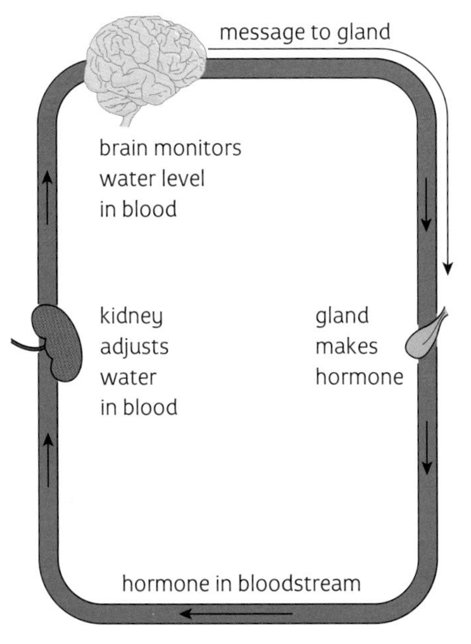

Keeping the right amount of water in the body

The body contains lots of water. We get water from food and drink. We lose water when we breathe out, when we sweat, and when we urinate. It is important that the amount of water in the body is carefully controlled if we are to stay healthy.

The amount of water in the blood is monitored as it passes through the brain. The brain sends a hormone to 'tell' the kidneys how much water to get rid of in the urine.

An adult's body contains about 40 litres of water. We feel thirsty if we lose only 2 litres. If we lose 10 litres we die.

Questions

1. Where are your kidneys?
2. Why is important to get rid of urea from the body?
3. What does the bladder do?
4. Explain how urine is kept in your bladder until you go to the toilet.
5. What are nephrons?
6. What happens in a nephron?
7. Describe what happens to urea from the time it enters the kidney in the blood to when it leaves the body in urine.
8. Explain how the amount of water in the blood is controlled.

2.26 The skin and temperature contr

Objectives

This spread should help you to

- describe what the skin does
- describe the structure of the skin
- describe what happens in your skin when you get too hot or too cold

What skin does

1 It protects your body against dirt and microorganisms.
2 It has sense organs for temperature, touch, and pain.
3 It loses water and mineral ions as sweat.
4 It helps keep your temperature at 37 °C.

The structure of skin

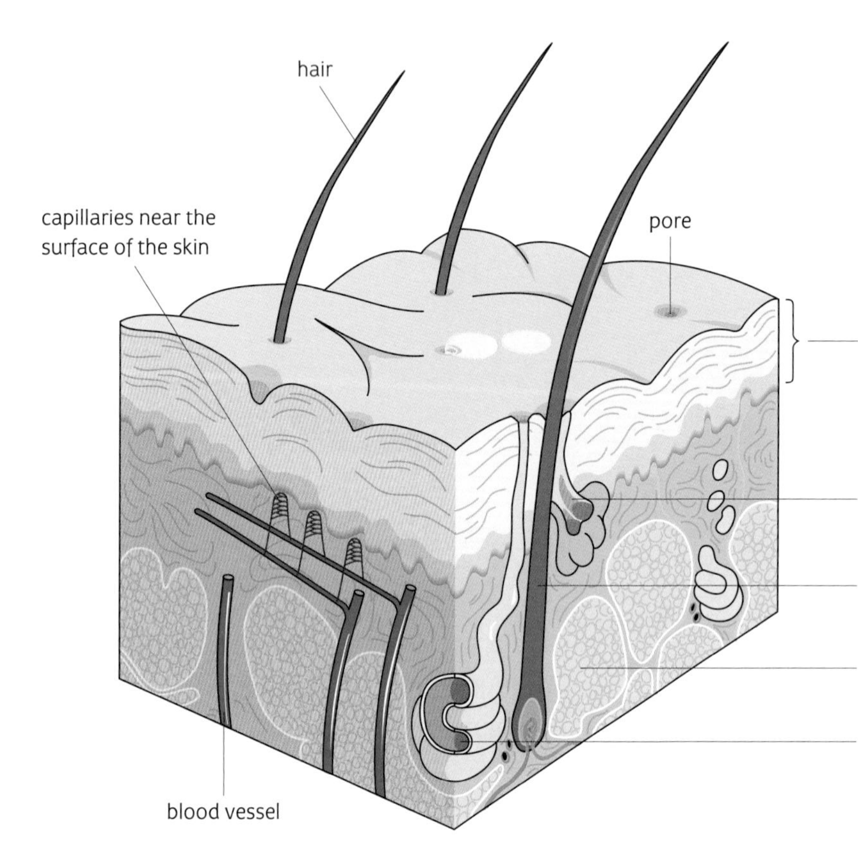

The outer layer is made of dead cells. As they rub off they are replaced by new cells from underneath.

Glands make an oily substance that keeps your skin supple and waterproof.

Hair protects your head from sunlight. It helps keep hairy animals warm.

This **fat layer** helps keep you warm.

Sweat glands make sweat. Sweating helps cool your body down.

Questions

1 List four things that the skin does.
2 What replaces dead skin cells as they rub off?
3 What keeps your skin waterproof?
4 Suggest why fat people get hot quicker than thin people.

This polar bear has lots of hair. This traps warm air round the body and helps to keep the animal warm.

When you get too hot:

1 You sweat a lot. Sweat evaporates and cools you down. You should drink more to replace the water lost in sweating.

2 Blood vessels in the skin expand. More blood gets near the surface – you go red! Heat goes from the blood into the air so you get cooler.

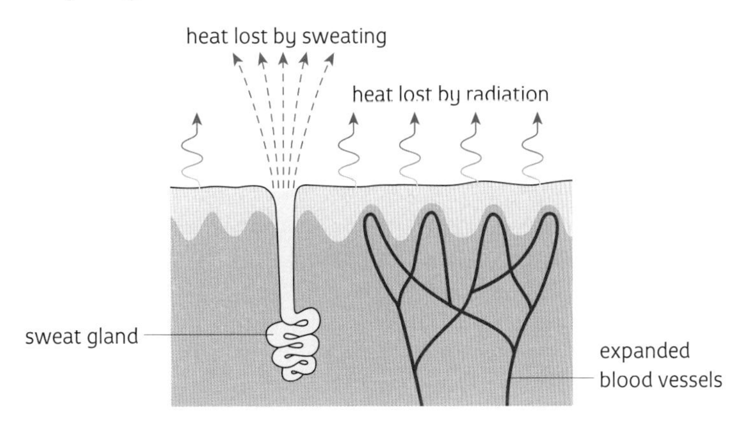

Did you know?

Only birds and mammals produce enough heat to maintain their body temperature. All other animals have a body temperature that goes up and down as the temperature of their surroundings goes up and down.

When you get too cold:

1 You don't sweat.

2 Blood vessels in the skin contract. Less blood gets near the surface – your skin looks paler. Heat stays in the body instead of being lost to the air.

3 You start shivering. This produces heat which warms you up.

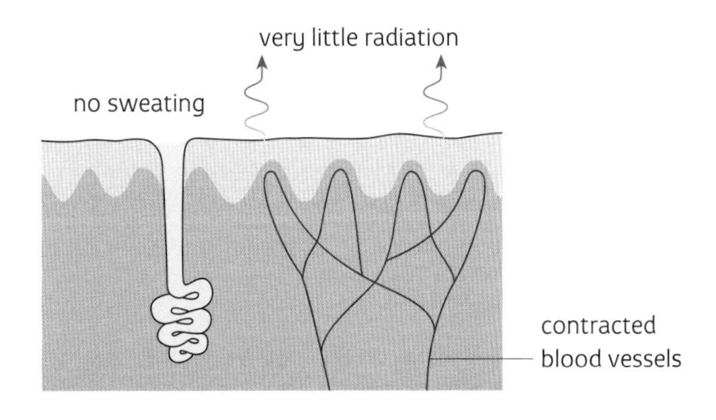

Questions

5 What is the normal body temperature of a human being?

6 Explain why it is important to drink lots of liquid when you get hot.

7 Describe the changes that happen in your skin when you get cold.

8 Explain why your face goes red when you are hot and paler when you are cold.

2.27 From sea to land – a problem?

Most plants and animals that live in the sea are not able to survive out of water. The sea and the land are very different environments. Nature has had to make some clever changes to plants and animals to enable them to live on land all the time.

A problem of support

Have you ever noticed that you feel lighter when you are swimming? This is because water gives you more support than air. Seaweed is all floppy when the tide is out. But when it is supported by water the seaweed is upright. On land even the smallest bodies need support to lift them above the ground. We have a skeleton for support inside our bodies. Insects have skeletons on the outside.

A problem of drying out

All living things contain large amounts of water – as much as 90%. Wet objects dry out quickly in the air, especially if it is warm. In the sea there is no problem. However, to avoid drying out on land, living things need a waterproof outer layer – at least on those bits that are exposed to the air.

A problem for reproduction

Sperms need something to swim in if they are going to successfully fertilize an egg. In the sea this is easy as there is plenty of water. On land, however, living things face more of a problem. Some, such as mosses, are restricted to living in damp places. Others have found some way of getting sperms straight into the female body. The penis is an excellent way of doing this.

A problem of temperature

Air warms up and cools down quicker than water does. During the day there is a greater temperature range on land than in the sea. This temperature range is even greater over a year. Animals in particular are able to regulate their body temperature to help them keep warm in cold weather and cool down in hot weather.

A problem of getting nutrients

The sea provides everything that animals and plants need – dissolved gases, nutrients, and light for photosynthesis. On land these things are available but in separate places. The soil contains water and nutrients, but it is dark. Above ground there is only light and some gases.

Animals can move to get the things they need. Plants, however, have to get water and nutrients from the soil, and light and gases from the air. To solve this problem, plants have root and shoot systems connected together by a transport system.

A tough, waterproof skin helps animals to survive on land.

Animals help to control their body temperature by sheltering from the heat of the Sun.

In water seaweed floats near the surface to get light. On land seaweed is all floppy.

Sperm need something to swim in. This is no problem in the sea. On land this is more difficult.

Plants living on land get their water and minerals from the soil and transport it to other parts.

Talking points

1. List the problems facing life on land.
2. Why is reproduction only successful in wet conditions?
3. Explain why an elephant has four legs but a bird has only two.
4. Explain why plants have a transport system.

Practice questions

1 The diagram shows a plant.

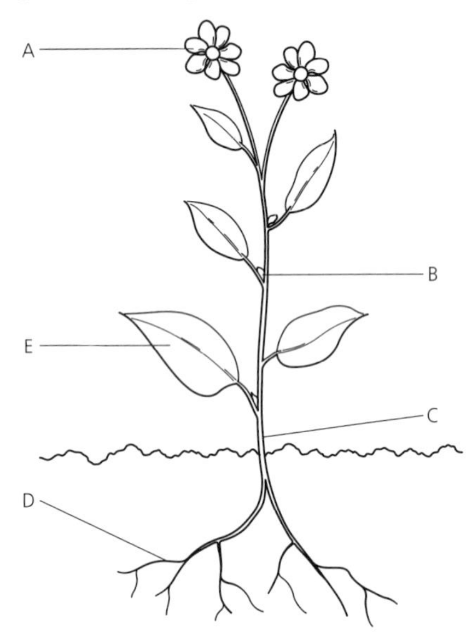

a List the letters A–E and write the correct label beside each one from the words below.

Parts: *bud, flower, leaf, root, stem*

b Copy the list of plant parts above and write the correct job alongside each one.

Jobs:

supports the plant

where reproduction takes place

anchors the plant in the soil

where photosynthesis takes place

where new growth takes place

2 The diagrams show a stem and a root.

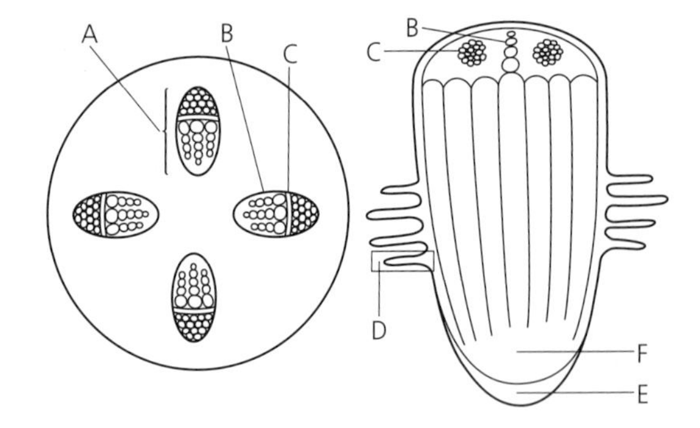

a List the letters A–F and write the correct label beside each one from the words below. (Some words may be used more than once.)

growing tip, phloem, root cap, root hair, vascular bundle, xylem

b Name two things that both roots and stems have.

c Which part:

i carries water up the plant

ii carries sugars around the plant

iii absorbs water and minerals from the soil

iv protects the growing root from damage?

3 The diagram shows the inside of a leaf.

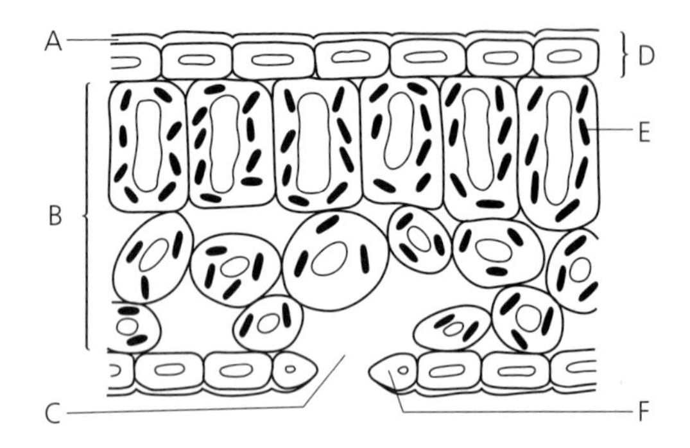

a List the letters A–F and write the correct label beside each one from the words below.

chloroplast, waxy cuticle, guard cell, leaf tissue, upper skin, stoma

b Copy the list of descriptions and write the correct part from the list above alongside each one.

Descriptions:

cells that make food by photosynthesis

makes the leaf waterproof

hole that lets gases in and out of the leaf

has chlorophyll inside

4 The diagram shows a leaf photosynthesizing.

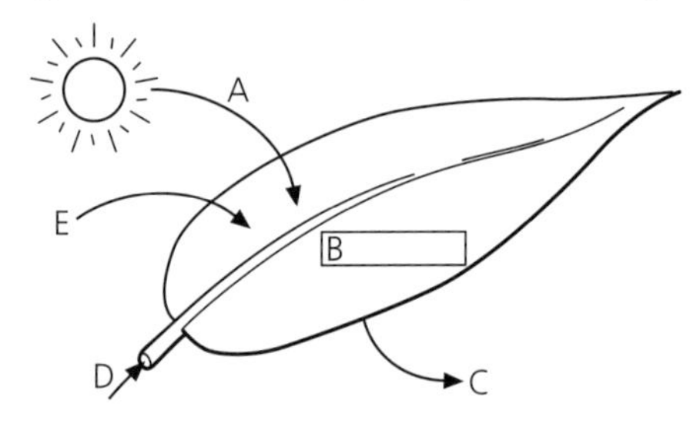

a List the letters A–E and write the correct label beside each one from the words below.

carbon dioxide, energy, glucose, oxygen, water

b Where does the energy for photosynthesis come from?

c What are the raw materials for photosynthesis?

d What are the products of photosynthesis?

e Write a word equation for photosynthesis.

5 The speed of photosynthesis depends on how much light and carbon dioxide there are, and also on the temperature.

The graph shows how the speed of photosynthesis is affected by carbon dioxide and light.

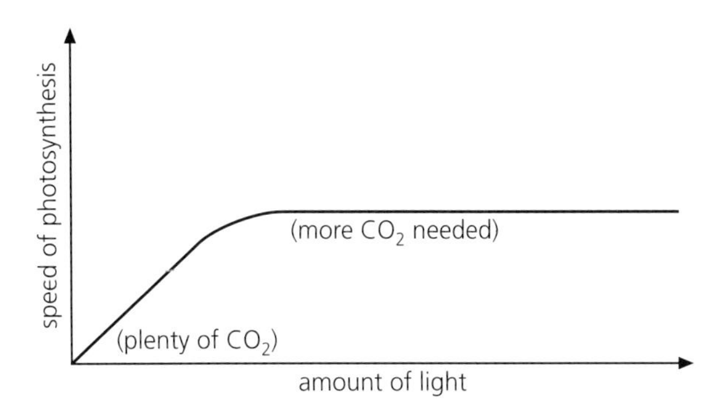

a Describe the change in the speed of photosynthesis.

b What causes this change?

c Suggest what would happen if there was more carbon dioxide.

This graph shows how the speed of photosynthesis is affected by light and temperature.

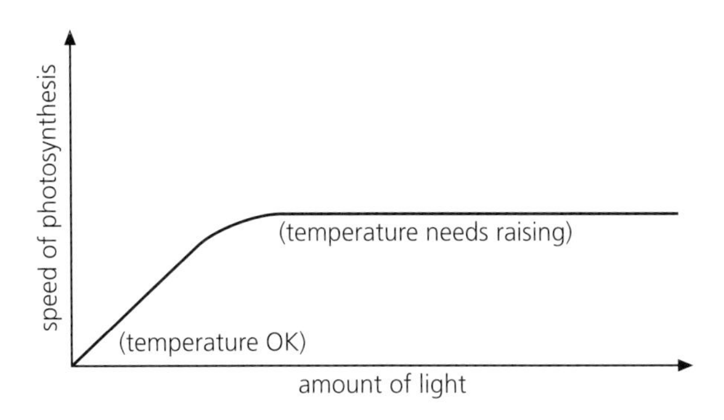

d What effect does temperature have on the speed of photosynthesis?

e Explain why crops grow better on warm sunny days.

6 The following diagrams show some plants. Each plant has been starved of one important mineral.

a Describe the appearance of a plant that has been starved of:

i magnesium **ii** phosphorus.

b What is the difference in the appearance of a plant starved of magnesium and a plant starved of nitrogen?

c Nitrogen is known as the 'leaf maker'. Explain why fertilizer containing nitrogen is spread onto fields where food crops are to be grown.

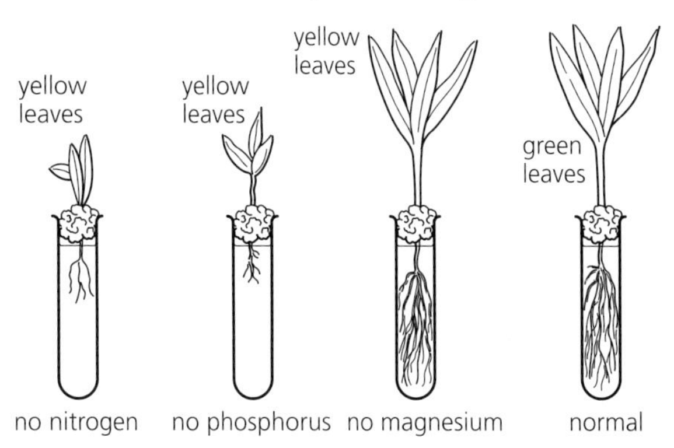

7 Two identical raw potato chips are measured. One potato chip is put in a beaker of water. The other chip is put in a beaker of strong sugar solution.

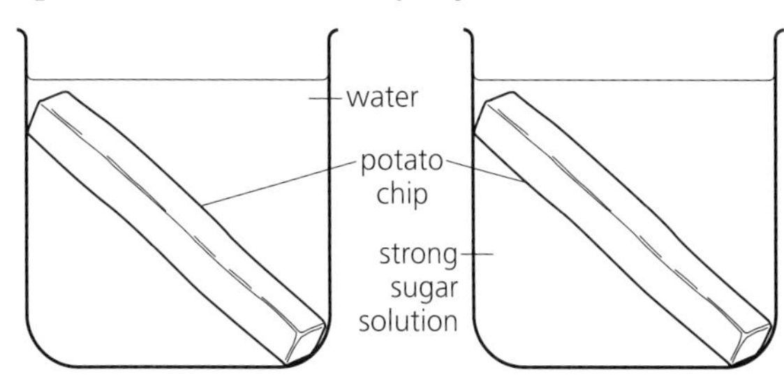

After one hour the chips are removed from the beakers and measured again. The results are in the table.

	Length of chip at start	Length of chip after one hour
chip in water	10 cm	11 cm
chip in sugar solution	10 cm	9 cm

a What has happened to the chip in:

i water **ii** sugar solution?

b What makes water move in and out of cells?

c Explain why the chip in water is firm and rigid but the chip in sugar solution is floppy.

d A plant that has plenty of water is upright and well supported. A plant that is short of water starts to wilt. Suggest a reason for this.

Practice questions

8 The diagram shows a piece of apparatus. It can be used to measure the speed of transpiration in a leafy shoot.

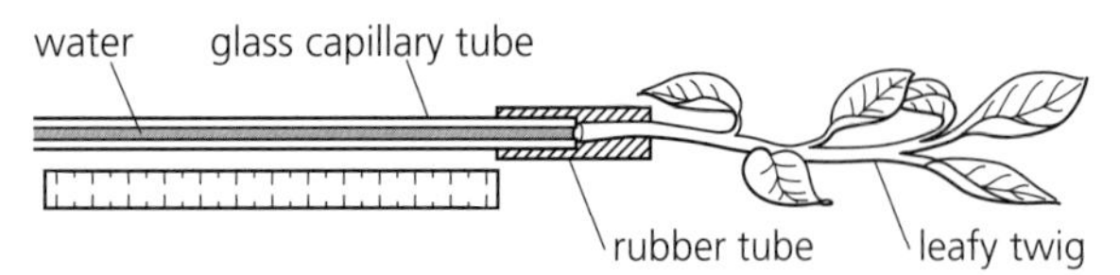

All you do is measure how far the water moves along the tube in a minute.

a What is transpiration?

b Why is it important to get an airtight seal between the leafy twig and the glass tube?

c Explain why it is important for the apparatus to be set up as soon as the leafy twig is cut from the plant.

d Suggest how you could use this apparatus to find out how transpiration is affected by different weather conditions.

e How does a plant control transpiration?

f Explain why transpiration is:

i useful to plants **ii** a nuisance to plants.

9 Some cress seeds are sown in a seed tray and watered. The tray is put into a box with a hole cut out of one end.

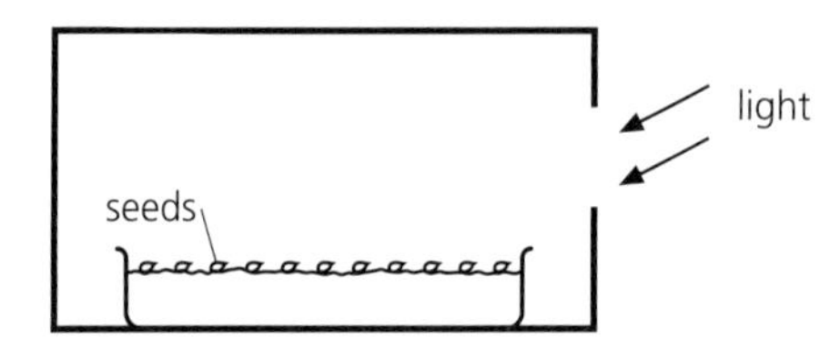

After a few days the seedlings looked like this.

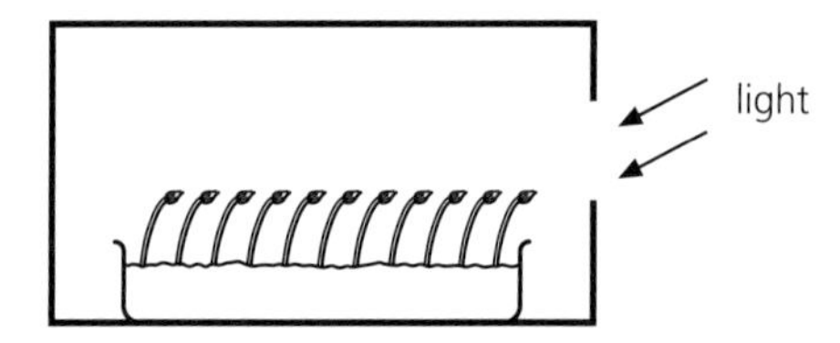

a i Describe what the seedlings have done.

ii Why have they done this?

b What do we call a plant's response to light?

c Describe how you would show that it is only the tip of the plant shoot that responds to light.

d i Name the plant hormone that causes this response.

ii Suggest what might happen if you gave a plant a lot of this hormone.

iii Give one example when it might be useful to give plants more of this hormone.

10 a The following diagram shows the parts of the eye.

List the letters A–K and write the correct label beside each one from the words below.

ciliary muscle, cornea, iris, jelly, liquid, optic nerve, lens, pupil, retina, sclera, suspensory ligament

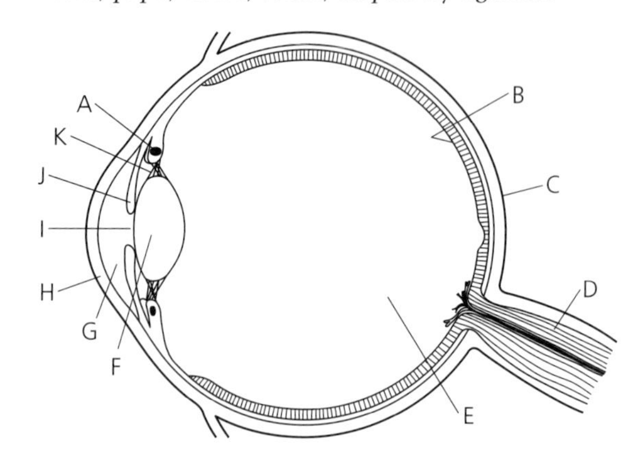

b Describe what happens to the shape of part F when the eye is focused on:

i a near object **ii** a far object.

c Describe what happens to part I when you walk into:

i a dark room **ii** a well lit room.

d What is the job of part D?

11 a Copy the list of sense organs and write the correct stimulus alongside each one.

Sense organs: *ears, eyes, nose, skin, tongue*

Stimulus: a bitter drink, colours of the rainbow, a fire alarm, a girl's perfume, a hot plate

b Which sense organ also helps us keep our balance?

c Suggest why you seem to lose your sense of taste when you have a cold.

12 a The diagram on the next page shows where hormones are made in the body.

List the letters A–E and write the correct label beside each one from the words below.

adrenal glands, ovaries, pancreas, pituitary gland, testes

b Copy this list of hormones and write the correct job alongside each one.

Hormones: *adrenaline, insulin, oestrogen, testosterone*

Jobs:

speeds up heartbeat and breathing

gives girls their female features

gives boys their male features

controls the amount of glucose in the blood

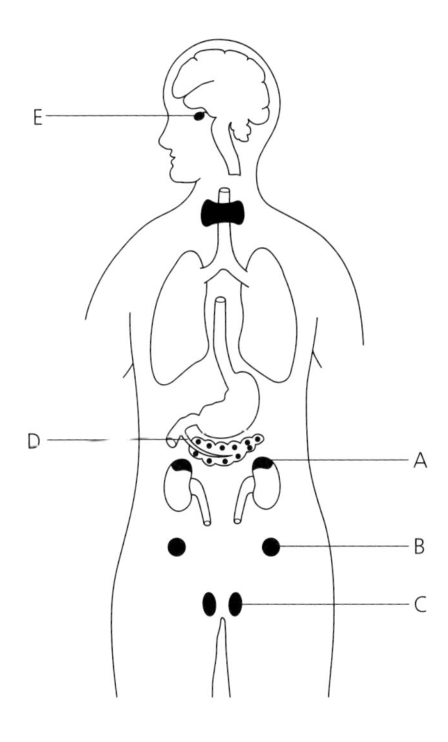

c Which hormone would be released if you were chased by a bull?

d A shortage of one of the hormones causes a disease called diabetes. Which one?

e How do hormones travel round the body?

f Explain why nervous 'messages' travel much faster than hormone 'messages'.

13 a Use this list of drugs to answer the questions.

alcohol, aspirin, caffeine, ecstasy

Name a drug that is:

i a depressant **ii** a stimulant

iii a painkiller **iv** a hallucinogen.

b Drugs can:

i become addictive. What does this mean?

ii have side effects. What does this mean?

14 The diagram shows part of the skin.

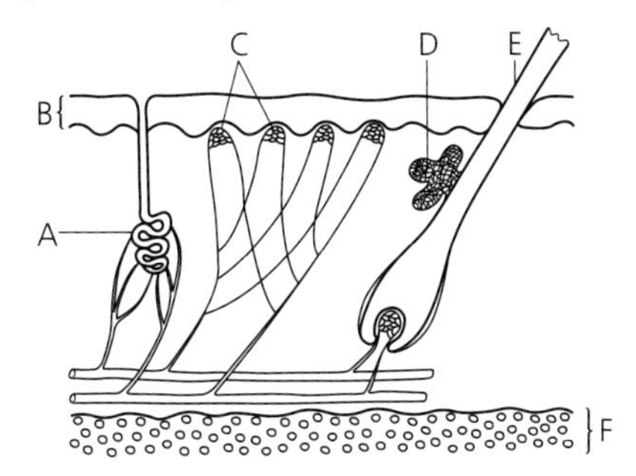

a List the letters A–F and write the correct label beside each one from the words below.

blood vessels, dead skin, fat layer, hair, sweat gland, oil gland

b Explain how:

i hair keeps hairy animals warm

ii sweating cools you down in hot weather.

c i Describe what happens to the blood vessels in the skin when you get cold.

ii How does this help stop heat loss from your body?

15 The diagrams show the windpipes of a smoker and a non-smoker.

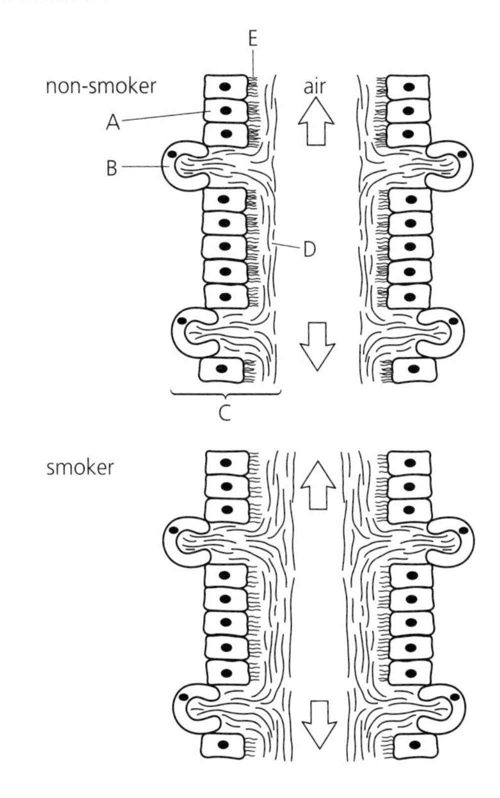

a List the letters A–E and write the correct label beside each one from the words below.

cells that make mucus, cells with cilia, cilia, mucus, windpipe wall

b Give two differences between the windpipe of a smoker and that of a non-smoker.

c Suggest how a smoker can try to get rid of extra mucus.

d Name the disease caused by bacteria getting into the lungs.

e i Name one other disease caused by smoking.

ii Explain how this disease is caused.

auxins hormones made in the tips of plant roots and shoots

brain the centre of control and coordination in the body

cancer a disease in which cells go on dividing out of control

carbon monoxide a poisonous gas which stops blood carrying oxygen

cell membrane a thin skin surrounding a cell; it controls what goes in and out of the cell

cell wall a rigid non-living layer around the outside of plant cells; it gives the cell support

central nervous system the brain and spinal cord together

chemical addiction the body gets used to a drug and needs it; without the drug the person suffers withdrawal symptoms

chlorophyll a green chemical that traps the sunlight energy needed for photosynthesis

chloroplasts green structures in plant cells, where photosynthesis happens

controlled drugs drugs with laws about them that say what they can and cannot be used for

coordinated body systems working together and at the right time

cytoplasm a jelly-like substance in a cell, where chemical reactions happen

depressant a drug that slows down nerve impulses

diabetes a disease caused when the pancreas doesn't produce enough insulin

drugs substances that change the way the body works

endocrine gland an organ where a hormone is made

excretory organs organs that get rid of poisons and unwanted substances from the blood

fat layer a layer of fat beneath the skin; it keeps animals warm

focus to make a clear image

geotropism plant growth movements in response to gravity

glands (in the skin) an organ that makes an oily substance to keep the skin supple and waterproof

guard cells two cells that surround a stoma

hallucinogen a drug that causes people to see things that aren't there

hormones chemicals made in endocrine glands that help coordinate the body

hydrotropism plant growth movements in response to water

impulses the way messages travel along nerve fibres

insulin a hormone that helps control blood sugar levels by changing glucose to glycogen

kidneys excretory organs that remove waste from the blood

leaves organs of photosynthesis

limiting factor something that slows down or stops photosynthesis, e.g. the amount of light

micrometre (μm) 1/1000 of a millimetre, a unit used to measure microscopic things

motor nerve cells (motor neurones) cells that carry impulses from the central nervous system to muscles and glands

nephrons tiny tubes in the kidneys which filter the blood and remove wastes

nerve endings parts of the skin that are sensitive to touch

neurones nerve cells

nicotine an addictive, poisonous drug in tobacco smoke

nucleus the control centre of a cell

osmosis the movement of water from an area of high water concentration to an area of low water concentration across a partially permeable membrane

partially permeable membrane a membrane which allows water through but not some other substances

photo-synthesis plants using light energy to make food from carbon dioxide and water

phototropism movements in response to light

pituitary gland a gland that makes hormones that control other endocrine glands

psychological addiction a person feels the need to keep taking a drug

reflex actions actions you don't have to think about before doing them

relay nerve cell a neurone that connects a sensory neurone with one or more motor neurones

response the way the body reacts to a stimulus

roots parts that anchor the plant in the ground and absorb water and mineral salts

sedatives drugs that calm the body down, e.g. sleeping pills

sensory nerve cells (sensory neurones) cells that carry impulses to the central nervous system from sense receptors

solvents the liquid part of glue, paint, and cleaning fluids

spinal cord bundle of nerve fibres running from the brain down the middle of the backbone

stereoscopic vision a view from each eye put together by the brain to give a 3-D effect

stimulant a drug which speeds up nerve impulses

stimulus something that triggers a sense receptor

synapse a microscopic gap between two nerve cells

tar (in tobacco smoke) a sticky substance containing chemicals which cause lung cancer

tissue a group of cells that are the same and do the same job

tranquillizer a drug which calms the body down

transpiration the loss of water vapour from the leaves of a plant

transpiration stream the flow of water in the xylem from roots to leaves

transport system tubes that carry water and sugar through the plant

tropisms growth movements in plants

unit (of alcohol) a measure of the amount of alcohol in an alcoholic drink

urea a waste substance made in the liver and excreted by the kidneys

urine a liquid containing urea and other wastes excreted by the kidneys

voluntary actions actions you have to think about before doing them

withdrawal symptoms unpleasant side effects when a person stops taking drugs

Animals and plants are usually well adapted to survive in their normal environment. Their populations depend on many things such as competition for food, avoiding being eaten, and avoiding being infected by diseases. By looking closely at the numbers and sizes of organisms in food chains, we can find out what happens to energy and biomass as it passes along a food chain.

All animals and plants eventually die. They produce waste when they are alive. Micro-organisms play an important part in decomposing this material so that it can be used again by plants. The same material is recycled over and over again.

The growth of the human population, and the increased use of natural resources, are upsetting the balance of natural ecosystems. Pollution of air, water, and land, along with the destruction of natural habitats, are reducing the number of different animals and plant species. With so many people in the world, there is a serious danger that we will cause permanent damage not just to local environments but to the whole world. Conservation is a way of keeping things at a steady level, but not everyone is prepared to give up a comfortable lifestyle to pay for it.

Module 3

3.01 The environment

Objectives

This spread should help you to

- explain what the environment is
- describe a habitat, a population, and a community
- describe two ecosystems

What does environment mean?

Environment is a scientific word for surroundings. You are probably reading this in your school environment or your home environment. Your environment provides you with air to breathe, water to drink, and a suitable temperature in which to live. These are the physical or non-living parts of your environment. Your life is also affected by other living things. These could be the people in your class, your family, your pets, and even bacteria in the air. Living things together with their physical environment form an **ecosystem**. In an ecosystem many different cycles happen. These help keep the environment the same as years goes by.

The word **habitat** is used to describe that part of the environment where an animal or a plant lives. Habitats contain lots of different animal and plant species.

Individuals of the same species living together are called a **population**. For example, a population of squirrels is made up of young and old squirrels living in a woodland habitat.

A **community** is made up of a number of different animal and plant populations. A wood is a good example of a community. It is made up of populations of trees, ferns, birds, primroses, squirrels, and many more different kinds of animal and plant.

A woodland ecosystem

Soil gives water and minerals for plants to grow. Animals depend on the plants for shelter and for food.

Questions

1. What does 'environment' mean?
2. What does 'ecosystem' mean?
3. What do these words mean? Give an example of each.
   a habitat **b** population
   c community
4. **a** Describe the environment where you are reading this book.
   b Give two non-living parts of this environment.
   c Give two living parts of this environment.
5. Name three animals that live in a woodland habitat.
6. One land ecosystem is shown on this page. Name two others.

A freshwater ecosystem

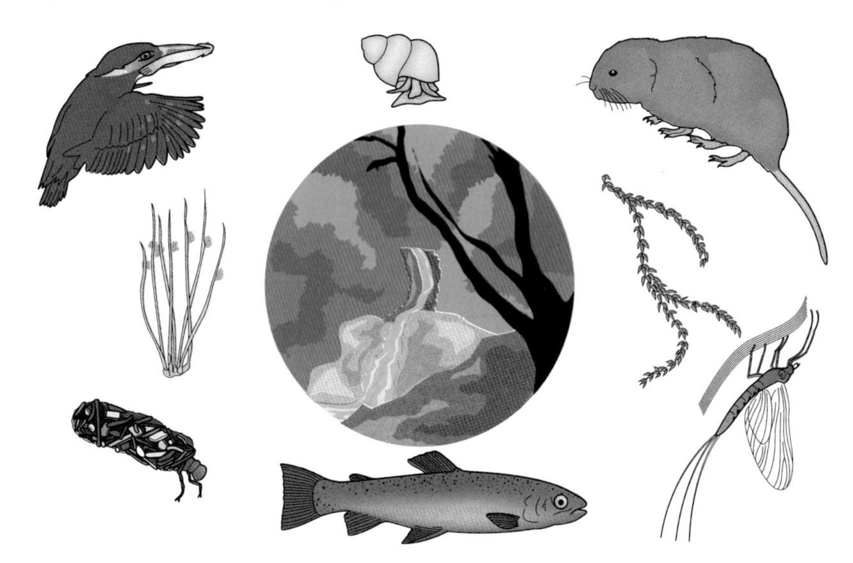

Each of these animals and plants affects the way the others live. They depend on each other.

A saltwater (marine) ecosystem

Animals and plants are adapted to survive in this ever changing environment.

Questions

7 Name two plants that live in a freshwater habitat.

8 Why do you think kingfishers live in a freshwater stream environment?

9 Name three animals that live in a saltwater environment.

10 Some seaweeds have air bladders on their leaves. How do you think this helps them?

11 Two water ecosystems are shown on this page. Name another one.

12 Describe the environment you would choose for an ideal summer holiday.

3.02 Adaptations

Objectives

This spread should help you to

- describe how the polar bear is adapted to live in very cold conditions
- describe how the camel and the cactus are adapted to live in the desert
- suggest how other living things are adapted to survive in their environments

Adapted for life

Every living thing is adapted to live in a certain way. Lions have strong teeth for tearing flesh and crushing bones. Butterflies have long tubular mouths for sucking nectar. Most flowers have colour and scent to attract insects for pollination.

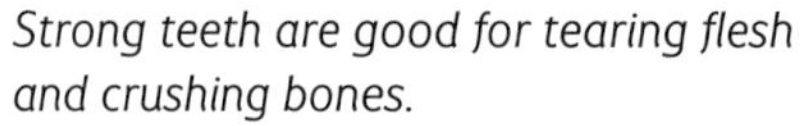

Strong teeth are good for tearing flesh and crushing bones.

Long mouthparts let the butterfly feed on nectar deep inside the flower.

Some living things have extra special adaptations to help them survive in very harsh environments.

The polar bear

The polar bear lives in **arctic** conditions where it is very cold, especially in winter. It has lots of special features which help it live in a very cold climate.

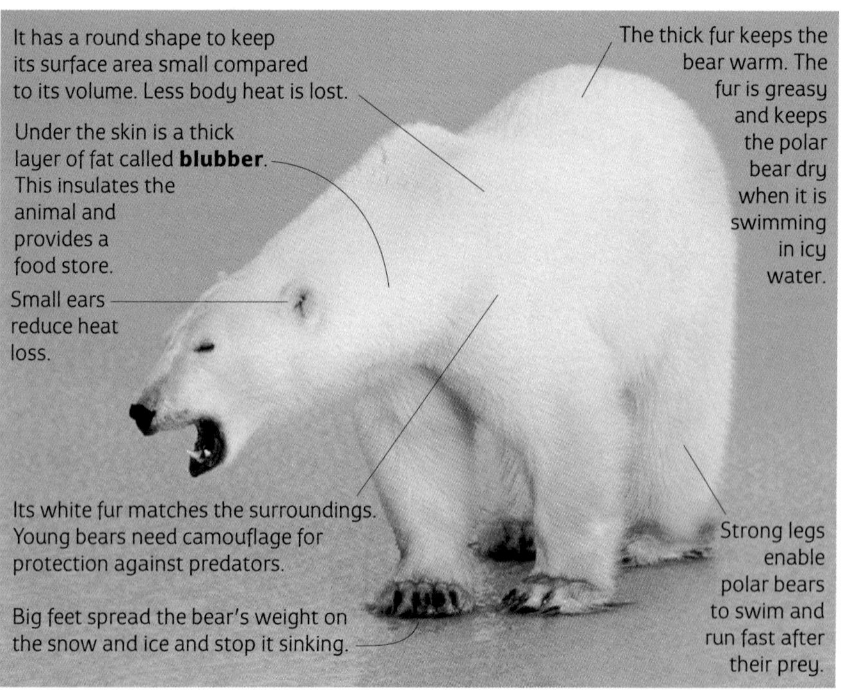

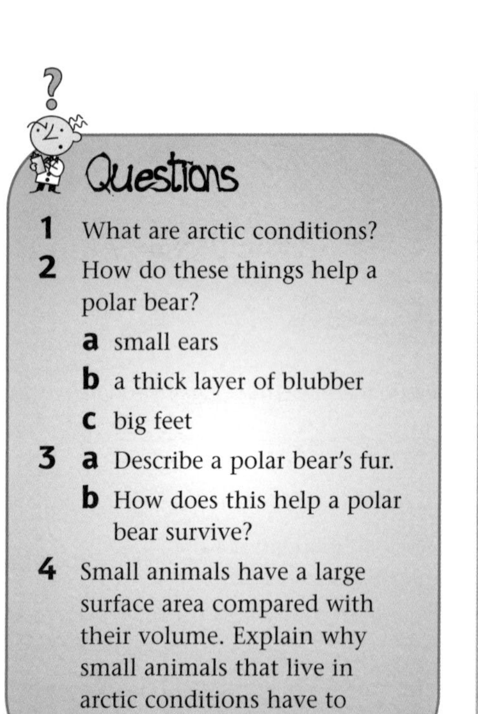

Questions

1 What are arctic conditions?

2 How do these things help a polar bear?
 a small ears
 b a thick layer of blubber
 c big feet

3 a Describe a polar bear's fur.
 b How does this help a polar bear survive?

4 Small animals have a large surface area compared with their volume. Explain why small animals that live in arctic conditions have to hibernate in winter.

The camel

Camels live in **deserts** where it is very hot and there isn't much water about. They are very well adapted for life in the desert.

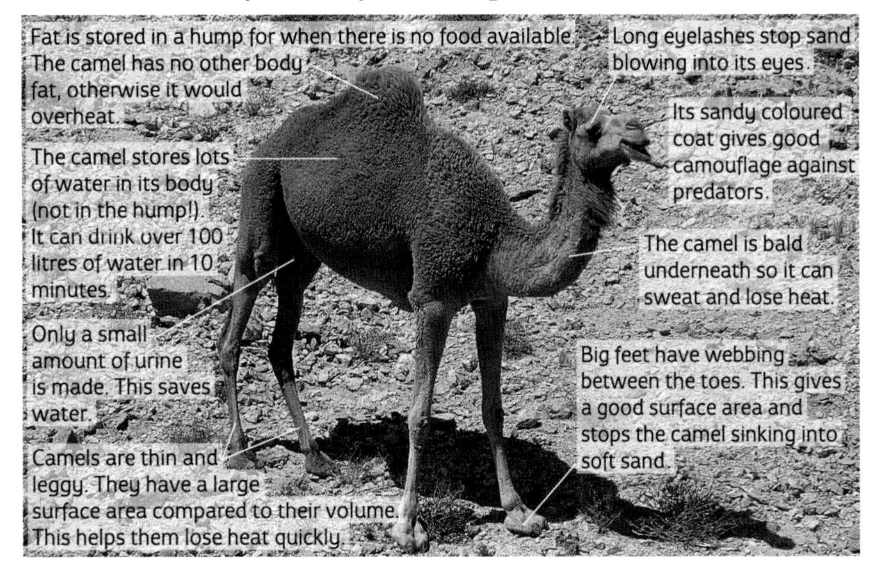

Did you know?

Fish that live in the Arctic produce an antifreeze for their blood to help them survive in the ice-cold water.

The cactus

Cacti also live in deserts. Like camels, they are well adapted to cope in a dry environment.

Questions

5 What are desert conditions?

6 How do these help a camel?

- **a** big feet
- **b** long eyelashes
- **c** a bald tummy
- **d** a hump

7 Explain how a camel can go for long periods without drinking water.

8 Describe the adaptations that help the cactus survive in desert conditions.

9 Look at the drawing of a fish below. Explain how each adaptation helps the fish move easily through the water. For example, fins help steer the fish in the right direction.

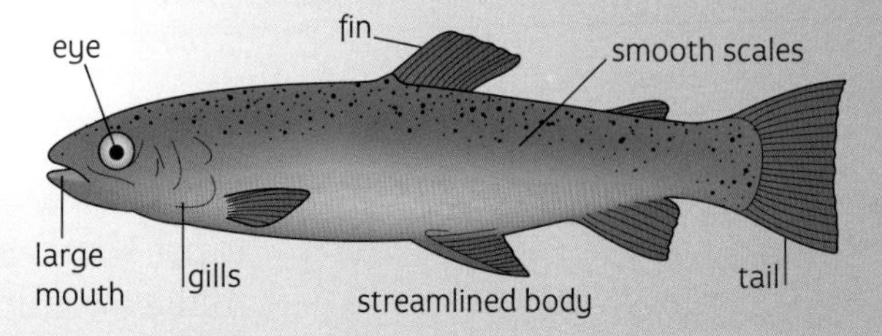

3.03 Competition

Objectives

This spread should help you to
- describe what animals compete for
- describe what plants compete for

In any ecosystem there are always more living things born than can ever survive. Only those best adapted to their environment will survive. This was one of Charles Darwin's important observations. It helped him come up with his theory of evolution.

Competition happens between animals and plants of the *same* species and also between animals and plants of *different* species.

Competition between birds of the same species and birds of different species.

What do animals compete for?

If they are going to survive, animals need food and water. The more food and water an animal can get, the better its chances of survival. Animals can move from one place to another, so they can hide from predators or shelter from bad weather.

For the survival of the species, individual animals must mate. The bigger, fitter males usually win the battle to mate with the females. Fighting is common between animals, especially males. If they are not fighting over a mate, animals will fight for space to live in.

In most animal species, males fight for a mate or to defend their territory.

Questions

1. List four things that animals compete for.
2. Why do animals fight?

What do plants compete for?

Plants make their own food using the energy from light. They must therefore try to get as much light as they can so they can make as much food as possible. The plants that grow fastest and tallest usually survive.

In a wood the faster growing trees win the race for light.

Plants need water and minerals from the soil. They can't make food without them. Plants with root systems that spread deeper and wider in the soil will be more likely to win the competition.

Bright, sweet-smelling flowers attract insects for pollination. The more attractive a flower is to insects, the better its chances of reproducing.

A good root system helps a plant compete for water and minerals.

Insects are attracted to the brightest, most colourful and sweetest smelling flowers.

Questions

3 Explain why it is an advantage for an animal to move from one place to another.

4 Give one example of competition between:

- **a** animals of the same species
- **b** animals of different species.

5 List four things plants compete for.

6 Explain how having a good root system helps a plant survive.

7 Plants can't move from place to place. Why is this a disadvantage for plants?

8 Straight tree trunks are useful for building. Explain why trees that are grown for building are planted very close together.

3.04 Wild populations

Objectives

This spread should help you to
- describe the four stages of a population growth curve
- describe some environmental pressures that affect populations of animals and plants

Population growth

If a healthy population of wild rabbits moves into a new area where there is plenty of food, space, and shelter, it will quickly get bigger. This happens because the extra food makes the rabbits more fertile and so many more rabbits are born than die. But soon space gets filled up, competition for food begins, and predators move in. Now fewer rabbits are born and more rabbits die. The rabbit population changes.

Foxes help to keep the rabbit population steady.

If the change in the number of rabbits is plotted on a graph, a **growth curve** is produced. There are four stages in a growth curve.

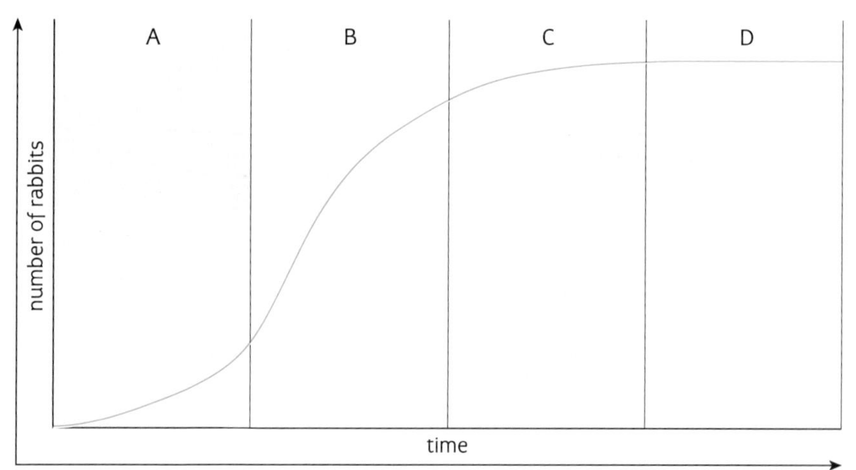

Questions

1. Explain why wild animal populations grow quickly when they move into a new area.
2. Give two things that help keep wild populations steady.
3. Why does the growth of a wild population start slowly?
4. Explain why the population grows rapidly in stage B.
5. Bacteria reproduce by dividing into two every 20 minutes. Starting with one bacterium in ideal conditions:
   - **a** Draw a table showing the population growth for five hours.
   - **b** Use the information to plot a growth curve. Put time on the horizontal axis.

Birds like the robin fight for their own space.

These lions have lots of choice.

Flooding destroys the homes of burrowing animals as well as this snail.

Few plants survive forest fires. Animals that can't escape are killed.

Environmental pressures

Things such as food shortages, which limit the growth of an animal or plant population, are called **environmental pressures**.

There are two main groups of environmental pressure:

1 Pressures due to the size of the population itself

- Shortage of food, water, or oxygen affects all living things. Nothing can survive without them.
- Without light, plant populations can't photosynthesize. Without photosynthesis, plants die.
- Lack of space causes overcrowding. Breeding is upset, especially in birds that like their own territory.
- The more animals or plants there are in a population, the easier it is for a predator to catch its prey.
- Disease spreads more easily if lots of individuals live closely together.

2 Pressures that the population has no control over

- These have the same effect whether the population is big or small.
- Sudden temperature changes can kill many living things. Lots of birds die in cold winters.
- Forest fires destroy the habitats of many woodland plants and animals.
- Storms and floods wash away the homes of burrowing animals like rabbits. Plants can't survive if they are covered with water for long time.

Questions

6 What are environmental pressures?

7 Explain the difference between the two groups of environmental pressures.

8 List three things other than shortage of food, water, and oxygen that can cause the size of a population to fall.

9 Suggest one other environmental pressure that a population has no control over.

10 In the 1950s the rabbit population in Britain was almost wiped out by the disease myxomatosis. Why do you think the disease spread?

3.05 Predators and prey

Objectives

This spread should help you to

- describe how predators are adapted to catch their prey
- describe how prey are adapted to avoid their predators
- describe a predator–prey relationship

Exploding and crashing

When the growth curve of a population levels off, the numbers don't usually stay the same for long. Quite often an animal population gets much bigger or smaller as time goes by. These are called **population explosions** or **population crashes**. More food or more space can cause a population explosion. Disease can cause a population crash.

In any ecosystem there are animals that eat other animals. These are **predators**. The animals they eat are called their **prey**.

Predators are well adapted for catching and eating their prey.

An eagle's beak is adapted for tearing flesh.

The cheetah can run very fast to catch its prey.

Predators have eyes that face forwards to give good stereoscopic vision.

Prey animals are well adapted for avoiding being caught and eaten.

Wildebeest stay in herds where there is protection in numbers.

Zebra are well camouflaged to avoid being seen.

Prey animals have eyes on the sides of their head to give all-round vision.

A successful predator gets its prey.

Predator–prey relationships

There will be predators and prey in any community of living things. The relationship between predators and their prey plays an important part in controlling populations. When a predator eats its prey, it takes individuals from the population. These are usually sick or old animals. Neither of these are useful to the species as a whole. Sick animals spread disease to the others and the old use up valuable food and space. This sounds a bit hard, but the future of a species depends on healthy animals breeding successfully.

But what happens if a predator eats young healthy animals? Obviously the prey population starts to fall. If it falls too far (crashes) the population of predators will go down as well. Fewer predators means that fewer prey will be eaten. This allows the prey population to rise again (explode).

The graph below shows how the populations of a predator and its prey change over a period of time.

Notice that the number of predators is always lower than the number of prey, and that the predator population explodes and crashes soon after the prey population.

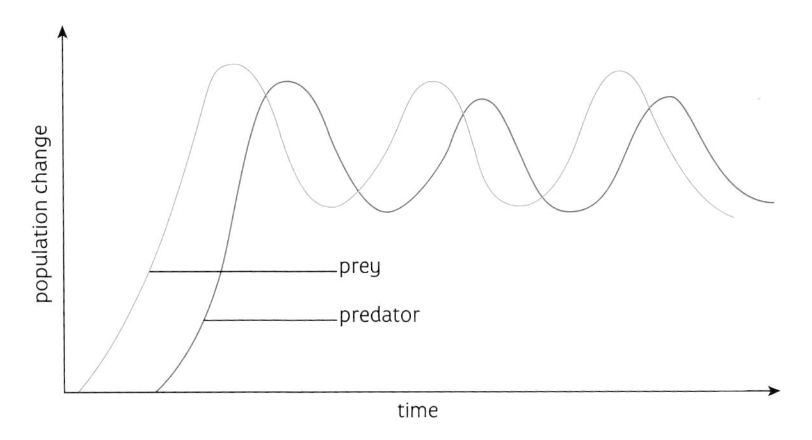

Questions

1. What is meant by: **a** a predator **b** prey?
2. Explain the difference between a population explosion and a population crash.
3. Give one example of something that might cause: **a** a population explosion **b** a population crash.
4. How are predators adapted to catch their prey?
5. **a** Which animals in a population does a predator usually catch? **b** Why is this good for the prey population?
6. Describe how the populations of predators and prey change over time.
7. Explain why predator–prey relationships are important in controlling populations.
8. Explain why the population of predators falls soon after the prey population falls.
9. Why do you think there are always more prey animals than predators?

3.06 Food chains

Objectives

This spread should help you to

- explain what a food chain is
- name the parts of a food chain

Passing energy along

Lots of things go on in an ecosystem, especially feeding. Food passes from plant to animal and from one animal to another along a **food chain**.

Green plants make their own food during photosynthesis. Energy in light is converted into the chemical energy in food. A living thing that makes its own food is called a **producer**. Green plants are producers.

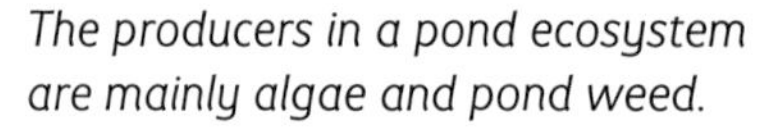

The producers in a pond ecosystem are mainly algae and pond weed.

The producers in a woodland ecosystem are more varied.

Animals however can't make their own food. They must get their food by eating plants or by eating other animals. Animals are called **consumers** because they eat or consume other living things.

Animals that eat plants are called **herbivores**. Rabbits are herbivores – they eat grass, cereals, lettuce, and lots of other plants. Animals that eat other animals are called **carnivores**. A fox is a carnivore. It feeds on animals such as insects, mice, birds, and rabbits.

Some animals feed on a diet that has both plants and other animals. These are called **omnivores**. Badgers are omnivores. They feed on things like grass, fruit, slugs, worms, and rabbits.

Questions

1 What is a food chain?

2 What is light energy converted into during photosynthesis?

3 What is:

a a producer **b** a herbivore

c a carnivore **d** an omnivore?

4 Give one example of a producer in:

a a pond ecosystem

b a woodland ecosystem.

5 Are you a herbivore, a carnivore, or an omnivore? Explain your answer.

A rabbit is a herbivore.

A fox is a carnivore.

Badgers are omnivores.

Links in the chain

Every plant and animal is a link in a food chain. Energy, in the form of food, passes along a food chain from producers to consumers.

Here is a typical food chain.

The arrows show which way the energy in the food is going. In this example the energy goes from the lettuce to the rabbit and on to the fox.

Notice there are three links in this food chain. Not all food chains have three links. Some have more, like this one.

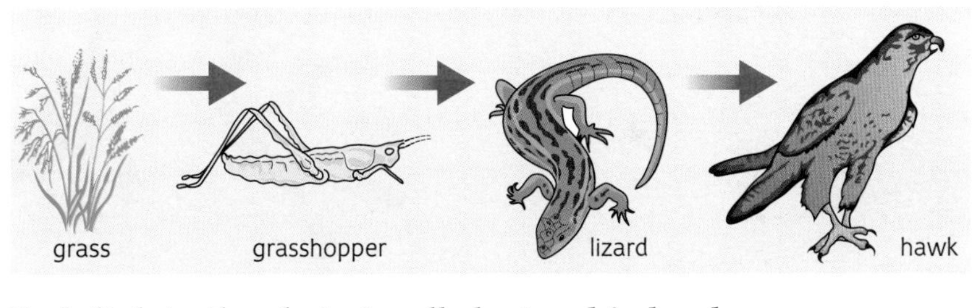

A field of producers and first consumers.

Each link in the chain is called a **trophic level**.

Herbivores that feed on producers are called **first consumers**. Mice, rabbits, horses, and sheep are first consumers. Carnivores that feed on first consumers are called **second consumers**. **Third consumers** feed on second consumers and so on up the food chain. Owls, foxes, cats, and dogs are carnivores that could be second or third consumers. It all depends on what they're eating at the time!

6 What do the arrows in a food chain mean?

7 What is a trophic level?

8 Explain the difference between a first, second, and third consumer.

9 Explain how a human could be both a first and a second consumer.

10 A student went to study a local pond. In the pond she noticed: water fleas eating algae; a heron with a perch in its beak; minnows eating water fleas; and a perch chasing minnows.

a What is the: **i** shortest **ii** longest food chain in the pond?

b How many consumers are there in the longest food chain?

3.07 Food webs

Objectives

This spread should help you to
- explain what a food web is
- describe how decomposers work

Joining up the food chains

Single food chains don't give us a full picture of the feeding relationships between plants and animals. Rabbits don't eat lettuce all the time and foxes eat other things besides rabbits.

If we were to trace every food chain involving lettuce, rabbits, and foxes we would end up with lots of interconnecting food chains. This is called a **food web**.

See how complicated it all can get even with only a few plants and animals in a food web!

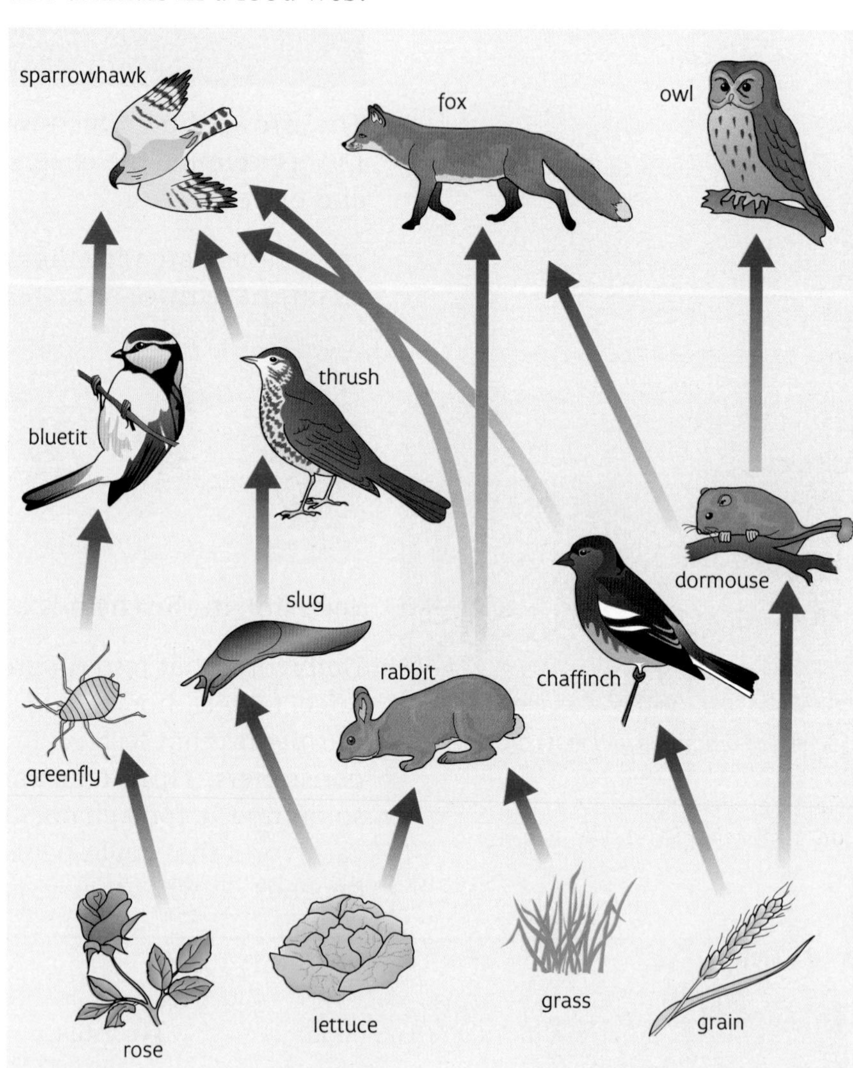

A North Sea food web

The producers in the food web on the opposite page are plant plankton. **Plankton** are microscopic organisms that live near to the surface of the sea. Animal plankton are the first consumers. There are lots of examples of second consumers such as the herring, and third consumers such as mackerel and cod.

Questions

1 What is a food web?

2 Look closely at the food web opposite.

a Write a food chain with:

i two links **ii** three links **iii** four links.

b What are the names of the herbivores?

c Which carnivore has the most varied diet? Explain your answer.

d Suggest one way that a sparrowhawk can:

i help **ii** hinder a gardener.

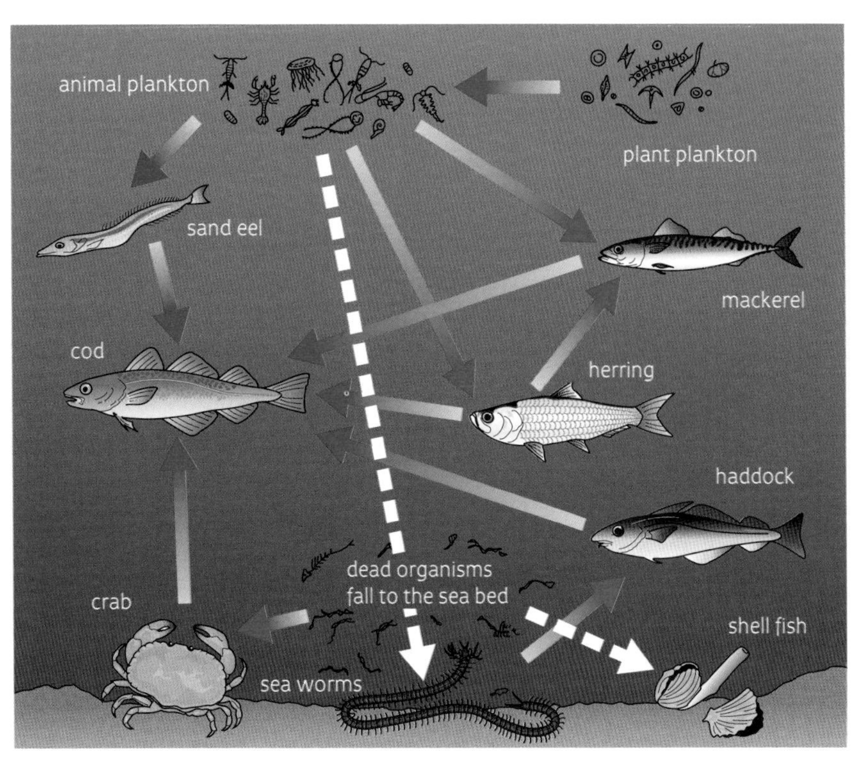

In compost heaps, decomposers turn waste plant material like potato peelings and garden weeds into fertilizer.

Fungi decomposing a dead tree. Minerals are being released into the soil.

Decomposers

Decomposers are bacteria and fungi. They feed on dead animals and plants, making them decompose or rot. They produce digestive enzymes which break down dead tissues. The digested material is then absorbed into the decomposer's cells.

Decomposers are important because they keep soil fertile. Dead animals and plants are removed and the minerals from their bodies are released into the ground. Decomposers work best where it is warm and damp and there is plenty of oxygen.

Producers take up the minerals and use them to make food. Animals feed on producers and the whole cycle repeats itself.

Decomposers are useful in breaking down human waste in **sewage works** and waste plant material in **compost heaps**.

Questions

3 Look closely at the North Sea food web above.

a Name: **i** the producers **ii** the first consumers **iii** a second consumer.

b Give one time when a mackerel is:
i a second consumer **ii** a third consumer.

c Why can the cod be called a fourth consumer?

4 What are decomposers?

5 What conditions do decomposers like best?

6 Explain why decomposers are important.

7 Explain how decomposers keep soil fertile.

3.08 Pyramids of numbers and biomass

Objectives

This spread should help you to

- describe a pyramid of numbers
- describe a pyramid of biomass
- explain the difference between pyramids of numbers and of biomass

Pyramids of numbers

In many ecosystems there are usually more producers than consumers. For example, in the food chain:

there are probably thousands of single grass plants, eaten by 15 rabbits which in turn may be eaten by one fox. Each time you go up one trophic level, the number of living things gets a lot less. It takes a lot of food from the level below to keep the animals above alive.

This information can be shown as a sort of bar chart. The length of each bar represents the number of living things in each link of the chain.

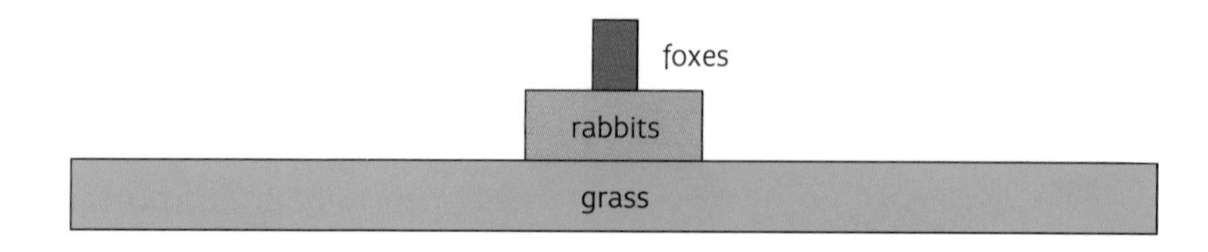

This is called a **pyramid of numbers**.

Not all pyramids are neat and tidy like the one shown above. Sometimes they look wrong. Suppose the producer in a food chain was a single oak tree. Feeding on the oak tree could be thousands of caterpillars. Small birds feed on the caterpillars. In turn hundreds of fleas could be feeding on the blood they suck from the birds. The 'pyramid' for this food chain is a very strange shape!

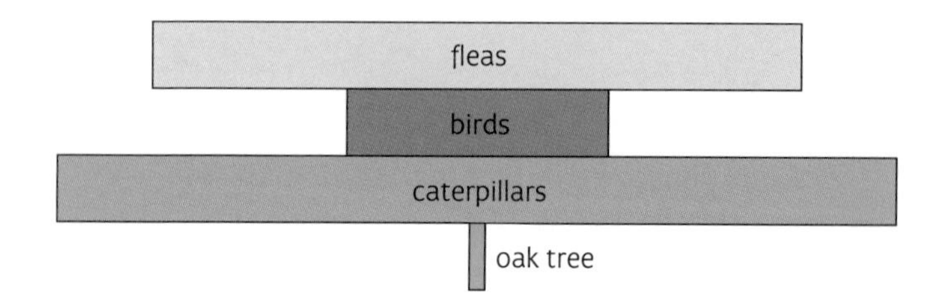

Questions

1 Why is the chart called a pyramid of numbers?

2 What does the length of each bar stand for in a pyramid of numbers?

3 Explain why pyramids of numbers can be different shapes.

4 Draw a pyramid of numbers for this information:

a 1 rose bush, 3000 greenfly, 1000 ladybirds, 2 bluetits

b 1000 wheat plants, 20 field mice, 1 cat, 200 fleas.

An oak tree has a big biomass.

Pyramids of biomass

Biomass is a word used to describe the mass of living material in an ecosystem. If you could collect up all the animals and plants in an ecosystem and weigh them, this would be the biomass for that ecosystem. So, using the food chain:

the grass would weigh about 500 kg, the rabbits would weigh 50 kg, and the fox would weigh about 10 kg. This information can also be shown as a bar chart.

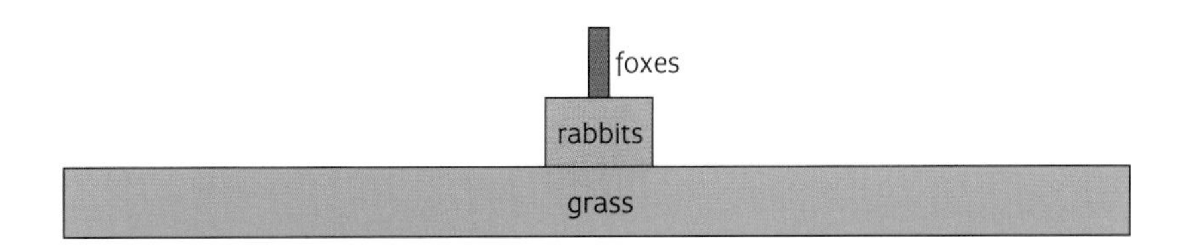

We get another pyramid – a **pyramid of biomass**.

If we draw a pyramid of biomass for the oak tree food chain we get a true pyramid shape this time.

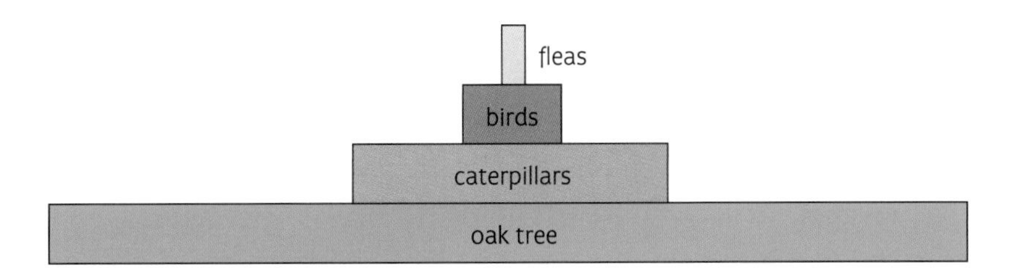

This is because a single oak tree weighs a lot more than the animals in the trophic levels above – even thousands of fleas!

Questions

5 What is biomass?

6 What is your biomass?

7 **a** Give one food chain where the pyramids of numbers and biomass are the same shape.

b Give one food chain where the pyramids of numbers and biomass are different shapes.

3.09 Energy flow through ecosystems

Objectives

This spread should help you to

- describe how energy flows through a food chain
- explain why energy is lost at each trophic level

Starting at the Sun

The Sun is the energy source for nearly all living things. Energy flows from one link to the next along a food chain. Producers convert the light energy from the Sun into chemical energy in food. Some of this energy is used by the plants in respiration. The rest of the food might be used to make protein for growth or stored as starch.

When consumers feed on plants they release the stored energy by digestion and respiration. They use this energy to move. Some energy is 'locked up' in the animal's body as protein and fat as it grows.

This goes on and on up the food chain until all of the energy is 'lost'. Energy is lost as heat to the environment.

Let's follow the flow of energy through a food chain which is important to humans.

grass → bullock → human

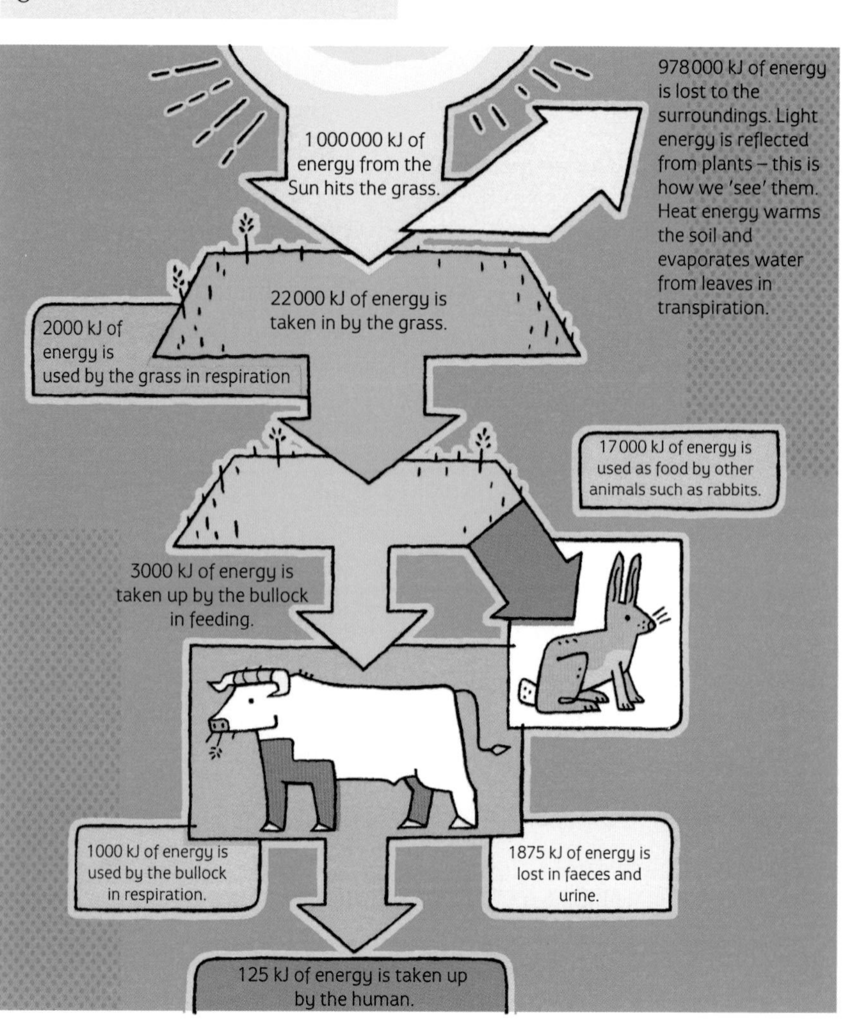

Questions

1. What is the source of energy for nearly all living things?
2. Give three things that plants use energy for.
3. What do we mean by saying energy is 'locked up' inside a body?
4. Where does energy go to when it has passed through a food chain?
5. Explain why a plant takes in only a tiny amount of the energy that hits it from the Sun.
6. Suggest where the energy might go to when a human eats meat from the bullock.

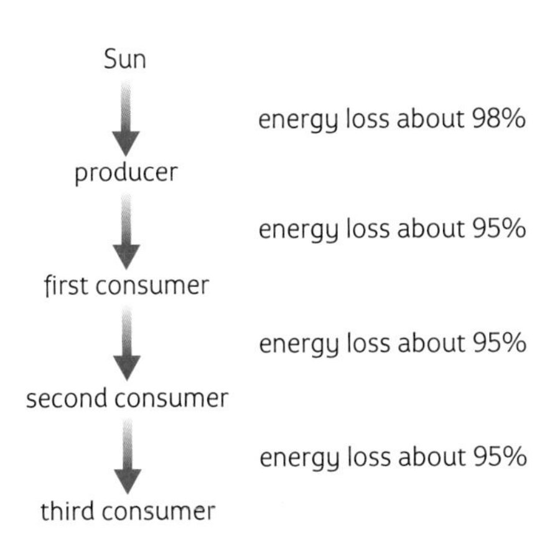

Losing energy

Energy flow through food chains is very **inefficient**. Lots of energy is lost as heat at each stage.

This energy loss could be reduced if we ate more food from lower down the food chain. Eating plants instead of the animals that feed on them makes better use of the energy that is available.

Of course, it is important to eat a balanced diet. Most people eat some meat but many people eat more meat than they need to. Land like moorland and hillsides, that isn't suitable for growing crops, can be used to rear animals such as sheep and deer. This is an efficient way of using the land we have available. Unfortunately, our demand for meat has resulted in animals being reared in intensive conditions. Intensive farming methods keep large numbers of animals in small spaces so they can't waste energy by moving about. This means they won't need as much food, so the meat costs less to produce.

More people can be fed by growing crops than by grazing animals like sheep and cattle.

These hens can't waste energy by moving around and keeping warm – they're too close together.

Wheat or meat?

An average person can get enough energy for one day by eating…

Questions

7 Why is energy flow through a food chain inefficient?

8 Roughly, how much energy is lost between:

a a producer and a first consumer

b a first consumer and a second consumer?

9 Explain why it makes sense to eat food from low down a food chain.

10 Explain why it is efficient to rear animals for food on moorland and hillsides.

11 a What is intensive farming?

b Explain why intensive farming produces meat at lower cost.

12 In some countries there is not enough food to go round.

a What should the people eat, bread or meat?

b Explain your answer.

3.10 More about decomposers

Objectives

This spread should help you to

- describe how microorganisms help in the disposal of sewage
- describe how microorganisms help in making compost

Decomposers are microorganisms such as bacteria and fungi. They are very useful in sewage works, where they break down human waste, and in compost heaps, where they break down waste plant material.

Sewage disposal

Sewage consists of urine and faeces from our bodies, water from washing, and some industrial wastes. It also contains harmful microorganisms that could spread disease. These things must be removed from the sewage so the water leaving the works is safe.

This is what happens in a sewage works.

1 Screening. Raw (untreated) sewage comes in through a metal grid. This strains out things like wood, rags, and paper, which are burned.

2 Grit tanks. Sewage contains grit which gets into the sewers from roads. The screened sewage goes slowly through tanks where the grit and stones sink to the bottom.

3 Settling tanks. The sewage is piped into tanks and left to settle. Faeces slowly sink to the bottom to form **sludge**. This is later used to make fertilizer and methane gas. The liquid above the sludge is called **liquor**. The liquor passes on to the next stage.

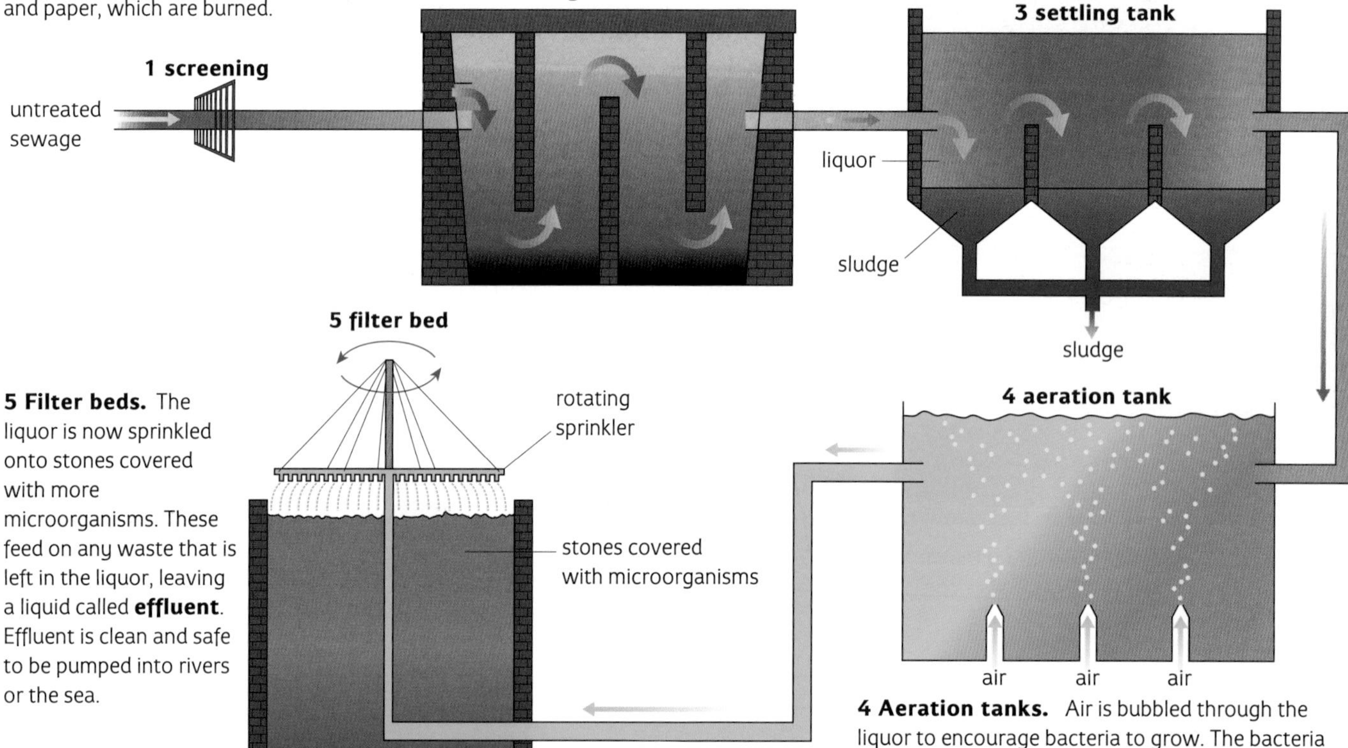

4 Aeration tanks. Air is bubbled through the liquor to encourage bacteria to grow. The bacteria feed on the waste in the liquor and produce carbon dioxide and other harmless substances.

5 Filter beds. The liquor is now sprinkled onto stones covered with more microorganisms. These feed on any waste that is left in the liquor, leaving a liquid called **effluent**. Effluent is clean and safe to be pumped into rivers or the sea.

Making compost

Every year we throw away thousands of tonnes of household and garden waste. Almost half of this can be made into compost.

All dead things are decomposed eventually. A compost heap is simply an environment where this natural process can be speeded up. To encourage a strong population of decomposer microorganisms, they must have the best conditions for growth.

Making compost reduces the amount of rubbish you throw away and provides free fertilizer for your garden.

These are:

1 plenty of food – microorganisms get this from the waste put into the heap. It is good to have a mixture of things that decompose fast, such as fruit and vegetables, and things that decompose slowly, like wood chippings and leaves. This gives the microorganisms a steady supply of food.
2 warmth – once the microorganisms get going they generate a lot of heat through respiration. The ideal temperature in a compost heap is about 60 °C. This is hot enough to kill weed seeds and the eggs of insects such as flies.
3 air – microorganisms need oxygen for aerobic respiration. This breaks the food down properly into simple substances that plants can use. A good compost heap should be well aerated.

Turning the compost regularly makes sure that all the compost gets enough air and stays at an even temperature. This will speed up decomposition.

Compost is ready when it has a crumbly appearance, an earthy smell, and when you can't recognize things in it!

What to put in

- garden waste, for example grass cuttings, flowers, leaves
- kitchen waste, for example fruit and vegetable peelings, eggshells
- shredded newspapers and tissues
- sawdust and wood shavings
- animal manure

What not to put in

- meat scraps
- dead animals
- conifer twigs and pine needles
- waste sprayed with pesticide
- cat and dog faeces

Questions

1 What is sewage?
2 What is the difference between sludge and liquor?
3 Give two uses for sludge.
4 Why is air bubbled though the liquor?
5 What happens on filter beds?
6 What do we call the clean liquid pumped into rivers from sewage works?
7 What three things do microorganisms need to grow?
8 List four things that:
a you can **b** you can't make compost from.
9 Describe a good compost bin. Explain why each feature is important.
10 Explain why compost takes longer to make in winter than in summer.
11 Give one way that sewage disposal and compost making are: **a** the same **b** different.

3.11 The carbon cycle

Objectives

This spread should help you to

- describe the carbon cycle
- explain how the amount of carbon dioxide in the air is kept the same

All living things need carbon. It is needed for protein, fats, and all the other things that living things are made of. The carbon comes from the carbon dioxide in the air.

Plants take in carbon dioxide from the air. They use it to make food by photosynthesis.

carbon dioxide + water ⟶ glucose + oxygen (ENERGY needed)

Animals get their carbon by eating plants (or each other).

The amount of carbon dioxide in the air stays roughly the same. This is because it is put back as fast as plants take it out. Animals and plants give out carbon dioxide when they respire. Bacteria and fungi respire and give off carbon dioxide when they are decomposing dead bodies in the soil.

glucose + oxygen ⟶ carbon dioxide + water (ENERGY released)

Wood, coal, gas, and petrol contain carbon. When they are burned the carbon joins up with oxygen in the air to make carbon dioxide.

carbon + oxygen ⟶ carbon dioxide

In our ecosystem, carbon goes from the environment into living things and then back again. This is called the **carbon cycle**, shown opposite.

Every time you light a gas cooker you are releasing carbon which was locked up millions of years ago.

A lot of carbon dioxide comes from car exhausts.

These trees help to keep the balance of carbon dioxide in the air.

Questions

1 **a** Why do living things need carbon?

b Where do plants get their carbon from?

c Where do animals get their carbon from?

2 Name one thing that:

a takes carbon dioxide from the air

b puts carbon dioxide back into the air.

3 Why is the carbon cycle a good name?

4 Explain why we should worry about the increased use of motor vehicles on our roads.

5 A large forest is cut down. The branches are then burned. Give two reasons why this may increase the amount of carbon dioxide in the air.

3.12 The human population

Objectives

This spread should help you to

- explain why the human population keeps on rising
- describe some of the effects that a rising human population is having on the environment

... humans can protect themselves from predators...

... agricultural techniques have improved so that many countries produce more food than they need.

The human population breaks the rules

Wild populations are controlled by predators, disease, climate, and food supply. Populations can explode but always crash when environmental pressures are put on them. But this doesn't apply to humans because...

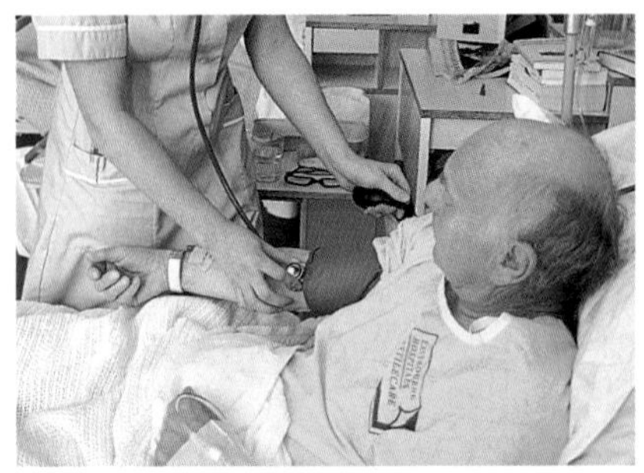

... they have the skill and knowledge to overcome many diseases...

... they build homes which protect them from the worst effects of changes in climate...

All this means that fewer babies die and old people live longer. The human population is growing bigger and bigger. It is growing by about 1.5 million every week – that's 150 a minute!

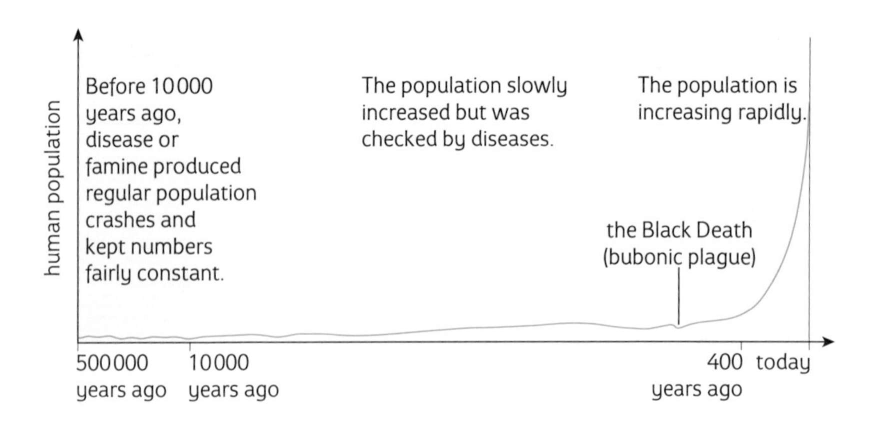

The table shows birth rate and death rate in some countries.

Area	Population in 1985 (in millions)	*Birth rate	*Death rate	% increase yearly	Doubling time	Population in 2000 (in millions)
Africa	551	45	16	+2.9	24 yrs	869
Asia	2829	28	10	+1.8	39 yrs	3562
Europe	492	13	10	+0.3	240 yrs	509
Latin America	406	31	8	+2.3	30 yrs	554
North America	264	15	8	+0.7	99 yrs	297
World	4845	27	11	+1.7	41 yrs	6135

*Births and deaths per 1000 of the population per year

In poor countries jobs are done by hand.

In rich countries jobs are done by machine.

The population in many poor, developing countries is rising quickest. The people in these countries have different problems from those living in richer parts of the world.

Poor countries

People are living longer in poor countries because of imported medical help. But the birth rates are still high, for many reasons.

- Children are needed to support the family. Most children are working by the age of 10.
- Most work is still done by hand, as people can't afford machines. So lots of people are needed to do the work.
- Having a son is seen as very important. Women have a low social status. Couples could have lots of daughters before they get a son.

Rich countries

Population growth in richer countries is slowing down and may soon stop because:

- Many couples are choosing not to have children. They are expensive!
- Most work is done by machines so fewer people are needed.
- Men and women have equal status.

Damage to the environment

The rapid rise in the human population could cause worldwide social, economic, and environmental disaster. As poorer countries become more industrialized, the demands on limited resources will increase rapidly. Pollution levels will go up as more fossil fuels are burned. Reserves of minerals will be used up to make things to improve people's lives. More trees will be cut down to make space to grow food and rear cattle.

We hope that developing countries will learn from the mistakes made by the richer ones. But who can blame them for wanting a better quality of life?

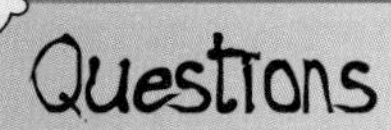

Questions

1 How fast is the human population growing?

2 Explain why the human population isn't affected by environmental pressure in the same way as wild populations are.

3 When did the human population start to go out of control?

4 Explain why the birth rate is higher in poorer countries than richer ones.

5 The population of some African countries could double in the next 25 years. Describe the environmental problems this will cause.

3.13 Pollution of air

Objectives

This spread should help you to

- describe what pollutes the air
- describe how the greenhouse effect is caused
- describe how acid rain is caused
- describe how the ozone layer is affected by CFCs

Any substance that changes the environment and harms living things is a **pollutant**. Pollutants can affect the air, water, and land.

The greenhouse effect

The Earth is surrounded by air. The air acts like a layer of insulation, keeping heat in, just like the glass in a greenhouse. Two gases, carbon dioxide and methane, are very good at keeping the heat in – this is why they are called **greenhouse gases**. They have been very useful in making the Earth warm enough for human life. But the amounts of carbon dioxide and methane in the air are increasing. The increased **greenhouse effect** is causing the Earth to slowly warm up. This is called **global warming**.

Methane is a greenhouse gas. It comes naturally from marshy areas such as where rice is grown...

... and from cattle as they belch and break wind!

The increase in carbon dioxide is mainly due to burning more fossil fuels. But, all over the world, we have been cutting down more trees to make space for farming and for building homes. This is called **deforestation**. Fewer trees means less photosynthesis. Less photosynthesis means less carbon dioxide is taken from the air.

Changes in climate could cause drought or flooding. A lot of food crops are grown in places such as Europe and North America. Global warming will make these areas drier and less fertile. The polar ice-caps will melt and the sea will expand as it warms up, so sea levels will rise. Low-lying areas such as parts of eastern England, Holland, and Florida will be flooded.

Acid rain

As well as carbon dioxide, sulphur dioxide and oxides of nitrogen are produced when fossil fuels are burned. These gases mix with clouds to form acids. These then fall as **acid rain**.

Acid rain makes lakes acidic. This has a serious effect on the ecosystem. The acid causes aluminium to be released from the soil into the water. Aluminium is poisonous to water life. It kills fish and the birds that eat them.

Acid rain causes the death of trees and other plants. Their leaves drop off and their roots are damaged so they can't absorb water.

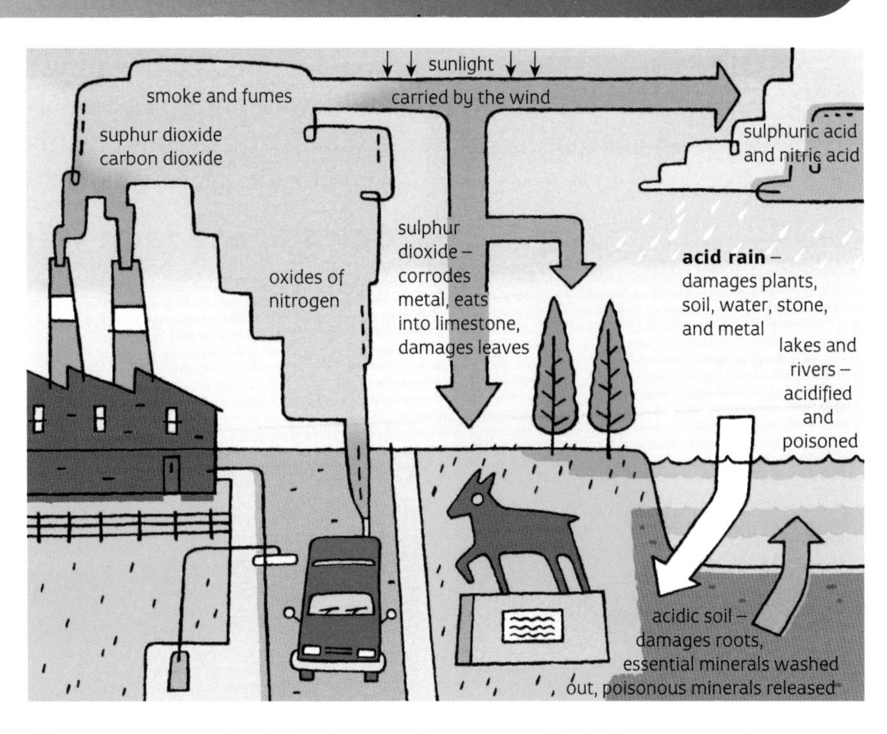

Harmful ultraviolet rays can cause skin cancer. The Slip! Slop! Slap! campaign in Australia encourages people to protect themselves.

The amount of acid rain can be controlled. Power stations now have '**scrubbers**' which take the harmful gases out of the smoke as the fossil fuels are burned. Modern cars and lorries are fitted with **catalytic converters** to clean up exhaust gases.

CFCs and the ozone layer

Ozone is a gas that forms a layer in the air high above the Earth. It is very important because it filters out harmful **ultraviolet** (UV) **radiation** from the Sun. CFC gases are used in some aerosol cans and fridges. Unfortunately when they get into the air they break up the ozone. The break-up of the ozone layer lets more UV rays get to Earth. In parts of the world, such as Australia and New Zealand, it is dangerous to go out in the sun because of the increased risk of skin cancer. CFCs are slowly being replaced by other gases but the damage done already will take a long time to repair.

Questions

1 What is a pollutant?

2 a What is air polluted by?
b Where do these pollutants come from?

3 a Name two greenhouse gases.
b Explain how they are causing global warming.

4 Explain how global warming can lead to flooding of low-lying areas.

5 Why should we be concerned about the amount of rice that is grown and cattle that are reared?

6 What gases cause acid rain?

7 Explain how acid rain is made.

8 Describe the effect of acid rain on limestone.

9 Explain how acid rain kills things living in lakes.

10 a What does ozone do?
b How is ozone affected by CFCs?
c Explain why it is a good idea to replace CFCs with other gases.

3.14 Pollution of water

Objectives

This spread should help you to

- describe what pollutes water
- explain the difference between biodegradable and non-biodegradable waste

Water is polluted by untreated sewage, chemicals from factories and farms, and spilled oil. Pollution affects the thousands of different animals and plants that live in water. Our health depends on a clean water supply.

Sewage and farm waste

Most of our sewage is treated in sewage works. But sometimes, especially after heavy rainfall, untreated sewage gets into rivers. Farmers put fertilizers onto their land to help their crops grow well. If too much fertilizer is used or if it rains after using it, the fertilizer gets washed from the soil into streams and rivers. Both sewage and fertilizers contain **nitrates**.

A sewage works uses bacteria to digest human waste.

A thick blanket of algae stops light getting to the water plants below.

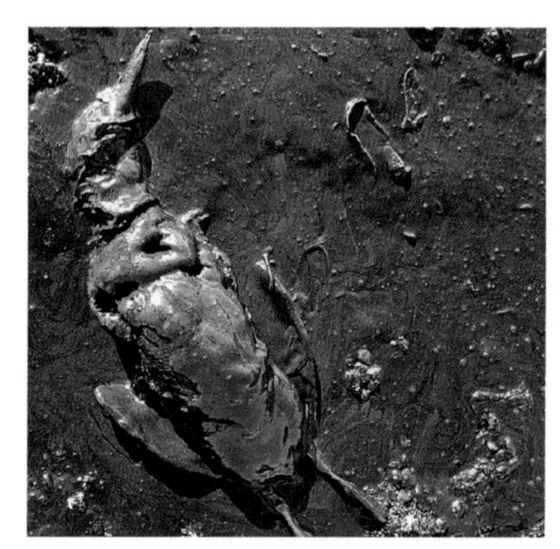

Oil kills sea birds.

Nitrates make plants grow quickly. Too much nitrate in a river makes the algae grow very fast. The algae forms a thick green blanket over the surface of the water. Light can't get through to the water plants below so they die. Bacteria decompose the dead plants, using up oxygen from the water. Less oxygen causes fish to die. With fewer fish to eat the algae, it grows even faster. When the algae eventually dies, bacteria decompose the remains. All the oxygen is used up as bacteria do this – the river is now dead! This whole process is called **eutrophication**.

Oil

Oil gets into the sea from wrecked tankers, tankers washing their tanks, oil rigs, and factories. Oil poisons sea birds and clogs their feathers so they can't fly. Oil, along with the chemicals that are used to break up oil spills, kills the plankton living near the sea surface.

Plant plankton plays a very important part in the carbon cycle. Like all green plants, it uses up carbon dioxide from the air and replaces it with oxygen during photosynthesis. Plant plankton is the start of all sea food chains – without it everything else dies.

Questions

1 **a** How does untreated sewage get into rivers?

b How do fertilizers get into rivers?

2 What chemical do sewage and fertilizer contain?

3 Explain how pollution by sewage and fertilizer leads to the death of a river.

4 Explain how oil pollution increases the amount of carbon dioxide in the air.

Pesticides are used to kill pests that damage crops.

Pesticides

Farmers use **pesticides** to kill pests that damage crops. Unfortunately the pesticides also kill useful animals like bees and beetles. This in turn means less food for birds.

Some pesticides are especially dangerous because they build up or **accumulate** in the bodies of animals. **DDT** is a pesticide which was developed in the 1940s. It was used to kill off not only those insects that attacked food crops, but also those that live off humans such as head lice. However, DDT doesn't decompose (it is **non-biodegradable**). Levels of DDT build up in the soil. From the soil it gets into rivers and into food chains. As it passes from one consumer to another it gets more and more concentrated. The animals at the tops of food chains are killed. The use of DDT is now banned in many countries.

Using unleaded petrol helps reduce the amount of lead in the environment.

This diagram shows how the osprey, a bird of prey, was almost wiped out by DDT.

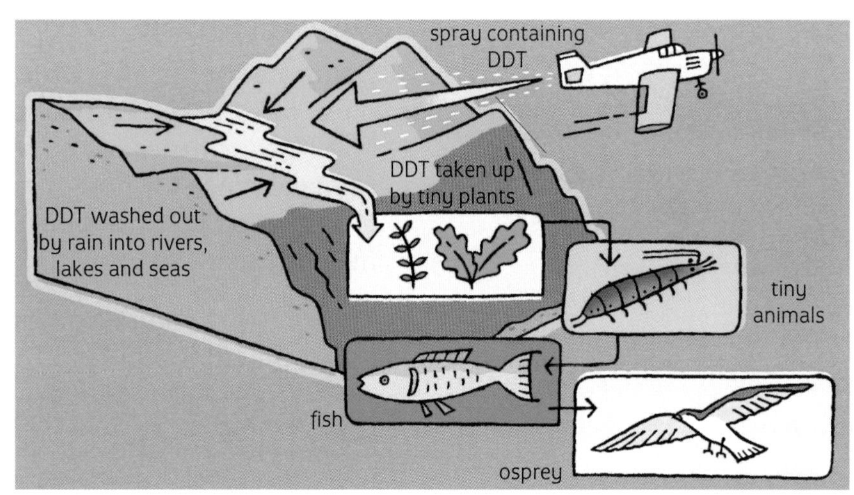

Other toxic chemicals

Industrial waste may contain **heavy metals** such as lead and mercury. These are not broken down in the environment or inside living things. They build up to high concentrations in soil, water, and living tissue. Lead and mercury severely damage the nervous system.

Questions

5 a What does 'non-biodegradable' mean?

b What does 'biodegradable' mean?

6 Explain why pesticides such as DDT affect animals at the tops of food chains.

7 Suggest why carnivores that live near water have higher levels of DDT in their bodies than carnivores that live and feed over land.

Pollution of land

Objectives

This spread should help you to

- describe what pollutes the land
- explain the link between land pollution and food production
- describe the problems associated with deforestation

As the human population grows bigger and bigger, people need more food, more homes, and more resources. They also make and need to get rid of more and more waste. The land is being seriously affected by these things.

Farming for food

Farming is important because it produces lots of food from less and less land. It is a big business giving us more choice at low cost. But the environment has had to pay the price. Huge fields have been made by cutting down hedges. Hay meadows have been ploughed up. The natural habitats of many animals and plants have been destroyed so that crops can be grown and harvested efficiently.

Over 90 per cent of our eggs come from hens kept in battery cages like these.

Over-use of fertilizers pollutes the land and the water that runs over it. Pesticides upset food chains. **Intensive farming** of hens, pigs, and calves may give us cheap eggs and meat but these animals have no space to move and rarely see daylight.

Deforestation

Deforestation means cutting down trees and not planting new ones. Most of the Earth's surface used to be covered by forests. The rich countries of the world have already cut theirs down to make space for homes, factories, and roads and to grow food. Poorer countries are now cutting their forests down for fuel and space to grow food as their populations rise.

This drawing shows how deforestation can cause serious environmental problems.

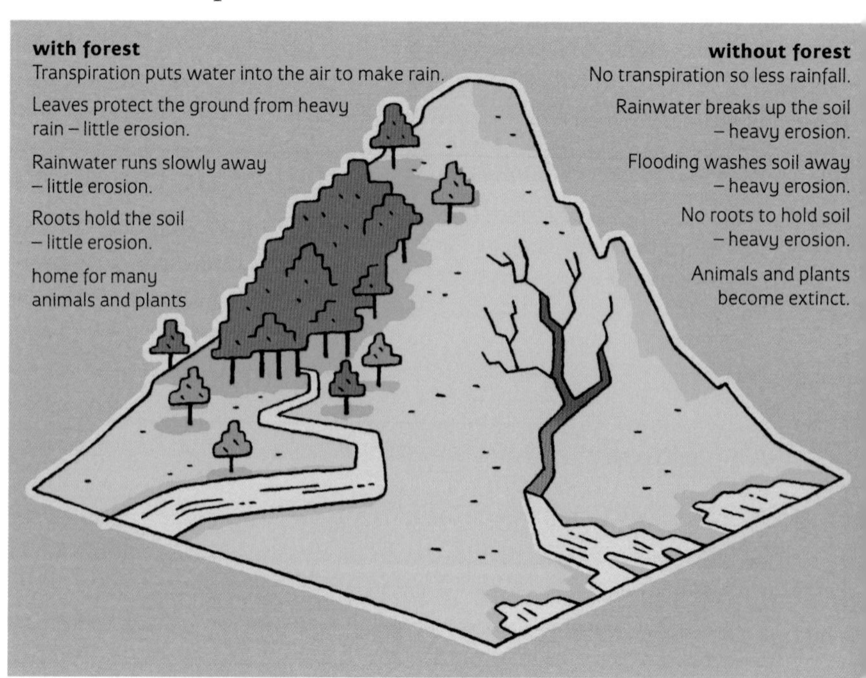

Questions

1 a Why is farming important to us?

b Give four ways that modern farming is affecting the environment.

2 a What is intensive farming?

b Give one advantage and one disadvantage of intensive farming.

c Why do you think most people buy eggs laid by hens kept in battery cages?

3 What is deforestation?

Other ways we pollute the land

Stone for house-building and gravel for roads comes from quarries. Quarrying leaves big scars on the land which will never go away.

We put our rubbish in landfill sites. These can leak poisons which get into our water supply. Bacteria decompose the rubbish producing poisonous (and explosive) methane gas. What happens when all the landfill sites are full?

Did you know?

It is estimated that 1% of tropical forest is lost each year.

Building more and more homes uses up more and more land. It costs more to clear old, derelict buildings than to use green fields to build on.

Questions

4 Why are poorer countries cutting down their forests?

5 Explain how deforestation affects rainfall.

6 Explain why forests are useful.

7 Why are new homes built on green field sites?

8 What are the disadvantages of putting rubbish in landfill sites?

9 List some of the pollutants of your school environment. Suggest how these could be controlled.

10 List some of the pollutants of your town environment. Suggest how these could be controlled.

3.16 Knocking out the competition

Objectives

This spread should help you to
- explain what pesticides are
- describe some of the problems of using pesticides
- describe biological control of pests
- explain how using sterile males can help reduce pests

Pesticides have to be used over and over again.

Using pesticides

The more people there are in the world, the more food is needed to feed them. As the world population goes up, so does the demand for efficient farming methods. Farmers must grow as many crops as they can.

Fertilizers are added to the soil to produce more food crop. But fertilizers don't just help the crops to grow – other plants such as thistles grow better as well. These other plants (weeds) compete with the food crop for light, water, and minerals. So, to give the food crop a better chance, chemicals are used to kill the weeds. These chemicals are called **selective weedkillers** or **herbicides**, because they kill the weeds without harming the crop.

Fungicides are used to kill any fungal competitors that might affect food crops. A fungus called wheat rust feeds on living wheat plants. This reduces the amount of wheat produced.

There are animal competitors too. Small mammals such as mice eat the crop before and after it is harvested. Insects like mealy bugs are a bigger problem. In some parts of the world one-third of all food crops are eaten by insects. Chemicals called **insecticides** have been developed to kill insects.

Herbicides, fungicides, and insecticides are all pesticides, because they kill pests.

There are problems with the use of pesticides:

- Some pests are resistant to them.
- They have to be used over and over again because the effect doesn't last long.
- Useful insects like bees and ladybirds are killed as well as the pests.
- People are getting concerned about pesticides on their food.
- Pesticides can cause pollution.

So what is the answer?

Biological control

The use of pesticides can be cut by encouraging the natural predators of insect pests. This is **biological control**.

Whitefly damage food crops such as tomatoes. They are a menace in greenhouses where conditions are ideal for their rapid growth – warm, moist, and plenty of food. *Encarsia* is a tiny wasp which lays its eggs inside whitefly. When the eggs hatch into larvae (grubs) they slowly feed on the whitefly from the inside, eventually killing it.

Questions

1 What are:
a herbicides
b fungicides
c insecticides?

2 Herbicides, fungicides, and insecticides are sometimes put together under one name, pesticides. Why?

3 Explain why it is important for farmers to 'knock out the competition'.

Whitefly can be kept under control by the grubs of this Encarsia wasp.

Plants can now be genetically engineered to make their own pesticide.

Biological control has advantages over pesticides.

- The pest doesn't become resistant to the predator.
- The predator usually feeds on only one kind of prey.
- There's no need to replace the predators – they breed themselves.

One disadvantage of biological control is that the predators might escape and upset natural food webs. Another is the time it takes to work. This graph shows that it takes about 1–2 years before the pest is controlled.

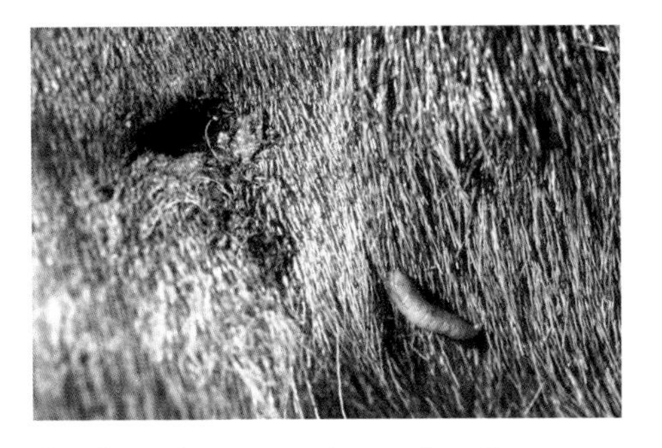

Sterile males are used to reduce the population of screw-worms (grubs) in Africa. Screw-worms infect the wounds of farm animals and can kill them.

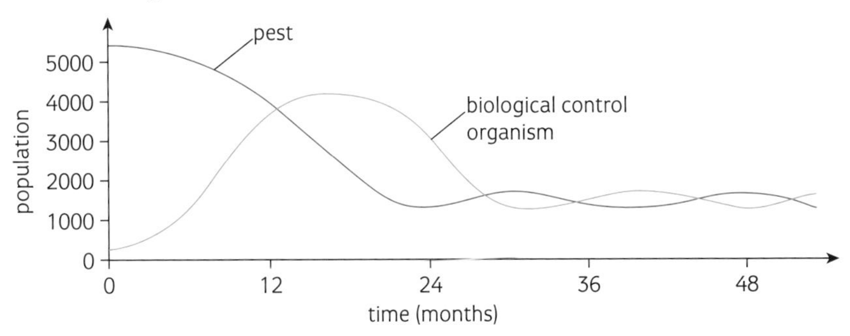

Sterile males

Insect pest populations can be reduced by breeding large numbers of sterile males in a laboratory. These are released into an area where the pest is a problem. If there are enough of them, the chances are that a 'wild' female will mate with a sterile male. The females lay infertile eggs and the pest population drops.

Questions

4 Name:

a a plant pest **b** a fungal pest **c** an insect pest.

5 Describe the problems of using pesticides.

6 What are the advantages of biological control over the use of pesticides?

7 What is a disadvantage of using biological control?

8 Does biological control kill all of the pests? Explain your answer.

9 Explain how whitefly can be controlled without using pesticides.

10 Explain how sterile males can be used to control pest numbers.

3.17 Managed ecosystems

Objectives

This spread should help you to

- explain the difference between a natural and a managed ecosystem
- describe some ways that humans manage ecosystems

Meeting energy demands

All living things depend on energy from the Sun. This energy is absorbed by green plants during photosynthesis, and then passed along food chains.

When humans first evolved, they got their energy by collecting wild grain, fruits, roots, and nuts, and by killing animals. Because the population was small, early humans had little effect on the environment. **Natural ecosystems** like woodland and rivers supplied the demand.

Today things have changed. There are more of us. Modern farming methods supply our food. In the year 1200, one hectare of land produced food for only five people. Today it could feed 50. Cows produce twice as much milk as they did 50 years ago. Chickens lay more eggs than ever before. We now have **managed ecosystems** to meet the demand.

A natural ecosystem.

Some pigs are kept in crates like these. They have food and water, but they can't move about.

Intensive farming

Some modern farms are like factories (**factory farms**). Most of our pigs and poultry are bred in **intensive units**. Light, heat, humidity, ventilation, and hygiene are carefully controlled. A large poultry unit can hold as many as 100 000 birds. Feeding, cleaning, watering, and egg collection are all done by machines. The animals are warm and well fed and production costs are low. But the animals are kept in crowded conditions and can't behave naturally.

Selective breeding and genetic engineering continue to increase crop yields. Hormones are used to control fertility in cattle and to increase milk production. Plant hormones control the ripening of fruit.

Questions

1. Give one example of a natural ecosystem.
2. Why were natural ecosystems enough to supply the needs of early humans?
3. Why do we have managed ecosystems?
4. **a** Explain what an intensive unit is.
   b Give one advantage of rearing animals in intensive units.
   c Give one disadvantage of rearing animals in intensive units.
5. Give one example of the use of hormones in the intensive farming of:
   a animals **b** plants.

Lettuce, tomatoes, and cucumbers are some of the crops that are grown under glass. Everything the plants need is carefully controlled to give the best crops.

Irrigation has turned this desert into productive land.

Natural fertilizers like manure and compost improve soil texture but release nutrients slowly. **Artificial fertilizers** are chemicals that are added to the soil. They can be taken up by plants straight away so crops grow quicker. Unfortunately the soil gets poorer and turns to dust which is easily blown or washed away.

Greenhouses

Greenhouses give protection from wind, rain, and also pests such as birds. They trap warmth from the Sun, which makes plants grow faster than they would if they were outside. In winter, extra heat can be provided by heaters.

Carbon dioxide, light, temperature, and water supply can be controlled easily. The best conditions for photosynthesis can be provided so the plants will grow quickly. Insect pests can be kept down by chemicals or biological control. Greenhouses also make it possible to grow plants from warm climates in cooler regions.

Irrigation

Crops can be damaged or killed if they don't get enough water. **Irrigation** is artificial watering. Dry, infertile soil can be turned into productive land.

There are various methods of irrigation. Water can be sprayed or sprinkled on to the soil or run in underground pipes straight to the roots of the plants. Water for irrigation is usually stored in reservoirs. These fill up when it rains.

Questions

6 **a** Explain the difference between natural and artificial fertilizers.
b Give one advantage of each.

7 Describe a greenhouse.

8 Why are greenhouses useful for growing plants?

9 List the things that are controlled in greenhouses.

10 **a** What is irrigation?
b Give one advantage of irrigation.

11 Salmon are bred in fish farms. The fish are kept in cages, given food, and are protected from predators such as seals. Waste drops out of the bottom of the cage to the sea bed.
Are fish farms natural or managed ecosystems? Explain your answer.

3.18 Conservation

Objectives

This spread should help you to
- describe how animals and plants are being destroyed
- explain why conservation is important
- describe some methods of conservation

We share the Earth with millions of other living things. But many of the things we do destroy the lives of the animals and plants around us. The number of different animal and plant species is falling. Every year more species become **extinct** – gone forever.

How animal and plant species are destroyed

- Habitats are destroyed or broken up when land is cleared for farming, housing estates, roads, factories, and quarries.
- Pollution and the tipping of waste destroy habitats and poison animals and plants.
- People who hunt for fun, and to get things like ivory and furs, which we can do without, put species at risk of extinction.

Clearing land for a new road destroys the plants and takes away the habitat of many mammals, birds, and insects.

A coat made from wool or synthetic fibre would keep her just as warm.

Why conserve animal and plant species?

Most humans would rather see this...

...than this.

But there are more reasons why we should stop species dying out.

- A fall in the number of different species may result in flooding or drought. Deforestation has shown this already.
- Less genetic variation in populations may mean they won't be able to survive changes in their environment.
- Many of our medicines have come from plants. There are many undiscovered plants which may provide the cure for deadly diseases.

Questions

1. What does 'extinct' mean?
2. Describe three ways that animal and plant species are destroyed.
3. Tropical rainforests are home to many undiscovered plants and animals. Explain why we should be concerned about the deforestation of tropical rainforests.
4. Why is genetic variation important?

Signs like these help conserve animal and plant species.

How animal and plant species can be conserved

We can help **conserve** species by:

- creating and maintaining national parks and nature reserves to keep habitats protected
- planned land use – where possible, houses and roads should be built with the interests of wildlife in mind as well as the interests of humans
- organic farming methods produce top quality food without polluting the land and water with chemicals that poison animals and plants
- sustainable forests – this means planting more trees as others are cut down
- passing laws to protect endangered species
- breeding endangered species in captivity. When numbers increase they can be reintroduced into the wild.

To make up for the habitats we destroy, we must find ways to make new ones. This abandoned quarry has been turned into a nature park.

Many farmers leave the corners of fields 'wild'. These areas make ideal habitats for wild animals and plants.

Questions

5 How can farmers conserve wildlife?

6 What can you do to conserve wildlife?

7 What are sustainable forests?

8 What is the point of breeding endangered species in captivity?

9 If you were able to pass one law about the conservation of animal and plant species, what would it be?

Living indicators of pollution

BOD

The more pollution there is, the more bacteria there will be. Bacteria use up the oxygen in water. The more bacteria there are, the faster the oxygen is used up. This is called the **biological oxygen demand** or **BOD** of the water.

Living indicators of air pollution

Simple plants called *Pleurococcus* live mainly on the north side of tree trunks where it is damper away from direct sunlight. Gases like sulphur dioxide from burning coal and oil slowly poison *Pleurococcus*.

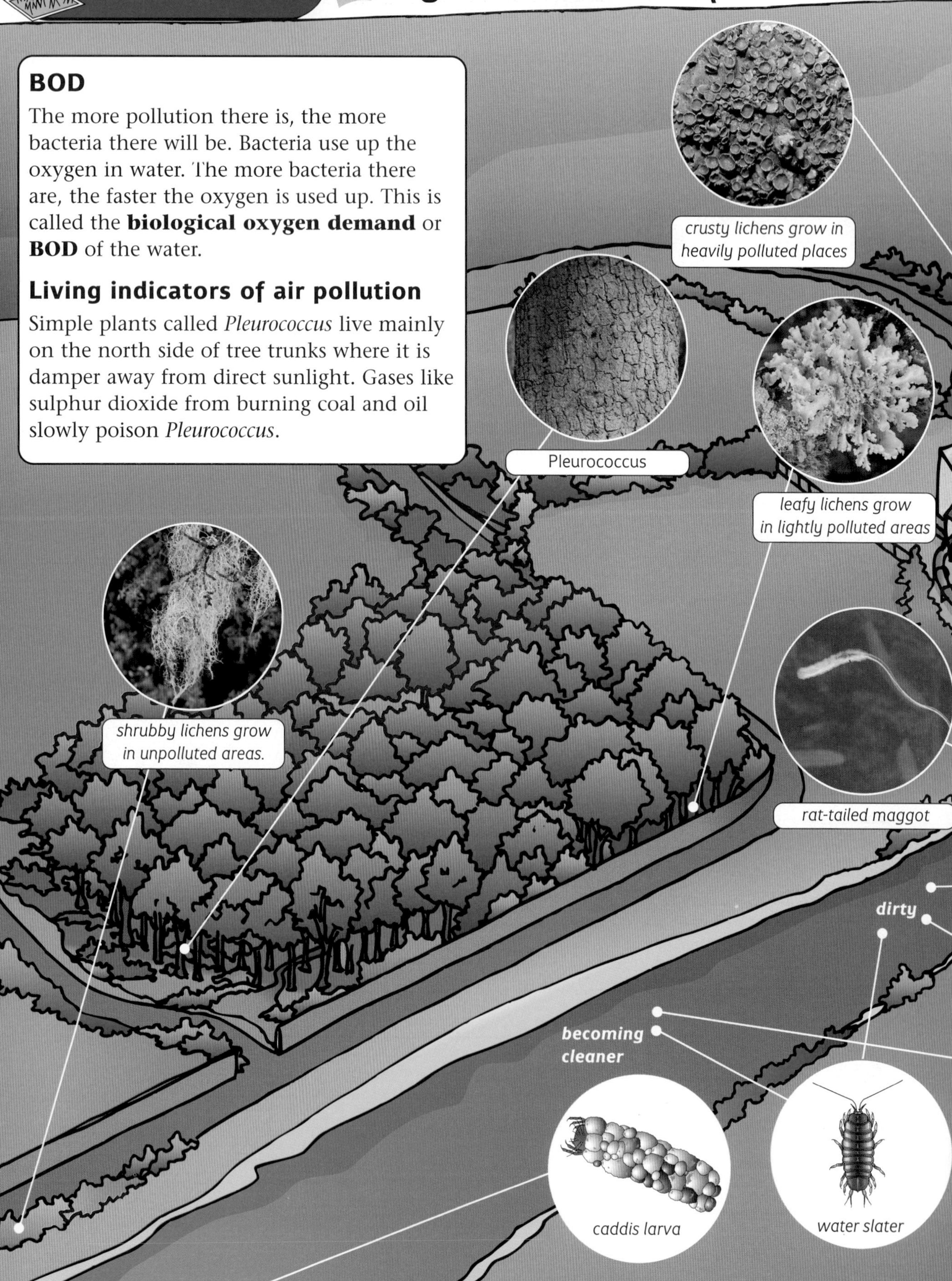

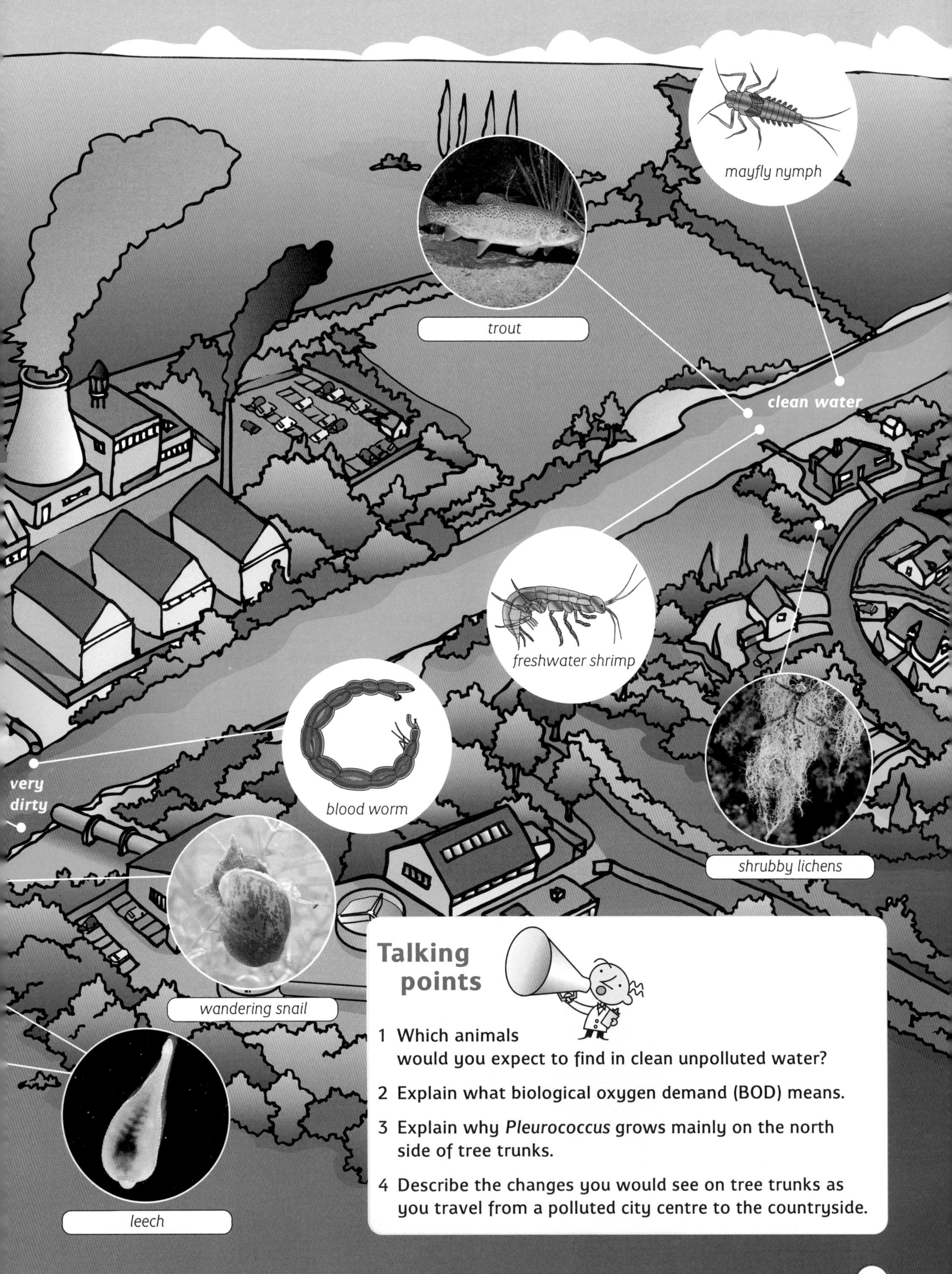

Talking points

1. Which animals would you expect to find in clean unpolluted water?
2. Explain what biological oxygen demand (BOD) means.
3. Explain why *Pleurococcus* grows mainly on the north side of tree trunks.
4. Describe the changes you would see on tree trunks as you travel from a polluted city centre to the countryside.

3.20

Asthma

What is asthma?

When someone has an asthma attack, the muscles in the walls of the tubes in the lungs contract making the air passages narrower. More mucus is produced in the walls of the airways, which then blocks the tubes. As a result, breathing becomes very difficult, usually with a characteristic wheezing sound. The sufferer usually gets very distressed and anxious, which makes the condition worse.

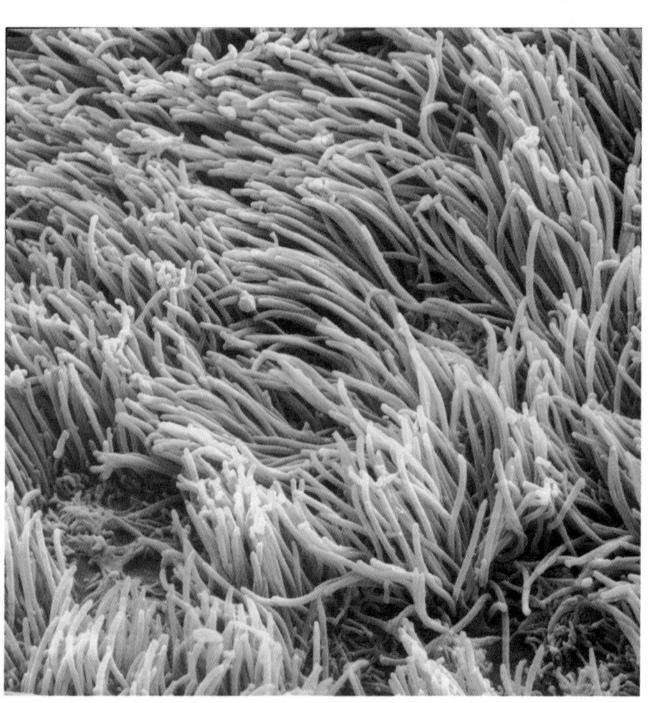

The airways to the lungs are lined with cells that produce mucus and cells with tiny cilia that help keep our lungs clean. These cells can be destroyed if an asthma attack goes on too long and isn't treated.

If the condition isn't treated, things can get even more serious. White blood cells are carried in the bloodstream and collect in the walls of the airways. They behave as though the person has suddenly been exposed to disease-causing microorganisms and the lungs are in danger of infection. These white blood cells produce chemicals designed to destroy invading microorganisms. Unfortunately the chemicals damage the cells lining the air passages.

What causes asthma?

Asthma is a disease that involves the body's immune (defence) system. If a child isn't exposed to many diseases, then the immune system develops a bit differently. Scientists have shown that in countries where diseases are still very common, fewer people have asthma.

Cigarette smoke can start an asthma attack.

Some asthmatics are very sensitive (allergic) to dust mite. Dust mites live in mattresses, pillow cases, and carpets. These should be cleaned regularly.

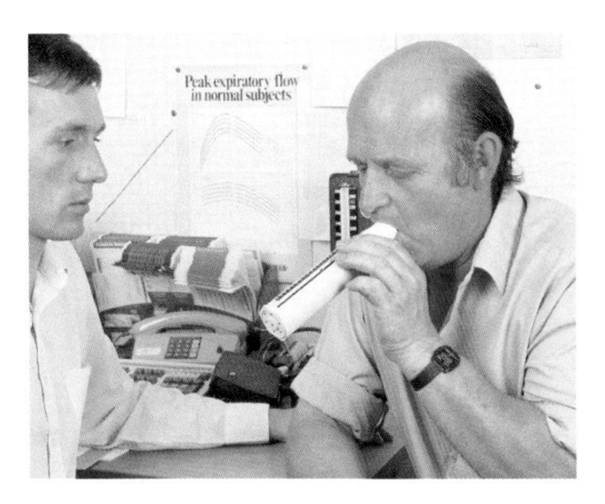

A peak flow meter measures lung volume. This helps doctors decide what treatment is best for an asthmatic.

Asthma attacks can be started by many different things. Exposure to pollen, cat or dog hair, exhaust gases from motor cars and lorries, viruses, and faeces of the dust mite have all been shown to set off asthma attacks. Asthma symptoms can also be triggered by colds, exercise, cold air, and tobacco smoke. People can have really bad asthma even though there has been nothing to set it off.

How can asthma be treated?

Asthma sufferers may carry an inhaler. This inhaler contains drugs which relax the muscles in the air passages allowing more air through to the lungs.

There is no cure for asthma.

Asthmatics can get fast relief by using their inhaler whenever they feel an attack coming on. The sufferer inhales a vapour containing a drug that relaxes the muscles around the air passages.

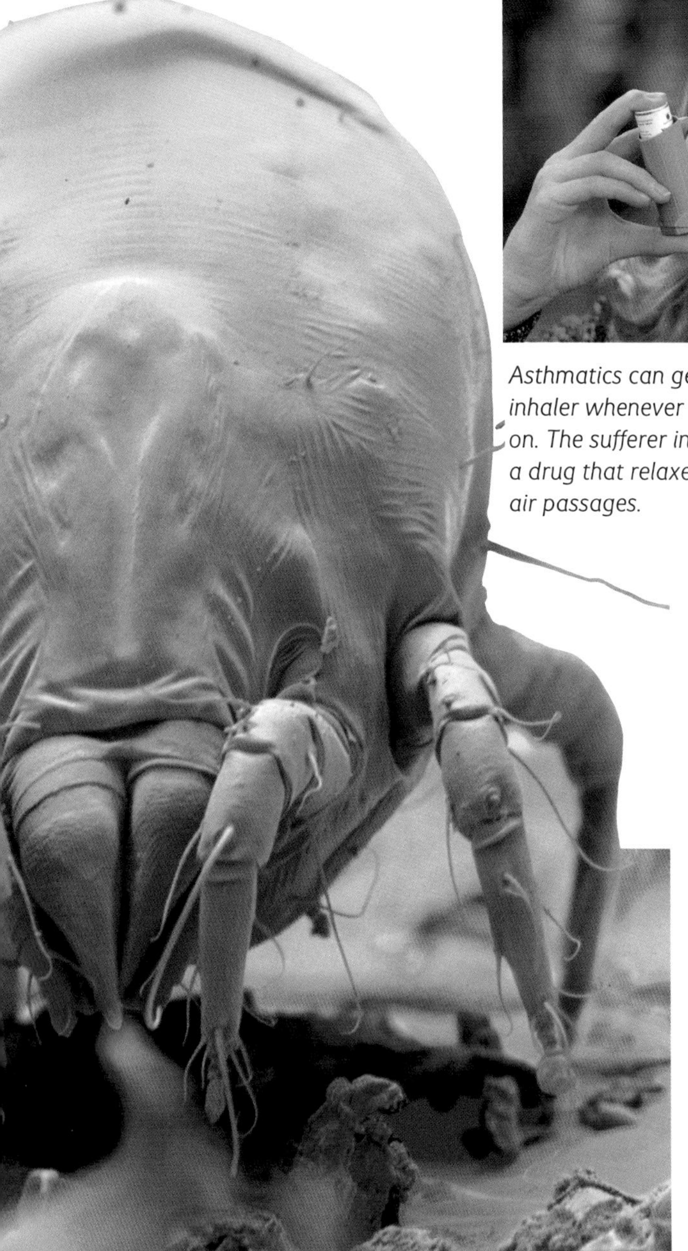

Talking points

1. Why do you think asthma sufferers get distressed when they have an attack?
2. Give four things that can start an asthma attack.
3. Explain how an inhaler gives relief for asthmatics.
4. Describe what happens to the walls of their air passages if an asthmatic isn't treated quickly.

Practice questions

1 The diagrams show two plants. One plant lives on land, the other lives in water.

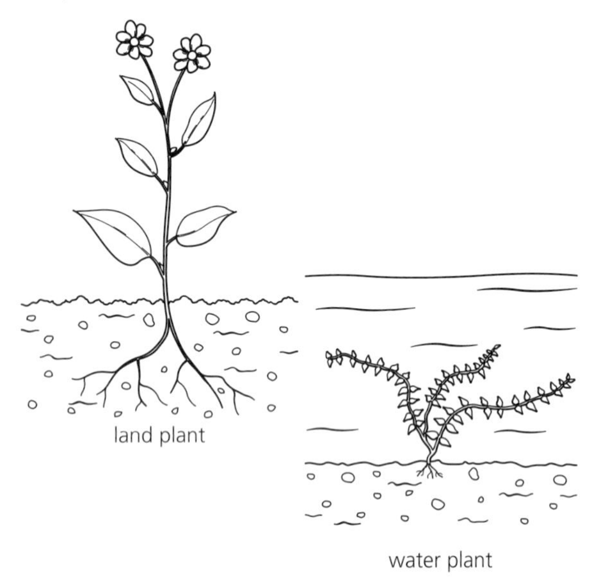

a **i** Name three features that both plants have.

ii What is the job of each feature to help the plant adapt to its surroundings?

b Explain why a land plant has a strong stem to keep it upright.

c Explain why the water plant is all floppy when it is taken out of water.

d Bubbles of air get trapped between the leaves of the water plant. Explain how this helps the plant survive in water.

2 Write these two headings:

Examples of competition **Examples of predation**

a Write these statements under the correct heading.

farm cats killing mice and stopping them eating cereal crops

two stags fighting for control over a herd of deer

trees in a forest growing upwards to get light

a seagull chasing off other birds from food in a garden

lions hunting a zebra for food

a spider catching a fly in its web

b Name two examples of prey from the statements above.

3 The diagrams show some living things.

a Describe how:

i the eagle is adapted to catch its food

ii the cactus is adapted to avoid losing water in the desert

iii the butterfly is adapted to suck nectar from flowers

iv the lion is adapted for tearing flesh and crushing bones.

b Gerbils live in the desert. They don't produce urine. They live in holes during the day and only come out at night.

i Explain how these adaptations help gerbils survive in the desert.

ii Suggest an adaptation that helps gerbils avoid being eaten.

4 The graph shows what happens when some *Paramecium* (single-celled organisms) are added to a population of yeast.

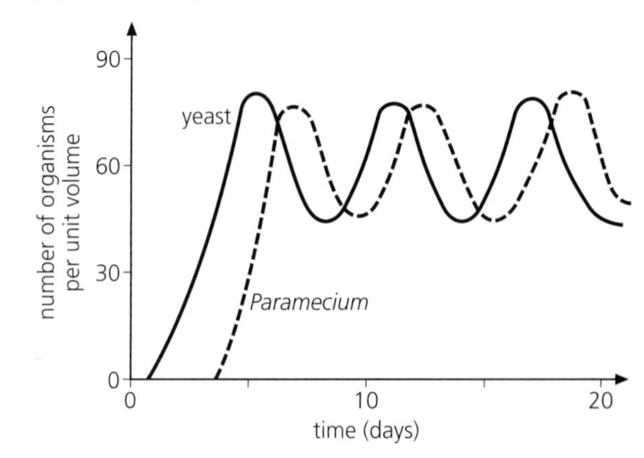

a **i** What happens to the yeast population after the *Paramecium* are added?

ii Explain why this happens.

b Suggest why the numbers of *Paramecium* go down soon afterwards.

c Explain why there is a time lag between the two curves on the graph.

d If the *Paramecium* were taken away, what would happen to the yeast population?

e Give two more examples of predator–prey relationships.

5 Anything that stops the growth of a population is called an environmental pressure. Copy this list of environmental pressures and write the best example alongside each one.

Environmental pressures: *shortage of light, overcrowding, forest fires, sudden fall in temperature, floods*

Examples:

rabbits' burrows washed away

birds die in bad winters

plants can't photosynthesize

pigs fight if kept together in large numbers

squirrels' homes destroyed

6 The graph shows human population growth.

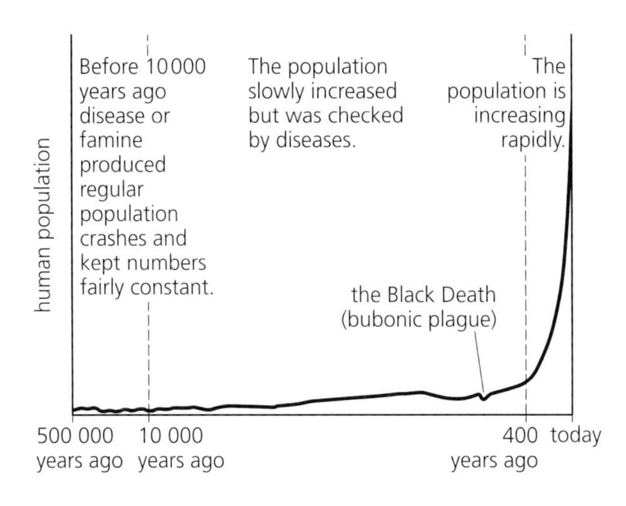

a When was there no real increase in the human population?

b What effect did the Black Death have on the population?

c Suggest why the population rose only slowly up to about 400 years ago.

d Give two reasons why the human population has got bigger and bigger in the last 400 years.

7 Write these two headings:

Natural ecosystems **Artificial ecosystems**

a Write each of these words under the correct heading:

woodland

tropical fish aquarium

field of wheat

garden

lake

grazing meadow

b i Explain why the Earth is described as a natural ecosystem.

ii Give one reason why it is important for humans to take care of the Earth ecosystem.

8 The diagram shows a map of a river.

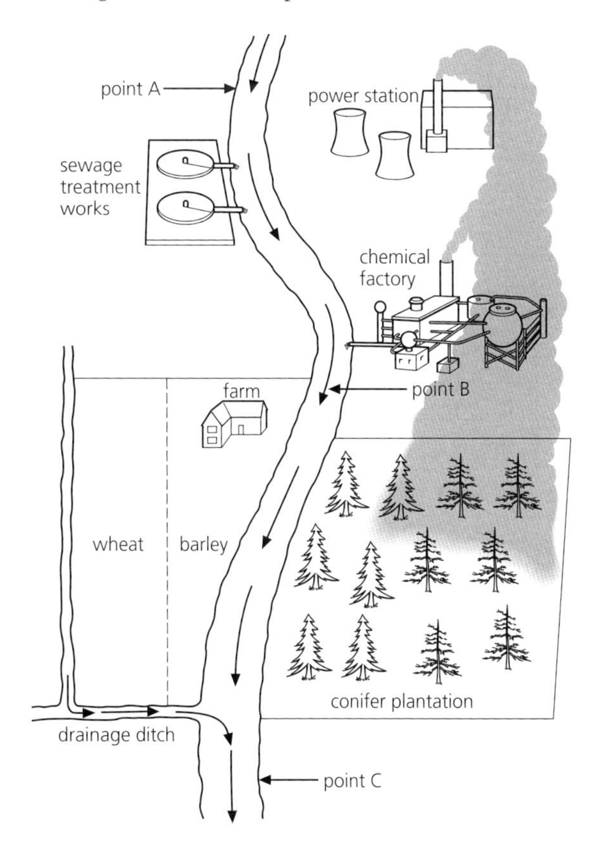

a Suggest why:

i there are more fish at point A than point C

ii fish-eating birds get ill when the farmer sprays his crops with pesticide

iii the amount of oxygen in the river would fall if untreated sewage was pumped into the river

iv trees in the conifer plantation are dying

v there is more water weed at point C after the farmer has put fertilizer on his wheat and barley.

9 a Copy this list of words and write the correct description alongside each one.

Words: *carnivore, omnivore, decomposer, herbivore, producer*

Descriptions:

plant that makes its own food

animal that eats plants

animal that eats other animals

animal that eats both plants and animals

animal that feeds on dead plants and animals

b The diagram shows a food chain.

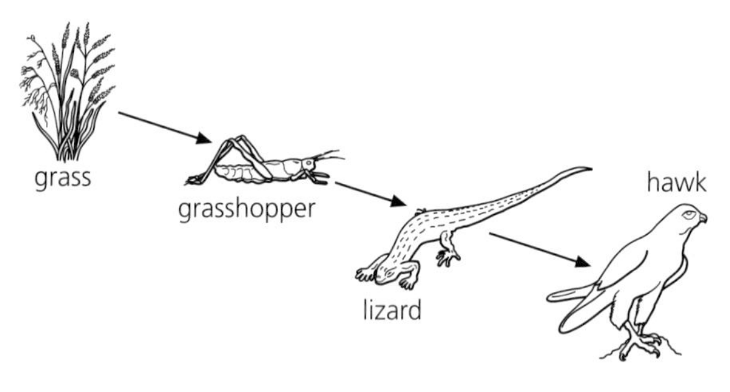

Name:

i a producer

ii a herbivore

iii a carnivore.

c Explain what the arrows mean in a food chain.

10 The following diagram shows a food web for a pond ecosystem.

a Write down two food chains ending at the water beetle.

b Name:

i a producer

ii a first consumer

iii a second consumer

iv a third consumer

v a decomposer.

c Suggest what might happen if all of the water beetles were taken out of the pond.

d Write down the longest food chain you can find in this food web.

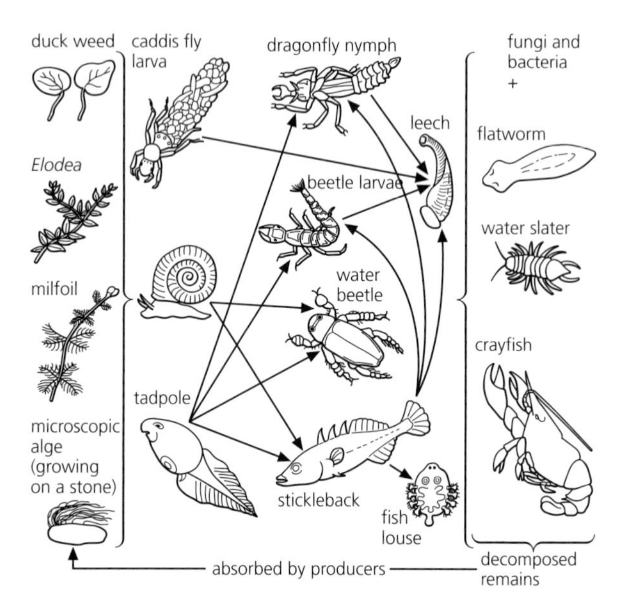

11 The diagrams show some pyramids of numbers.

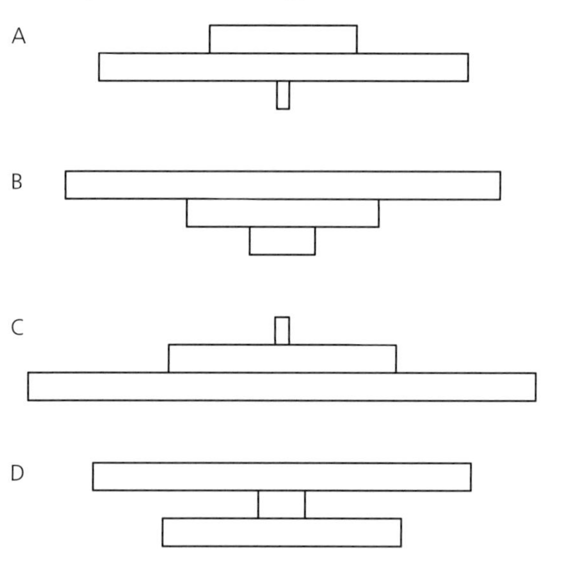

a Explain what a pyramid of numbers is.

b Copy these food chains and write the letter of the matching pyramid alongside each one.

grass → antelopes → lion

oak tree → caterpillars → sparrows

rabbits → fox → fleas

tomatoes → whitefly → parasites (microscopic grubs)

c What is a pyramid of biomass?

d Draw pyramids of biomass for each of the food chains in part **b**.

12 The diagram shows how energy flows through an ecosystem.

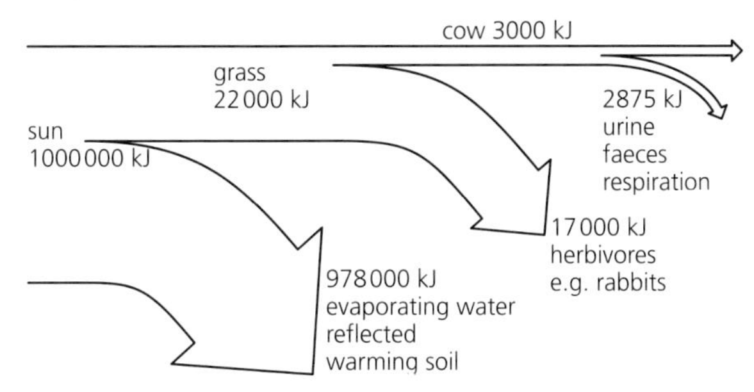

a **i** What form of energy from the Sun is used by plants?

ii What form of energy do plants turn this into?

b Describe what happens to the amount of energy as it goes through a food chain.

c Give two ways that energy is lost from the food chain:

grass → cow → human

d Of the 3000 kJ of energy taken in by the cow, how much is left for the human?

e Suggest why farmers try to keep rabbits off their fields.

13 The diagram shows the carbon cycle.

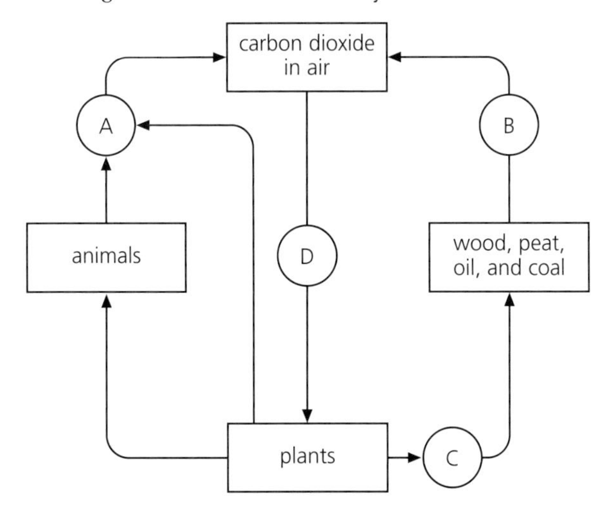

a Copy the diagram and use these words to label parts A–D.

burning, decay, photosynthesis, respiration

b What gas is needed for coal to burn?

c What gas is produced when coal burns?

d **i** Write a word equation for respiration.

ii Write a word equation for photosynthesis.

iii What do you notice about these two equations?

14 When living things die, they are broken down by microorganisms.

a What is the name for this breakdown?

b Name one type of microorganism that carries out this breakdown.

c Give three things that microorganisms need to carry out this breakdown.

Microorganisms are used in sewage works to remove organic waste from water before it goes into our rivers.

d **i** What is organic waste?

ii Explain why organic waste from our homes can't be put straight into a river.

e Suggest how we can recycle organic waste like vegetable peelings and rotten fruit instead of throwing it away.

15 In South America many people are very poor. They cut down the trees of the jungle to sell the wood. In their place they plant a crop, such as coffee, which they can sell to richer countries. With the money they buy food for their families.

a Explain why people in South America need to cut down the jungle.

b Give three reasons why cutting down forests and jungles damages the world environment.

c Suggest two things that might help solve the problem.

acid rain a mixture of sulphur dioxide, oxides of nitrogen, and rainwater

artificial fertilizer a mixture of the mineral elements that plants need, which is put directly onto the soil; artificial fertilizers can spoil the structure of soil

biological control using natural predators to control pests

biological oxygen demand (BOD) the speed that oxygen is used up in water ecosystems

biomass the mass of living material

carbon cycle the movement of carbon from the environment into living things and back again

carnivore an animal that eats other animals

catalytic converter a device that removes harmful gases from car exhaust fumes

CFCs gases in aerosols which destroy the ozone layer

community lots of different animal and plant populations living together

compost heap a pile of decomposing plant material

conserve to preserve and carefully manage natural resources

consumer an animal that eats other living things

crusty lichens types of lichen that can tolerate air pollution

DDT a non-biodegradable insecticide

decomposer bacteria and fungi that decompose dead things

deforestation the cutting down of lots of trees

ecosystem living things and their environment

effluent water that has been treated at a sewage works and is safe to put into rivers or the sea

environment surroundings

environmental pressure things that affect the growth of a population

eutrophication how life in a river is killed by fertilizer pollution

extinct describes a species that is dead and gone forever

factory farms farms with high productivity

first consumer an animal that eats plants

food chain food passing from plant to animal to animal, etc.

food web interconnecting food chains

fossil fuel a fuel produced by the decay of organisms that lived millions of years ago, e.g. oil

fungicides chemicals that kill fungi

global warming a rise in the temperatures on Earth

greenhouse effect warming of the Earth because heat is kept in the atmosphere; the increased greenhouse effect is causing global warming

greenhouse gas a gas in the atmosphere which stops heat escaping from Earth

growth curve a graph showing changes in numbers in a population over a long time

habitat the place where an animal or plant lives

heavy metals metals that cause pollution and build up in living things, e.g. lead

herbicides chemicals that kill plants

herbivore an animal that eats plants

indicator species organisms that are sensitive to pollution

inefficient (energy flow) lots of energy is lost at each trophic level

insecticides chemicals that kill insects

intensive farming a type of farming that produces as much of one product as possible, e.g. eggs

intensive units lots of animals or plants in a small space where everything is carefully controlled

irrigation adding water to the land, e.g. by spraying

leafy lichens types of lichen that can tolerate a little air pollution

lichen mutualistic relationship between a fungus and a green algae

managed ecosystems ecosystems that are affected by human activity, e.g. farms

natural ecosystems ecosystems that are not affected by human activity

natural fertilizer manure and compost that improve soil and release nutrients slowly

nitrates chemicals containing nitrogen

non-bio-degradable describes something that can't be decomposed

omnivore an animal that eats plants and animals

ozone a gas which forms a layer high above the Earth and filters UV radiation

pesticide a chemical that kills pests

plankton microscopic plants and animals living near the surface of the sea

pollutant a substance that harms living things

population a group of animals and plants of the same species

population crash a sudden, fast decrease in numbers in a population

population explosion a sudden, fast increase in numbers in a population

predator an animal that eats other animals

prey an animal eaten by another animal

producer an organism that makes its own food, e.g. a green plant

pyramid of biomass a diagram showing the mass of organisms at each trophic level in an ecosystem

pyramid of numbers a diagram showing numbers of organisms at each trophic level in an ecosystem

scrubbers devices that remove harmful gases from burning fossil fuels

selective weedkillers chemicals that kill only certain kinds of plants

shrubby lichens types of lichen that can't tolerate air pollution

third consumer an animal that eats second consumers

trophic level a link in a food chain

ultraviolet (UV) radiation the harmful radiation from the Sun which can cause skin cancer

Inheritance and selection

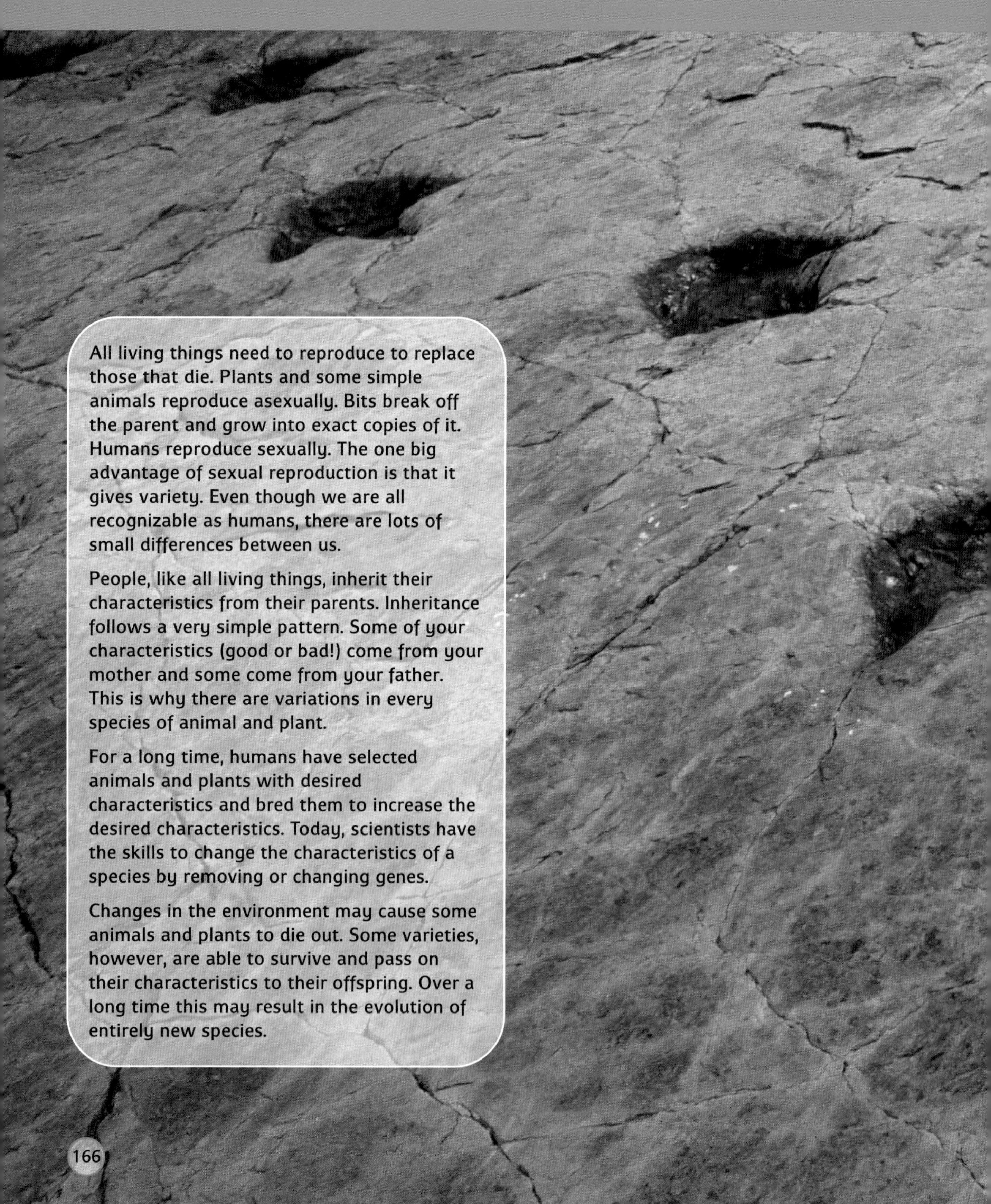

All living things need to reproduce to replace those that die. Plants and some simple animals reproduce asexually. Bits break off the parent and grow into exact copies of it. Humans reproduce sexually. The one big advantage of sexual reproduction is that it gives variety. Even though we are all recognizable as humans, there are lots of small differences between us.

People, like all living things, inherit their characteristics from their parents. Inheritance follows a very simple pattern. Some of your characteristics (good or bad!) come from your mother and some come from your father. This is why there are variations in every species of animal and plant.

For a long time, humans have selected animals and plants with desired characteristics and bred them to increase the desired characteristics. Today, scientists have the skills to change the characteristics of a species by removing or changing genes.

Changes in the environment may cause some animals and plants to die out. Some varieties, however, are able to survive and pass on their characteristics to their offspring. Over a long time this may result in the evolution of entirely new species.

Module 4

4.01 Variation

Objectives

This spread should help you to

- describe two types of genetic variation
- explain the difference between genetic and environmental variation

No two people are exactly the same. Even identical twins are different in some ways. People have different heights and weights. Their hair, eyes, and skin have different colours. These differences are examples of **variation**. Variation is caused by genetics or by changes in the environment.

Genetic variation

Humans, like all living things, **inherit** genes – they are passed on from their parents. Genes are the code inside the body that determines all of your **characteristics** or features. The different characteristics we all inherit cause **genetic variation**.

Characteristics that are distinct, such as being able to roll your tongue or having ear lobes, are examples of **discontinuous variation**. You either have the characteristic or you haven't.

Other characteristics are not so easily separated. If you measure and record the height of everyone in your class, you will find that they form a complete range of heights. This is an example of **continuous variation**.

No two people are the same.

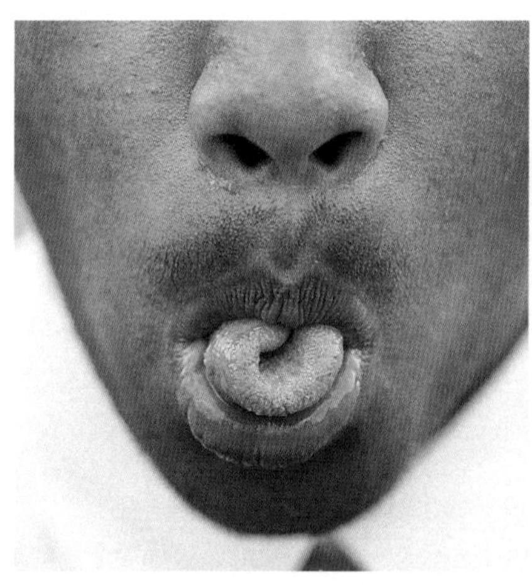

Are you a tongue roller or not? Either you can or you can't.

Height shows continuous variation.

Questions

1 **a** What is variation?
  b What causes variation?

2 Make a list of six inherited characteristics.

3 Explain the difference between continuous and discontinuous variation. Give one example of each in your answer.

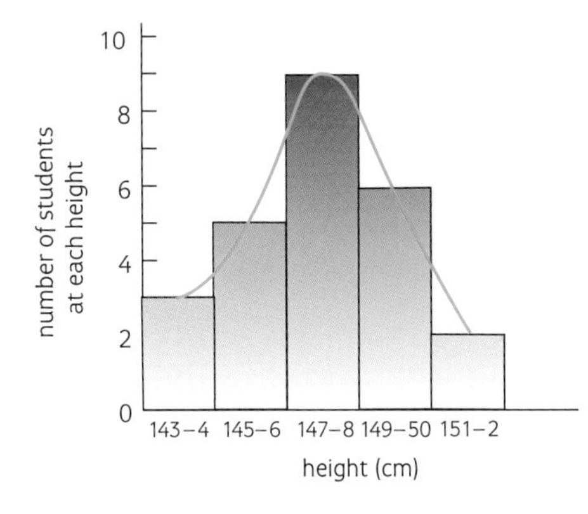

In your class you are likely to find a few very short people and a few very tall ones. But most will be bunched around the average height. If you draw a graph of the information it will look something like this.

This shape of graph is called a **normal distribution**.

Environmental variation

The amount of food you eat or the amount of exercise you do can affect your height and your weight. Your skin colour changes depending on how much sunlight you are exposed to. Thesc variations are not due to genetics. They are the result of what you do and how you are brought up – they are called **environmental variations**.

Plants are affected by environmental variation a lot. A plant will grow very fast and very tall if it has lots of light, a suitable temperature, and a good moist soil. Think how often you have to mow the lawn in the summer!

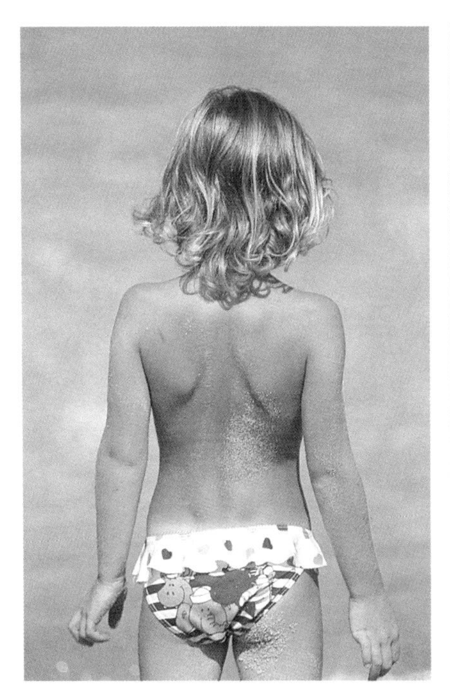

A warm, sunny climate soon changes this girl's appearance.

The shade from the tree affects the growth of the crop.

Questions

4 Intelligence shows a normal distribution. Explain what this means.

5 **a** What is environmental variation?
b Give two examples of environmental variation.

6 Explain why a farmer who fertilizes his soil gets better crops than a farmer who doesn't.

7 The leaves on a tree are usually bigger at the top than at the bottom. Give a reason for this.

Chromosomes and genes

Objectives

This spread should help you to

- describe what happens to chromosomes and genes at fertilization
- explain the difference between dominant and recessive gene alleles
- explain the difference between homozygous and heterozygous

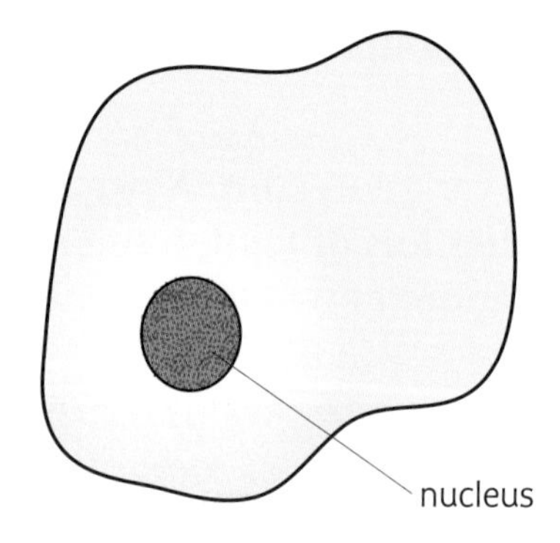

A cell from your body.

Chromosomes

Chromosomes are inside the nucleus of every one of your cells. They are fine threads that carry 'bits' of information about what you are like. These bits of information are called **genes**. Chromosomes and genes are made of a chemical called **DNA**. Each of your genes controls one or more of your characteristics that you inherited from your parents.

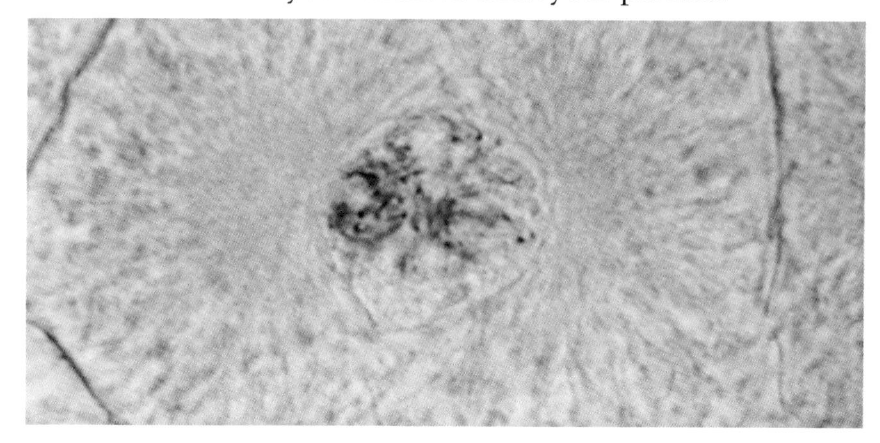

The chromosomes are the dark threads you can see in the nucleus.

Fertilization

A human sperm and egg each have 23 chromosomes. During fertilization, the nucleus of the sperm and nucleus of the egg join up. Each chromosome from the sperm pairs up with its 'partner' from the egg, making 23 pairs altogether. These are called **homologous pairs** of chromosomes. This brings the two sets of genes together. Genes for hair colour pair up, genes for skin colour pair up, and so on.

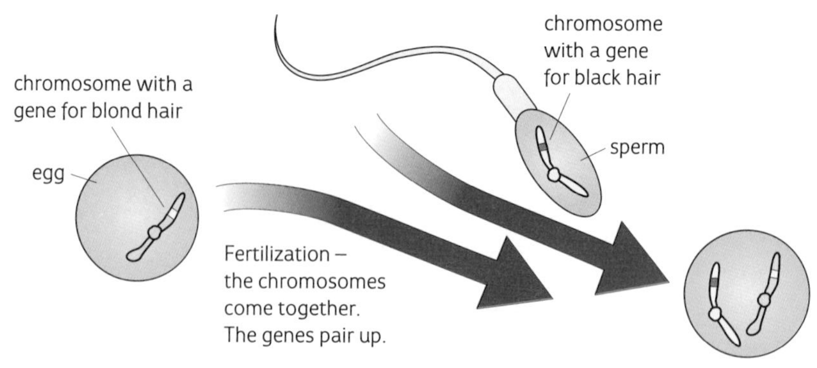

Dominant and recessive gene alleles

The genes in a pair may carry the same message. But sometimes one of the genes carries a different message from the other. These are called gene **alleles** – different forms of the gene. For example, a person has two alleles for hair colour (one pair). One allele says 'have blond hair', the other allele says 'have black hair'. These alleles are in competition with each other.

Questions

1. What are:
   - **a** chromosomes
   - **b** genes?
2. What are genes made of?
3. What are homologous pairs?
4. Explain how a human gets 46 chromosomes.
5. **a** What is an allele?
   b Give an example of an allele.

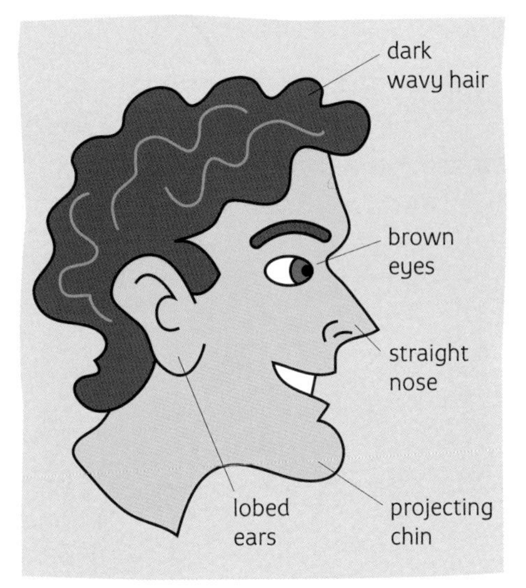

Other dominant characteristics.

The person with these alleles will have black hair because the allele for black hair is more powerful than the allele for blond hair. The allele for black hair is more **dominant** and produces the final hair colour. The allele for blond hair is **recessive**. Whenever a dominant allele pairs up with a recessive allele, the dominant allele produces the final effect.

Homozygous and heterozygous

A person who has two identical gene alleles for a characteristic is pure bred or **homozygous** for that characteristic.

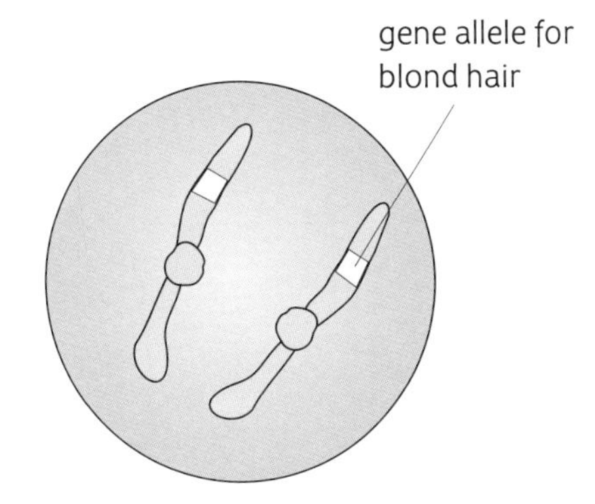

Both alleles say 'have blond hair'.

This girl is homozygous for hair colour.

On the other hand, a person with two different alleles (one dominant and one recessive) for a characteristic is **hybrid** or **heterozygous** for that characteristic.

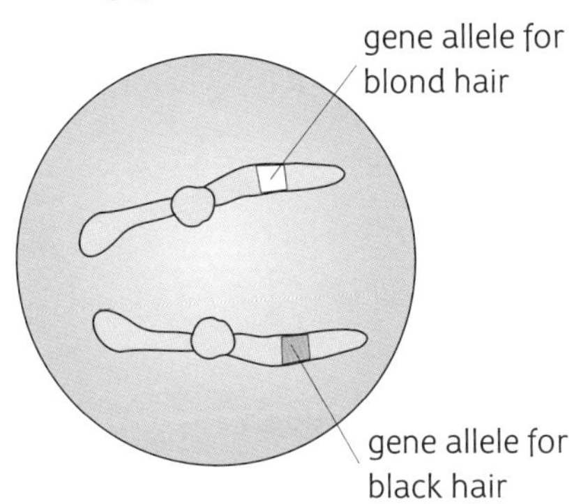

One allele says 'have blond hair' and one allele says 'have black hair'.

This girl is heterozygous for hair colour.

Questions

6 Explain the difference between a dominant and recessive gene allele.

7 Explain the difference between homozygous and heterozygous.

8 What does hybrid mean?

9 If a sperm carrying a gene allele for blue eyes fertilizes an egg carrying an allele for brown eyes, will the person be homozygous or heterozygous for eye colour?

10 Explain how a person with black hair could be either homozygous or heterozygous.

4.03 DNA

Objectives

This spread should help you to

- describe the structure of DNA
- describe what happens to DNA when cells divide

What is DNA?

Chromosomes and genes are made of DNA. DNA stands for ***d**eoxyribo**n**ucleic **a**cid*. Every chromosome contains one long DNA molecule. The genes are short lengths of DNA. There may be up to 4000 genes on one chromosome.

DNA is a sort of plan that determines how the body is made up. Every cell carries a copy of the plan.

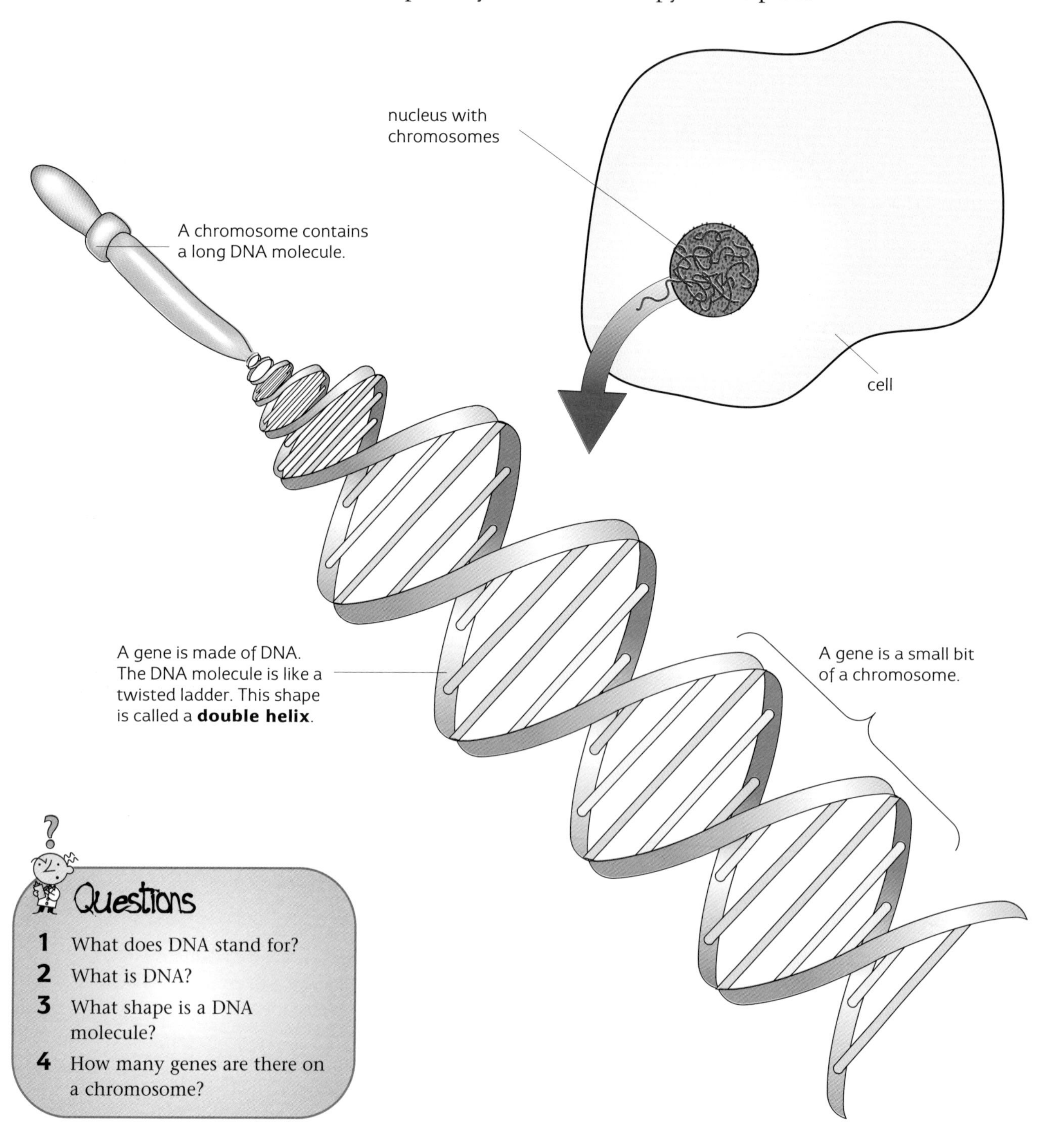

Questions

1. What does DNA stand for?
2. What is DNA?
3. What shape is a DNA molecule?
4. How many genes are there on a chromosome?

When cells divide

The rungs of the DNA 'ladder' are made from pairs of **bases.** There are four kinds of base. They have complicated names so we can use their initials instead. Bases can only fit together in one way, like this:

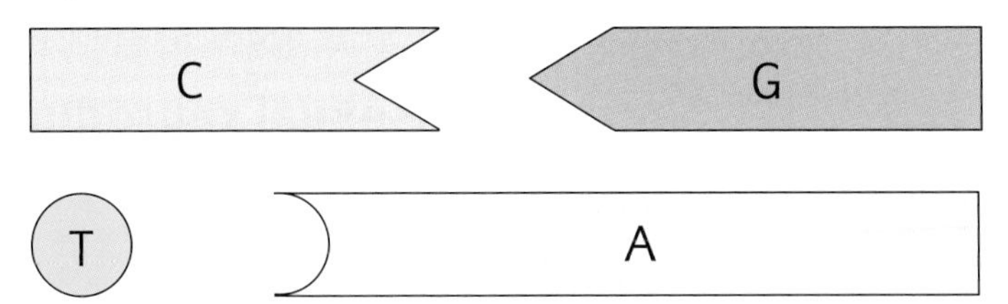

During cell division (mitosis) chromosomes split into two. This means that the DNA must divide. The DNA molecules must make exact copies of themselves. They do it like this:

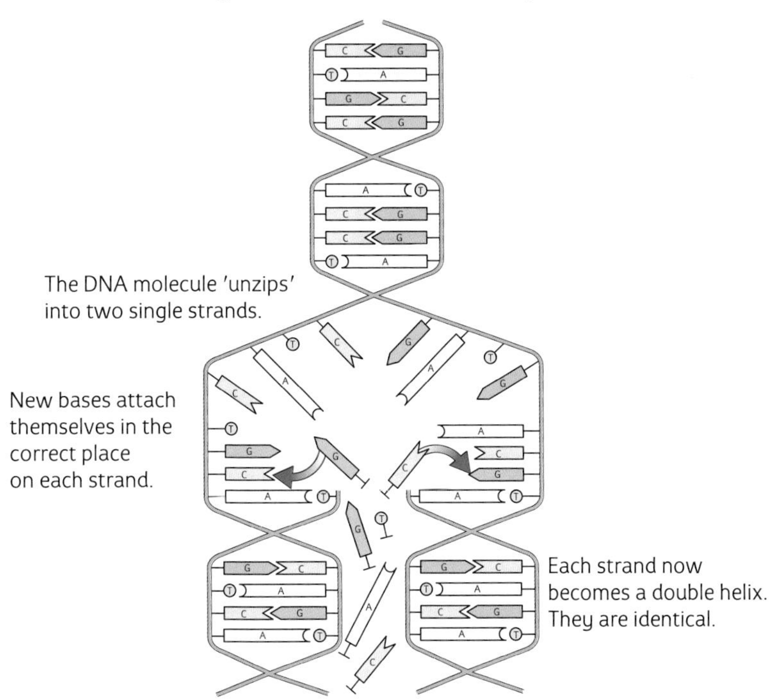

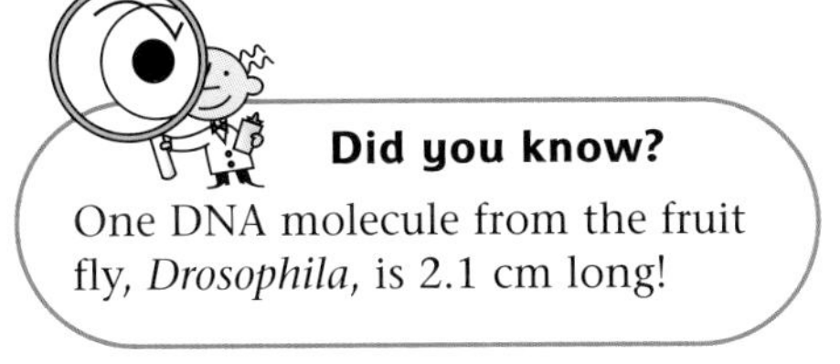

Did you know?

One DNA molecule from the fruit fly, *Drosophila*, is 2.1 cm long!

Sometimes mistakes are made in the copying of DNA. This changes the instructions carried by a gene. This is one way that **mutations** can happen.

Questions

5 a How many kinds of base are there on a DNA molecule?

b What letters are used for the bases?

6 Why do you think a large base is always paired with a small base?

7 a When do DNA molecules make exact copies of themselves?

b Describe how DNA molecules make exact copies of themselves.

8 Explain how a mutation might happen.

4.04 Two kinds of reproduction

Objectives

This spread should help you to

- explain the difference between asexual and sexual reproduction
- give some examples of asexual reproduction in simple animals
- describe how cells divide

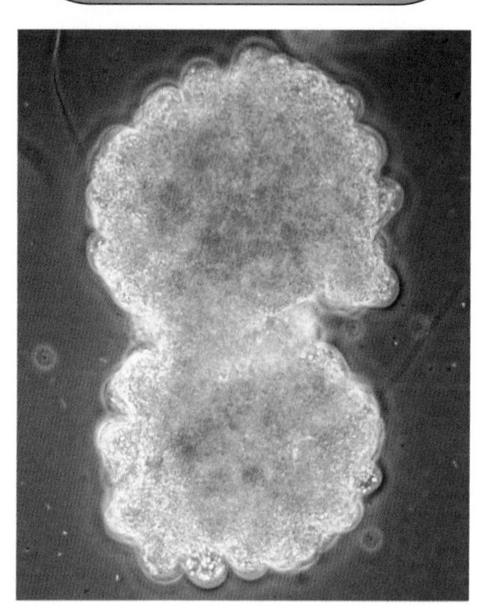

Amoeba takes about an hour to divide. The two new cells grow and each one divides again.

All living things must reproduce to replace those that die. There are two ways that reproduction can happen – by **asexual reproduction** or by **sexual reproduction.**

Asexual reproduction doesn't involve sex cells, while sexual reproduction does. Offspring from sexual reproduction show more variation than those from asexual reproduction.

Asexual reproduction

In asexual reproduction, there is only one parent and its offspring are exact copies of itself. The parent and offspring are genetically identical because they all come from one set of genes. Genetically identical organisms are called **clones.**

Single-celled organisms and simple animals reproduce asexually. Asexual reproduction is common in plants.

Asexual reproduction in *Amoeba*

Amoeba is a single-celled organism which lives in water, especially in stagnant ponds and ditches. It reproduces by dividing into two. Lots of other single-celled organisms reproduce in this way. It happens like this.

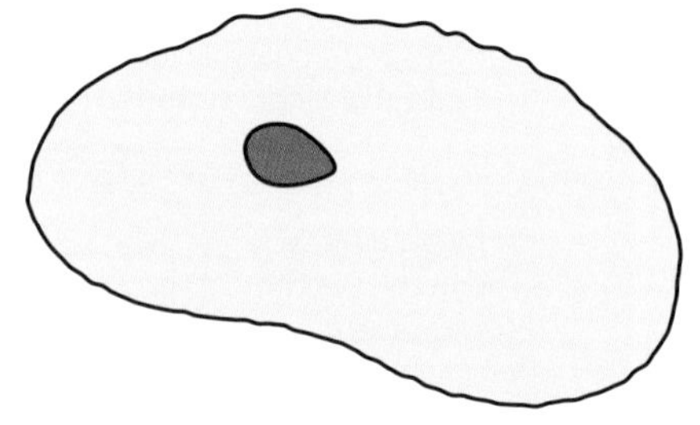

The genetic information inside the nucleus is doubled.

The nucleus divides. Each half contains the same genetic information.

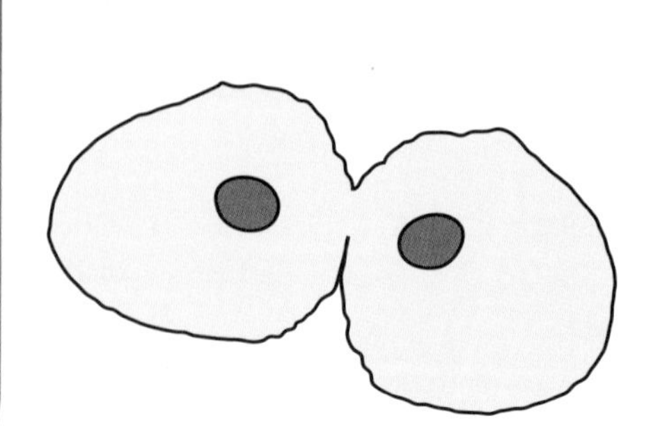

The cytoplasm divides into two.

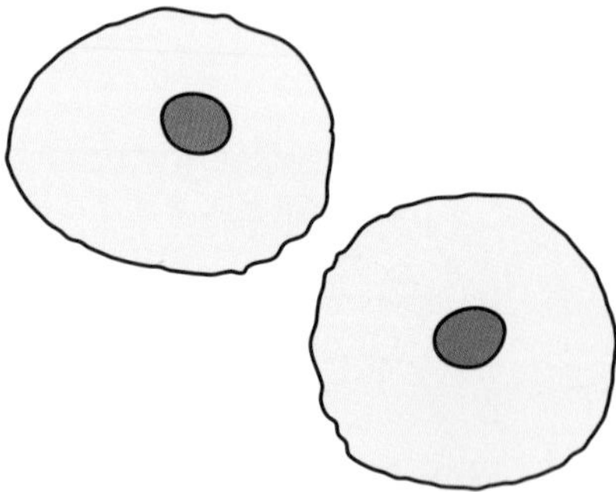

Two new *Amoeba* are produced.

Questions

1. Give two differences between sexual and asexual reproduction.
2. What do we call genetically identical offspring?
3. Explain why the offspring produced by asexual reproduction are genetically identical.
4. Where would you find an *Amoeba*?

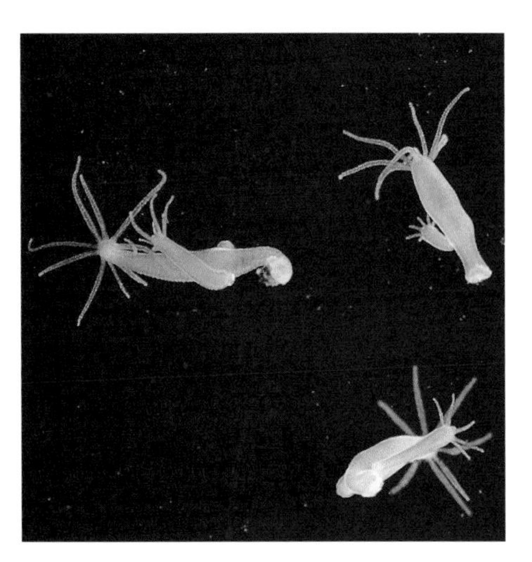

Hydra can grow more than one bud at the same time.

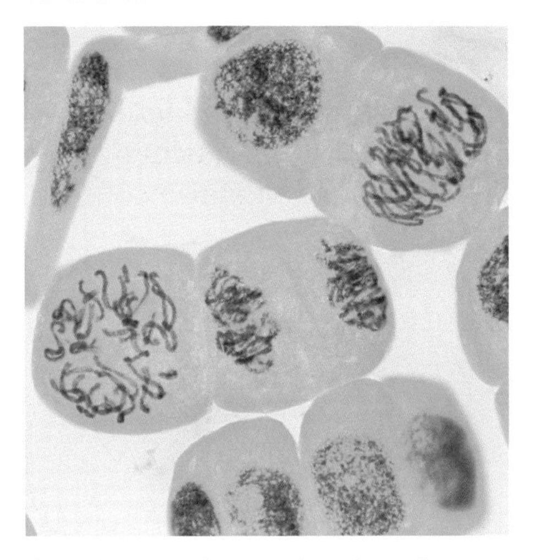

This photograph was taken through a microscope. It shows mitosis happening at a plant root tip. The cells are magnified 540 times.

Asexual reproduction in *Hydra*

Hydra is a simple multicellular animal. It is about 5–10 mm long and lives in ponds attached to pond weed. It traps smaller animals like water fleas with its tentacles, then swallows and digests them. *Hydra* can reproduce sexually but it has an asexual method as well. It happens like this.

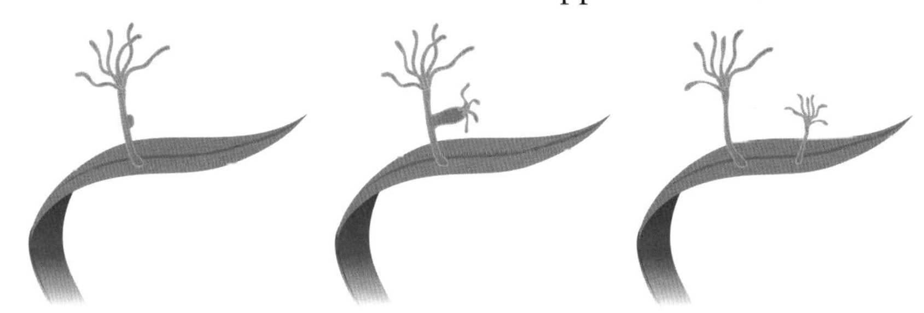

Cells on the side of the *Hydra* divide repeatedly to make a 'bud'. All of the bud's cells have the same genetic information as its parent.

The bud grows and develops tentacles.

The new *Hydra* breaks free and grows into an adult.

How cells divide

Before a cell divides, an exact copy of the genetic information in the nucleus is made. The two new cells will therefore have the same number of chromosomes as their parent and as each other. This kind of cell division is called **mitosis**.

Mitosis works like this

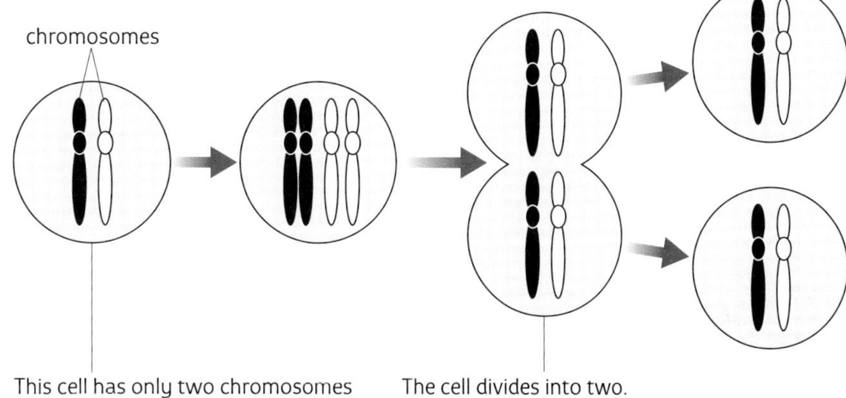

This cell has only two chromosomes – one pair. Each chromosome splits into two, making two pairs.

The cell divides into two. Each new cell gets a pair of chromosomes just like the original cell.

Questions

5 a What does *Hydra* feed on?

b How does *Hydra* catch its prey?

6 Describe one difference between the offspring of *Amoeba* and the offspring of *Hydra*.

7 a What happens to the number of chromosomes before a cell divides by mitosis?

b Explain why this happens.

4.05 Sexual reproduction in humans

Objectives

This spread should help you to

- describe the changes that happen in the body at puberty
- describe the male and female sex organs

Body changes

Puberty is the time when boys and girls become sexually mature. It usually starts around the age of 11–13 in girls and about 12–14 in boys. People vary a lot so there is nothing wrong with you if puberty comes earlier or later in your life.

During puberty we grow fast and get **secondary sexual characteristics**.

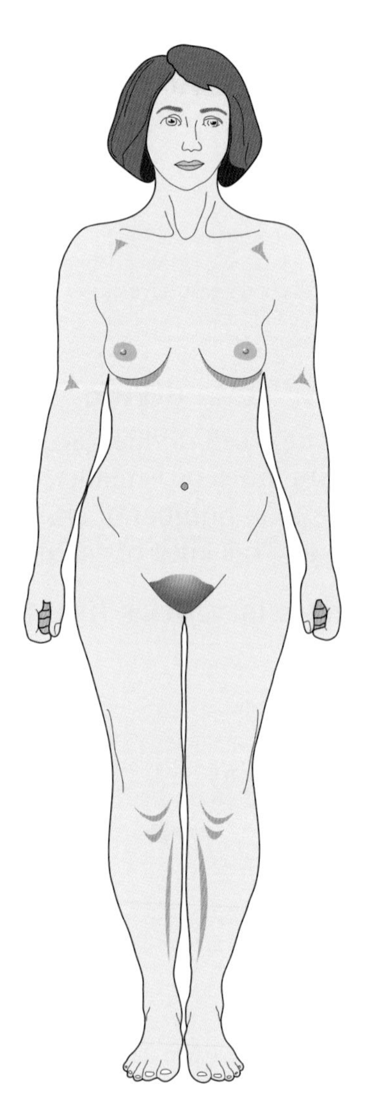

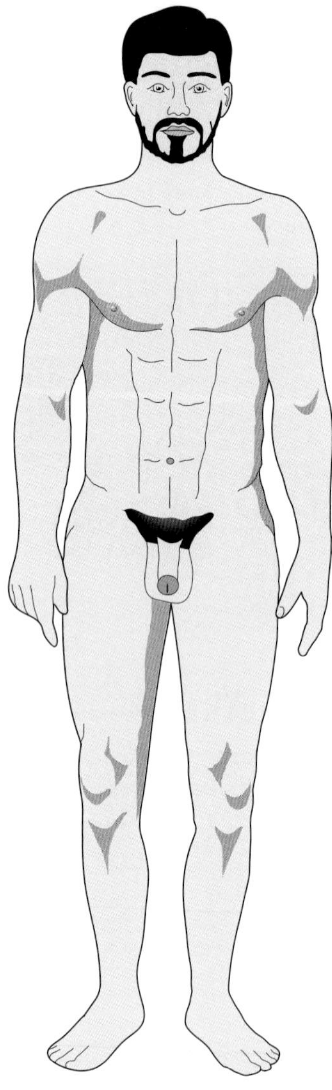

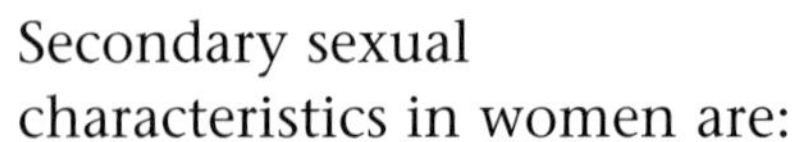

Secondary sexual characteristics in women are:

- breasts develop
- hips get bigger
- hair grows under the arms and around the vagina
- ovaries start producing **eggs** (female sex cells).

Secondary sexual characteristics in men are:

- voice gets deeper
- body becomes muscular
- hair grows on the face and body
- testes start producing **sperm** (male sex cells).

Questions

1 What is puberty?

2 Roughly when does puberty happen in:

a girls **b** boys?

3 Describe the changes that happen at puberty in the body of a:

a girl **b** boy.

Female sex organs

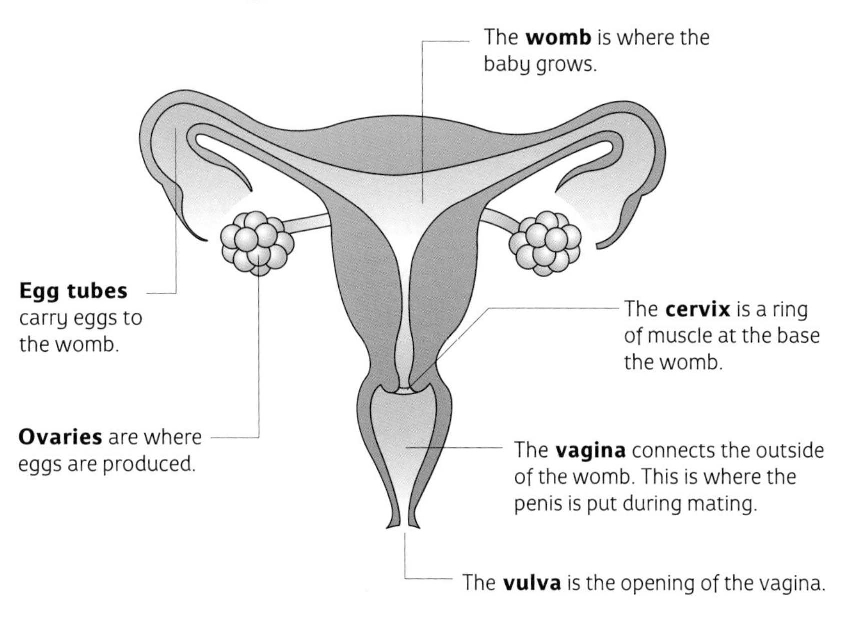

Male sex organs

Sperm tubes carry sperm from the testes to the penis. Sperm are in a liquid called **semen**.

These **glands** make the liquid part of semen.

Erectile tissue fills with blood and makes the penis hard and erect.

A **testis** is where sperms are made. A man has two testes.

Sperms go through the **penis** during mating.

Questions

4 Explain why a boy doesn't usually have to shave until he is in his teens.

5 Girls can sing high notes all through their lives but boys can't after the age of about 14. Explain this.

6 Which part of the body produces:

a eggs **b** sperm?

7 Where does a baby grow?

8 **a** Name the liquid that contains sperm.

b Where is this liquid made?

9 Explain how the penis gets hard.

4.06 The menstrual cycle

Objectives

This spread should help you to

- describe what happens in the menstrual cycle
- describe the job done by hormones in the menstrual cycle
- describe how these hormones can be used in fertility treatment and as a contraceptive

Monthly changes

Every month changes happen in a woman's body. These changes are called the **menstrual cycle**. The menstrual cycle starts at puberty and stops at the **menopause** when a woman is about 45–50. Each cycle takes about 28 days, though this can vary.

There are four stages in the menstrual cycle.

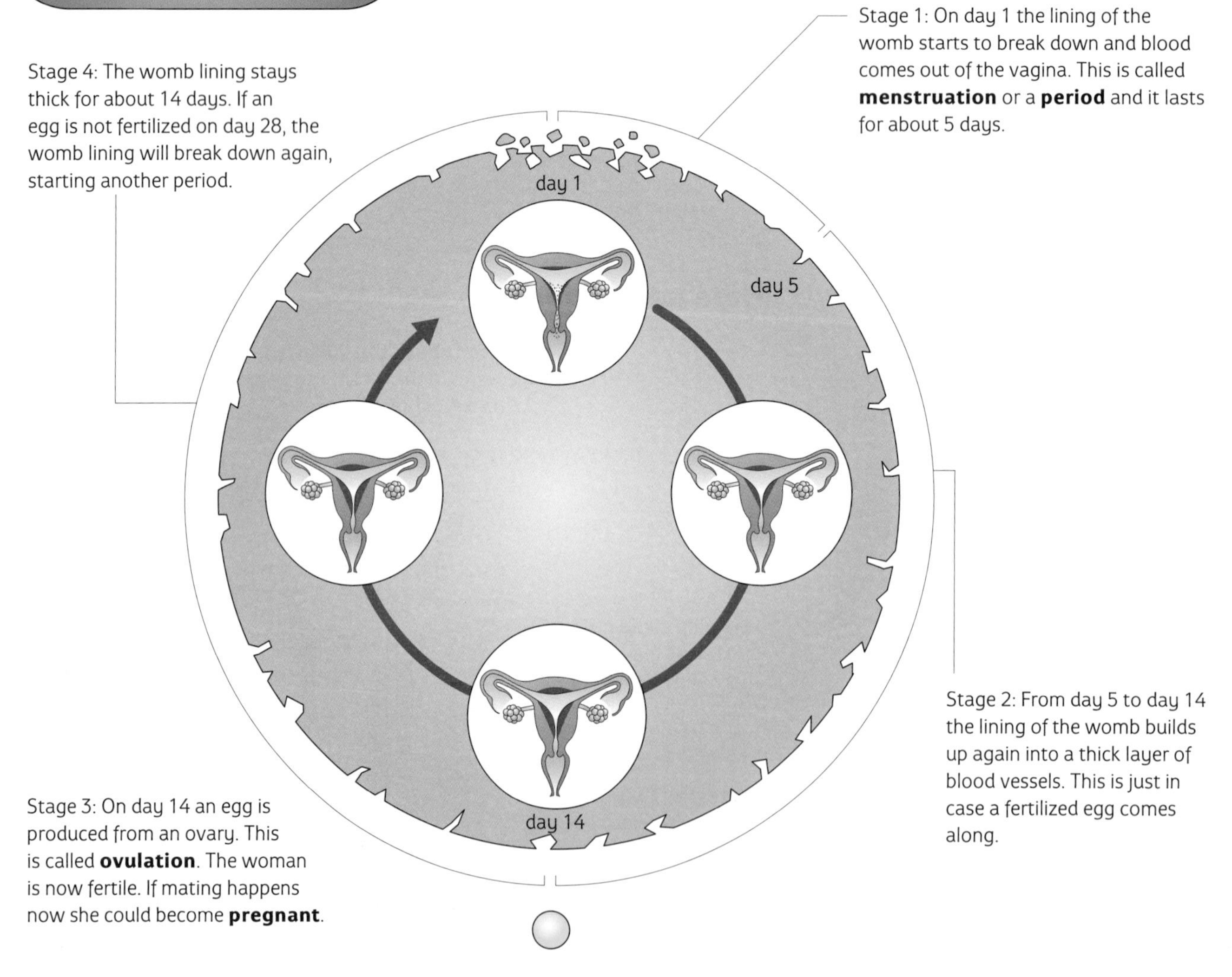

Hormones and the menstrual cycle

Two hormones control the menstrual cycle. They are called **oestrogen** and **progesterone**. Both hormones are made in the ovaries. Oestrogen makes the lining of the womb thicken after a period. It also makes one of the ovaries release an egg on day 14. Progesterone keeps the lining of the womb thick with blood vessels. When the level of progesterone falls, a period starts.

Questions

1. What is the menstrual cycle?
2. How many stages are there in the menstrual cycle?

These graphs show how oestrogen and progesterone work together to control the menstrual cycle.

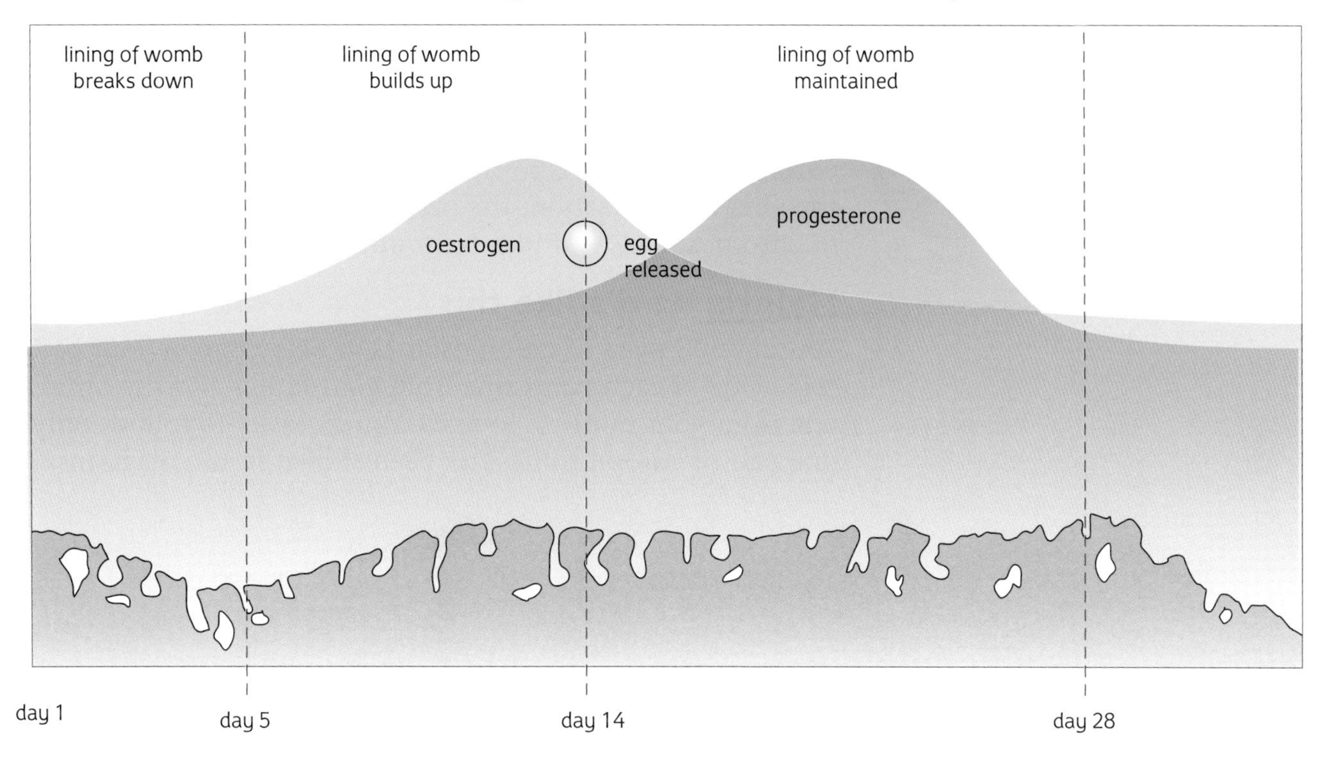

Fertility treatment and 'the pill'

The hormones that control the menstrual cycle can be used to treat women who are **infertile** (who don't produce eggs naturally). The woman is given a hormone that causes her ovaries to produce eggs and to make more oestrogen, so the eggs are released. This is called **fertility treatment**. Doctors have to be careful as too much hormone will cause too many eggs to be produced. This can lead to **multiple births**.

The **contraceptive pill** contains both oestrogen and progesterone. By taking the pill regularly, the amount of both hormones in the body is kept high. The body 'thinks' it is pregnant, and when a woman is pregnant she doesn't produce eggs. If no eggs are produced, she can't get pregnant.

Questions

3 How long does the menstrual cycle usually take?

4 What is a period?

5 If a woman starts to produce eggs when she is 13 and stops when she is 48, roughly how many eggs will she produce during her life?

6 Name the two hormones that control the menstrual cycle, and describe their jobs.

7 Why do some women have fertility treatment?

8 Explain why fertility treatment can lead to multiple births.

9 Explain how the contraceptive pill works.

4.07 Sex cells, mating, fertilization

Objectives

This spread should help you to
- describe how sex cells are made
- describe what happens during mating
- describe how fertilization happens

Making sex cells

Sex cells, or **gametes**, are made in the ovaries of a woman and the testes of a man. They are made by a special kind of cell division called **meiosis**.

Meiosis produces cells which have half the usual number of chromosomes. This is so that when a sperm and an egg come together at fertilization, the baby will have the right number of chromosomes, just like every other human.

Meiosis works like this

You have 23 pairs of chromosomes in every one of your body cells. One of each pair came from your father and the other one from your mother. To make things easier to follow, only one pair of chromosomes has been shown in the diagrams.

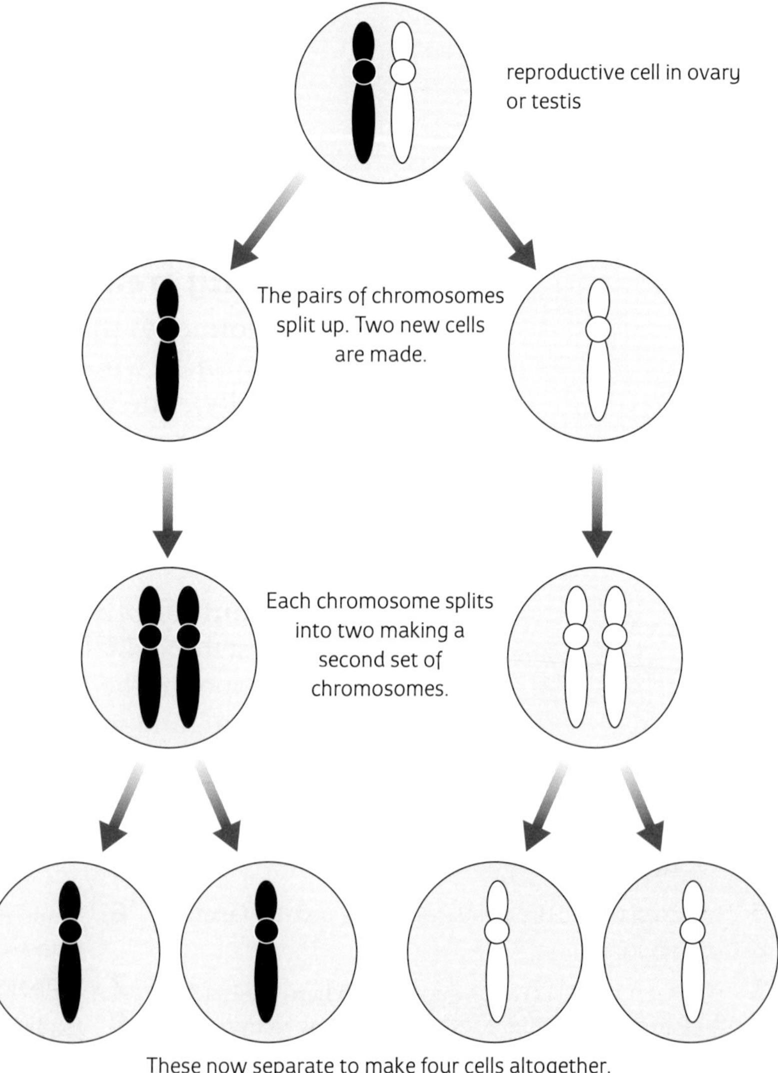

These now separate to make four cells altogether.
These are gametes. Each gamete has half the usual number of chromosomes.

Questions

1. Where are sex cells made?
2. What is the scientific name for a sex cell?
3. How many sex cells are made from one reproductive cell?
4. How many chromosomes are there in a normal human body cell?

Mating

If a sperm is to meet an egg, **mating** or **sexual intercourse** must happen. When a man gets sexually excited he gets an erection. He moves his penis up and down inside the woman's vagina. This causes semen to pump from the testes, through his penis, into the woman's body. This is called an **ejaculation**. A man feels a good sensation (an **orgasm**) when he ejaculates. A woman may also have an orgasm during mating.

Did you know?

A healthy human male will release between 3 to 400 000 000 sperm in one ejaculation. Only one is needed to fertilize an egg.

Healthy sperm swim at 4 mm per minute.

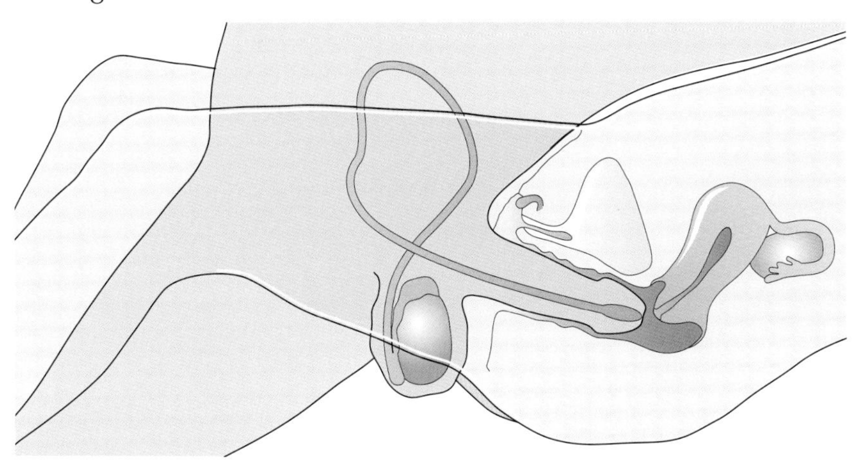

Fertilization

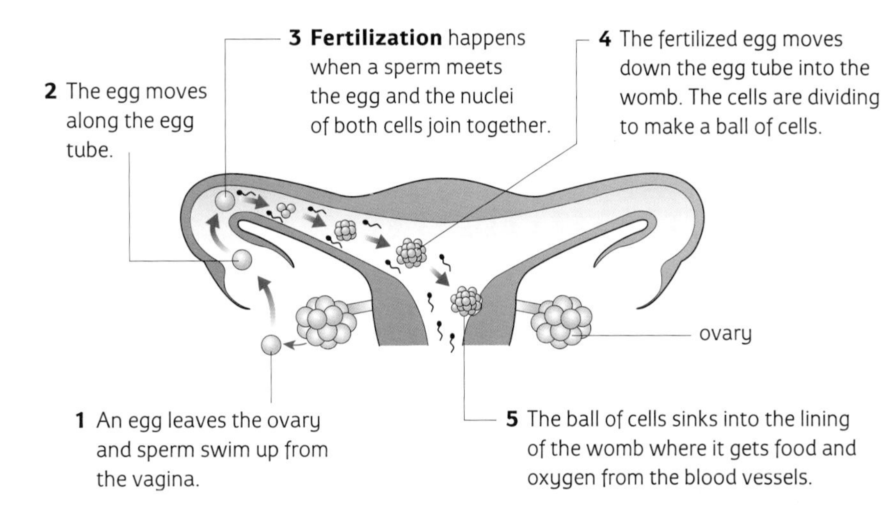

Questions

5 **a** What happens to the number of chromosomes at meiosis?

b Explain why this has to happen.

6 What is the scientific name for mating?

7 What do these words mean?

a erection **b** ejaculation **c** orgasm

8 Where does fertilization happen?

9 Describe what happens to an egg after it is fertilized.

4.08 Development and birth

Objectives

This spread should help you to

- describe how a baby develops from a fertilized egg
- describe the job done by the placenta
- describe the birth of a baby

Looking after the developing baby

The ball of cells in the womb lining slowly grows into tissues and organs. During the next 9 months the developing baby grows inside its mother.

The baby feeds, breathes, and gets rid of wastes inside its mother's body. The mother's blood brings food and oxygen and carries the wastes away. But the mother's blood does not mix with her baby's blood. Food, oxygen, and wastes are exchanged at the **placenta**.

The placenta lies against the womb wall. It is connected to the baby by the **umbilical cord**.

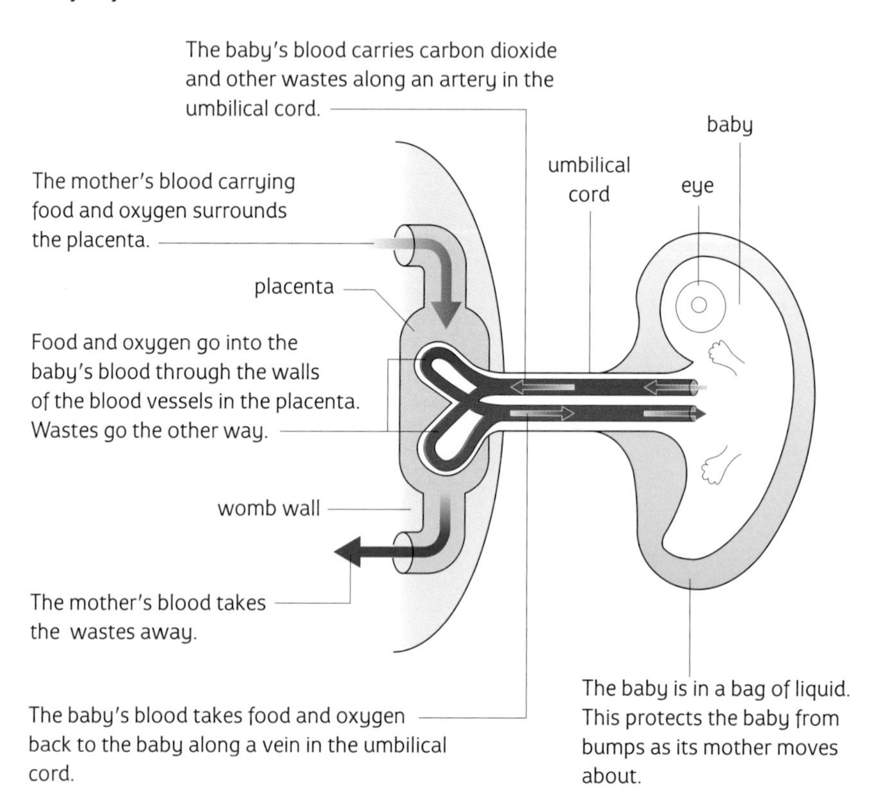

A pregnant mother shouldn't:

- drink alcohol. Alcohol can cause premature birth (born early) and damage the baby's brain.
- smoke. Smoking can cause smaller babies to be born.
- take drugs unless they are given by a doctor. Drugs can seriously affect her baby's development.

Growing on

A fertilized egg is no bigger than a full stop on this page. But in 9 months it will grow into a baby half a metre long and weighing about 3 kg.

Questions

1. What is the placenta for?
2. What is inside the umbilical cord?
3. What moves across the placenta:
   a from mother to baby
   b from baby to mother?
4. What is the job of the bag of fluid that surrounds the baby?
5. Explain why a pregnant woman should not drink alcohol, smoke, or take drugs.

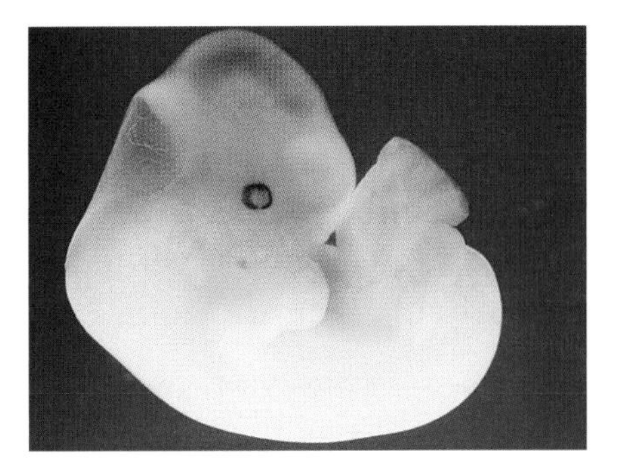

At 4–5 weeks the baby is about 0.5 cm long. Eyes and ears develop. Arms and legs are just bumps.

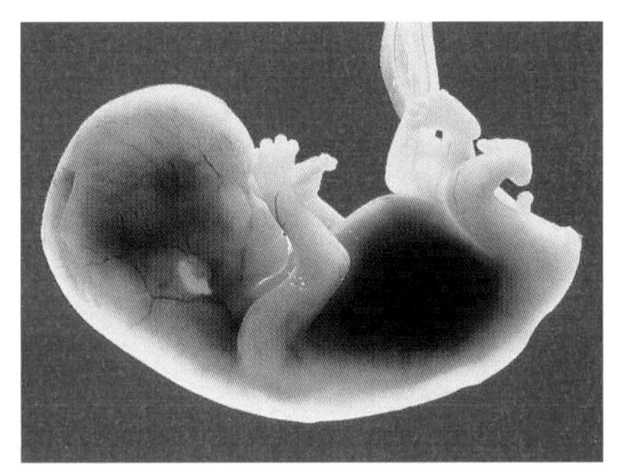

At 8–10 weeks it is now 4 cm long. It has a face, arms with fingers, and legs with toes. Its heart is beating.

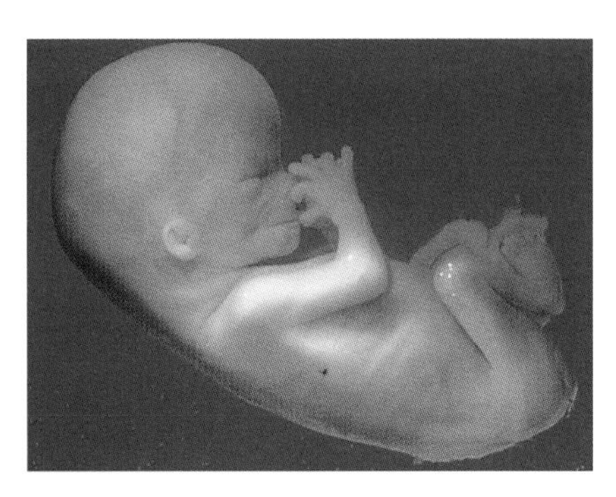

At 16 weeks the baby is 16 cm long. It now has all its organs and its mother will feel it moving inside her.

Birth

A few weeks before birth the baby turns upside down so its head lies just above the cervix.

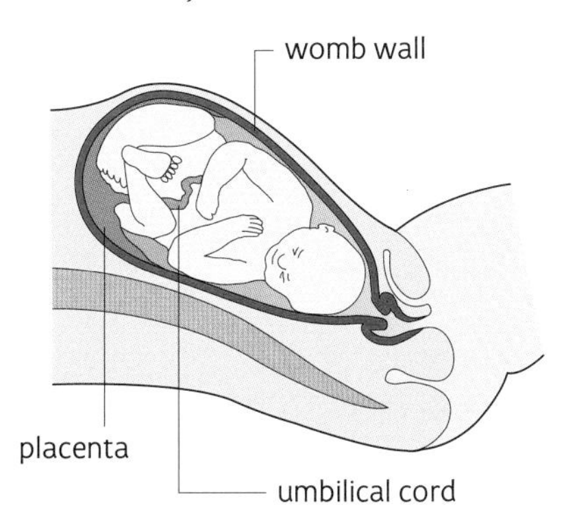

The baby ready for birth.

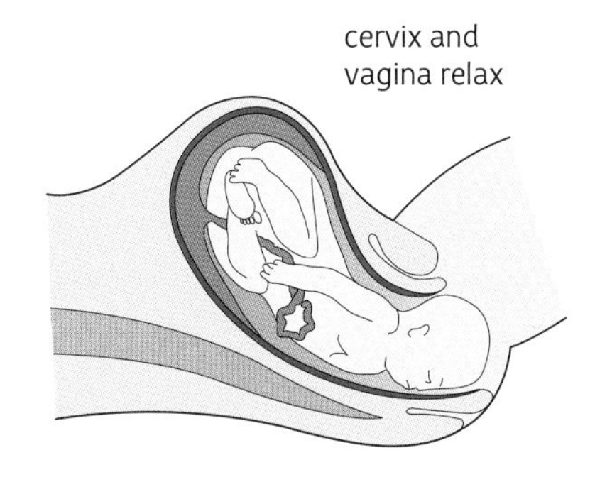

The birth.

During its first months the baby will get all the food it needs from its mother. It suckles milk from ***mammary glands*** *inside its mother's breasts.*

Birth starts with **labour pains**. The muscles in the womb wall start to contract rhythmically – slowly at first, then quicker and stronger. Muscular contractions of the womb burst the bag of liquid surrounding the baby. The cervix relaxes to let the baby's head through. The mother uses her tummy muscles to help the muscles in the womb to push the baby out into the world. The umbilical cord is tied and cut.

A few minutes after the baby is born, the womb muscles push the placenta and remains of the umbilical cord out of the womb. These are the **afterbirth**.

Questions

6 How long does pregnancy take in humans?

7 How long does it take for the baby to get its fingers and toes?

8 What does the baby do a few weeks before birth?

9 What are labour pains?

10 When does a mother use her tummy muscles?

11 What is the afterbirth?

12 Where are mammary glands?

4.09 Contraceptives

Objectives

This spread should help you to

- describe some methods of contraception

Birth control

Sexual intercourse is pleasurable and is one way that a man and a woman can show their love for each other. However, if a couple do not want to have a child they must somehow stop a sperm from fertilizing an egg. Fertilization can be prevented by using **birth control** or a method of **contraception**.

Some people believe it is wrong to use any form of birth control except the **rhythm method**. The couple simply avoid mating when there is a chance that an egg is being released from an ovary. The problem is that it is very hard to know exactly when an egg might be released. This is why this method is not very reliable.

Did you know?

No contraceptive is 100% safe.

Strange facts about contraception

Did you know that ancient Egyptian women used contraceptives? They put a piece of straw into their womb to stop them getting pregnant. This was an early form of IUD.

Some women put dried crocodile dung into their vagina before having sexual intercourse. The dried dung acted like a sponge, absorbing the sperm.

Rumour has it that other women tied cat's testicles around their neck. The male hormones were absorbed through their skin and upset the menstrual cycle. This method worked a bit like the contraceptive pill.

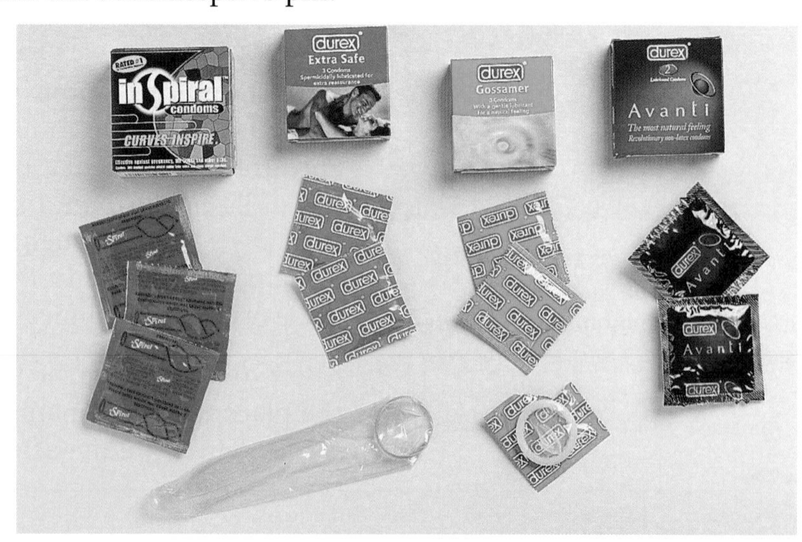

*The **condom** is a thin rubber sheath that is rolled over the erect penis before mating. Ejaculated sperms are collected in a small bulb at the end. Condoms are quite reliable and have no side effects. They are even safer to use if the woman uses a spermicide. The woman puts the spermicide into her vagina before mating. Condoms are the only method of contraception that give protection from sexually transmitted diseases (STDs).*

Questions

1. Why do people use birth control?
2. Describe the advantages of using a condom.
3. Why must a diaphragm be left in place for hours after mating?
4. **a** What is a spermicide?

   b Why is it a good idea to use a spermicide with a condom or a diaphragm?

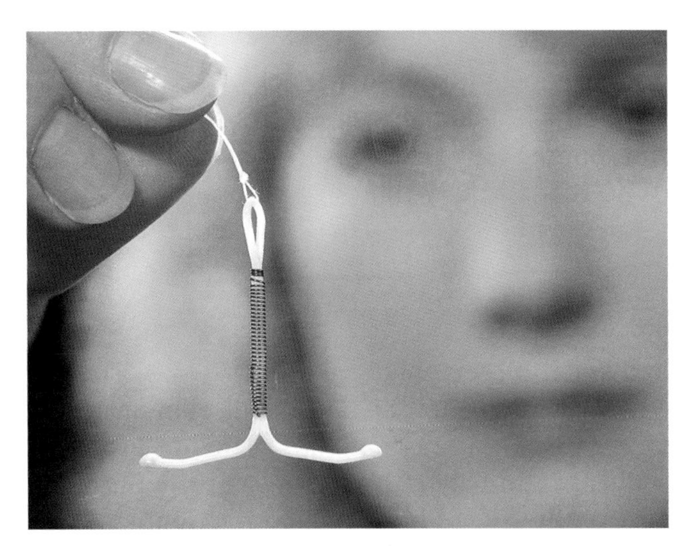

An ***IUD* (*intrauterine device*)** *is a small coil or loop of plastic. A doctor fits the IUD into a woman's womb where it stays for many months or even years. The presence of a 'strange' object in the womb stops a fertilized egg fixing itself to the lining of the womb and growing into a baby.*

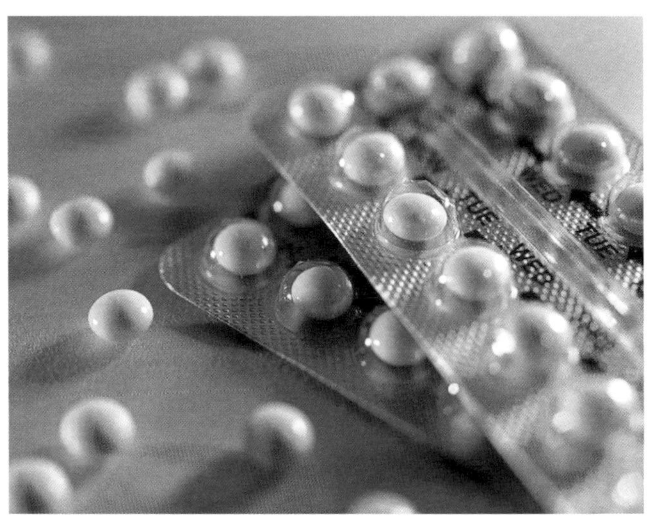

'The pill' *contains female hormones which stop the ovaries producing eggs. Contraceptive pills are prescribed by a doctor. They are very reliable but may have side effects. A woman with diabetes should not take the pill. Women who smoke whilst on the pill may get blood clots, migraines, or heart problems. Some women may also get breast cancer.*

The ***diaphragm*** *(or cap) is a circle of rubber with a plastic or metal spring round it. Before mating, the woman smears spermicide over it and puts it in her vagina to cover her cervix. The diaphragm stops sperm getting into the womb. It should be left in place for at least eight hours to make sure all the sperm are killed.*

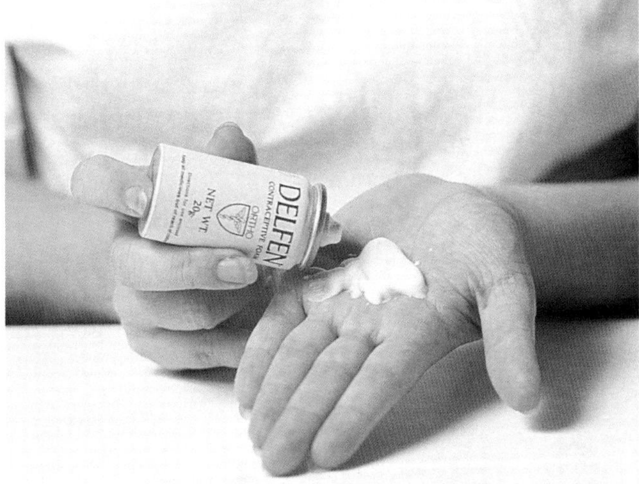

A ***spermicide*** *is a chemical that kills sperm. The chemical is usually in a cream or jelly. Spermicide is not meant to be used on its own. It makes using a condom or a diaphragm a lot safer.*

Questions

5 Explain why a woman must think carefully before going on the pill.

6 Explain how an IUD works.

7 A contraceptive is meant to stop fertilization. What makes an IUD different from all other kinds of contraceptive?

8 Explain why the rhythm method is not very reliable.

4.10 Sexually transmitted diseases

Objectives

This spread should help you to

- describe some sexually transmitted diseases

Sexually transmitted diseases (STDs) or **venereal diseases (VD)** are diseases passed on during sexual activity. They affect the reproductive organs but can also spread to other parts of the body. The microorganisms that cause STDs can't live long outside the body so it is unlikely that they can be caught in any way other than by sexual intercourse. Some of the diseases are very dangerous.

Crab lice live in pubic hair.

Lice or 'crabs'

Crab lice are very small insects that live in pubic hair. They feed on human blood which they suck from the skin around the reproductive organs. This causes itching. The lice lay tiny white eggs (called 'nits') and stick them firmly to the pubic hairs. Crab lice are easily spread by sexual contact with another person. Special shampoos which kill crab lice can be bought at a chemist's.

Gonorrhoea

Gonorrhoea is common and more people are catching it every year. Both men and woman can get gonorrhoea, but they can be cured. It is about ten days before someone knows they've got gonorrhoea. When they go to the toilet it is painful to urinate. The urine is followed by a greenish yellow discharge. If the disease is not treated, arthritis and damage to the eyes can happen.

Unfortunately most women don't have any symptoms at all, so the only way they know they have got gonorrhoea is when their partner gets it. If a pregnant women gets gonorrhoea, she can give the disease to her baby as it goes through the vagina during childbirth.

Chlamydia

Like gonorrhoea, more men and woman are catching **chlamydia** every year. Chlamydia can be easily treated and cured. But most infected people don't know they have the disease until it is too late. In women the egg tubes become inflamed, causing infertility. If a woman gets chlamydia while she is pregnant, the baby can get eye infections. There is also a greater chance of the baby being born too early (premature birth). In men the symptoms are like those for gonorrhoea.

Syphilis

Syphilis isn't as common as gonorrhoea or chlamydia but it is on the increase. Syphilis affects men more than women. In its early stages the disease can be cured, but because it doesn't hurt, people often don't know they have got it until it is too late.

Questions

1 What do these stand for?
  a STD **b** VD

2 How are STDs caught?

3 What are the symptoms that a person has:
  a 'crabs' **b** gonorrhoea
  c chlamydia **d** syphilis?

4 Explain why a woman with several sexual partners should have regular checks for gonorrhoea and chlamydia.

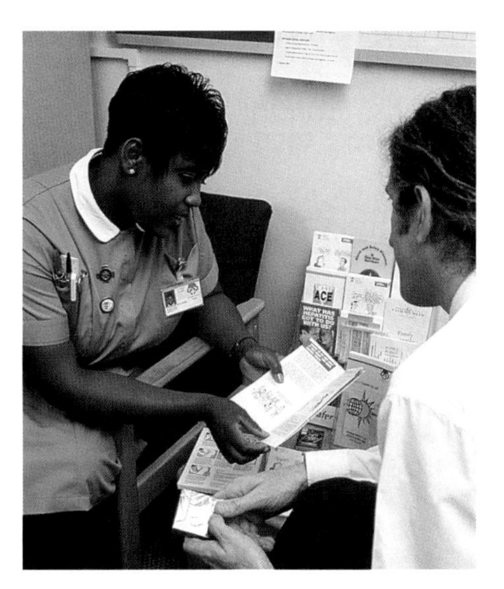

There are clinics in most towns and cities which give help and advice about sexually transmitted diseases.

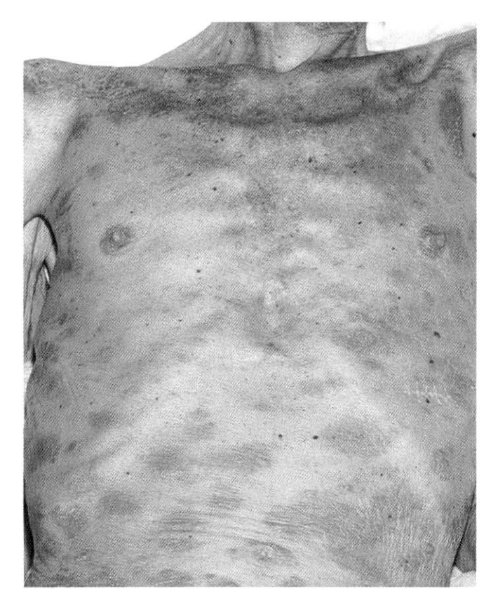

AIDS victims are unable to defend themselves against infection.

First, sores appear round the sex organs. They don't hurt and soon disappear. Next, a painless rash appears for a while. Finally, without treatment, the person could get heart disease or go blind or insane. Syphilis can pass from a pregnant woman to her baby which often dies in the womb or very soon after being born.

AIDS (acquired immune deficiency syndrome)

AIDS is caused by **HIV** (the **human immunodeficiency virus**). HIV attacks the body's immune system by stopping white blood cells making antibodies. People with AIDS can't fight off the infections that healthy people fight off easily. AIDS sufferers die from diseases such as pneumonia, diarrhoea, fungus infections in the lungs, skin cancer, and swelling of the brain.

AIDS is not easy to catch. HIV can't live for long outside the body so AIDS can't be caught by coughing, sneezing, or drinking from the same cup. Neither can you get it from touching, being in the same room as, or swimming in the same pool as someone who has AIDS. HIV is passed on when blood, semen, or vaginal fluids are mixed. This may happen during sexual intercourse or when injecting drugs with used needles or syringes.

There is as yet no known cure for AIDS. It is important to follow a few simple rules if you are to reduce the risk of infection.

- Have few sexual partners. Unless you are sure of your partner never have sex without using a condom.
- People who use drugs should never share syringes or needles.
- If you have your ears pierced, or body tattooed, or even have acupuncture, make sure that the equipment is sterilized first.
- Don't share a razor or a toothbrush with someone you don't know (gums can bleed when they are brushed hard).

Questions

5 Explain why it is important that syphilis is treated in its early stages.

6 Explain why the early detection of syphilis is important for a pregnant woman.

7 What do these stand for?

a AIDS **b** HIV

8 How do condoms reduce the chances of getting AIDS?

9 Why should a first-aider wear rubber gloves when helping someone who is bleeding?

10 Explain why you shouldn't share a toothbrush with someone who has AIDS.

11 Explain why it is safe to shake hands with someone who has AIDS.

12 What is unprotected sex?

4.11 Sexual reproduction in plants

Objectives

This spread should help you to
- describe the structure of a flower
- describe the reproductive organs of a flowering plant

Flowers and sex

Flowers contain a plant's male and female reproductive organs. Sex cells (gametes) are made here by meiosis.

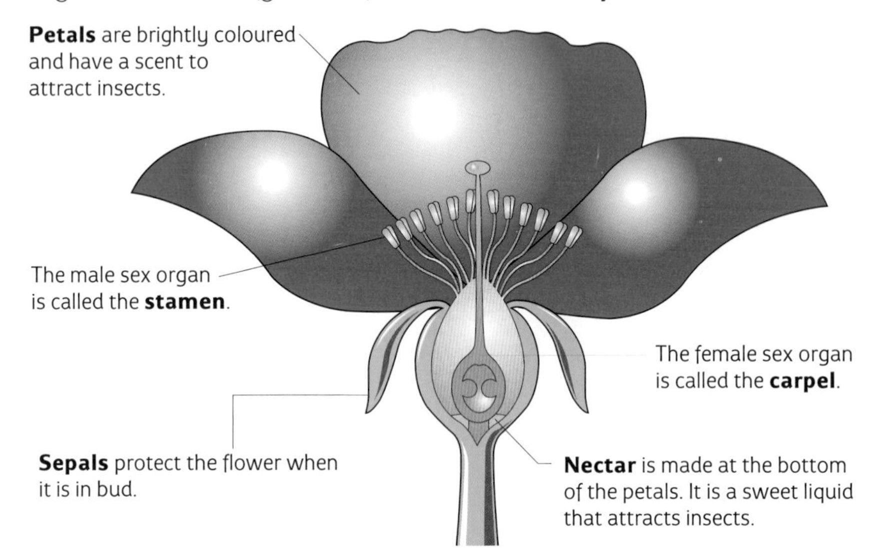

The **anther** is where **pollen** is made. Pollen grains contain the male sex cells.

The stalk is called a **filament**.

Parts of a stamen.

The **stigma** is sticky. Pollen sticks to this during pollination.

The **style** is a stalk that holds the stigma in the air to collect pollen.

The **ovary** protects the ovules.

An **ovule** (**egg**) contains the female sex cell.

Parts of a carpel.

The stamens are in a ring round the carpel.

Questions

1 Why are flowers important to a plant?

2 What is the name of:

a the male sex organ of a flowering plant

b the female sex organ of a flowering plant?

3 What do pollen grains contain?

4 **a** What is nectar?

b Where is nectar made?

5 What is the job of the:

a petals **b** sepals

c ovary **d** style?

Flowers come in all sorts of shapes, sizes, colours, and smells.

Apple blossom flowers have a sweet smell and lots of nectar to attract insects.

Grass flowers have very big stamens. See how the anthers hang outside each tiny flower so wind can carry the pollen away.

This daisy flower is really lots of tiny flowers bundled together.

Gorse flowers are completely enclosed. You can't see the reproductive organs at all.

Questions

6 Where does meiosis happen in flowering plants?

7 Explain how grass flowers get pollen into the air.

8 Explain why insects are attracted to apple blossom.

4.12 Pollination and fertilization

Objectives

This spread should help you to

- explain the difference between self and cross pollination
- describe wind and insect pollination
- describe how fertilization happens

Moving pollen grains

Pollination is the first step in sexual reproduction in plants. It happens when a pollen grain is carried from an anther to a stigma.

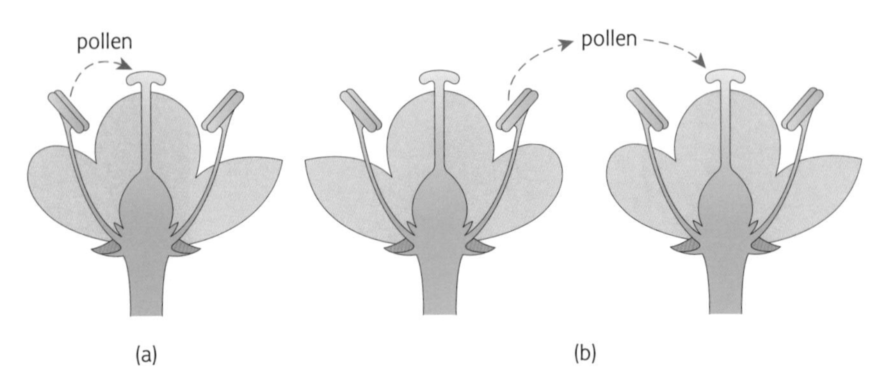

Pollen can be carried to a stigma on the same flower. This is called ***self pollination****.*

Cross pollination *happens when pollen is carried to a stigma of another flower.*

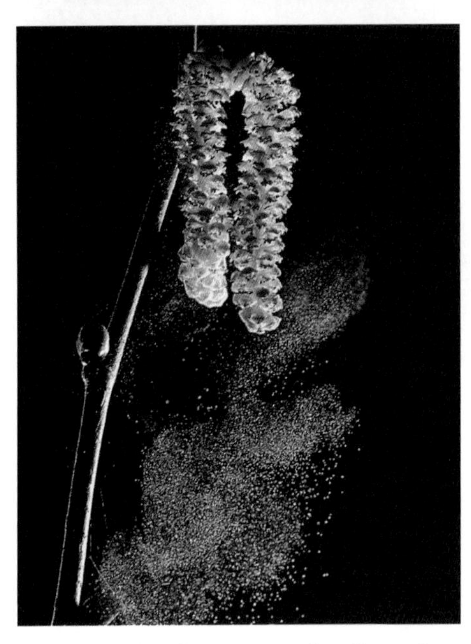

These hazel catkins are wind-pollinated flowers.

Pollination by wind

The flowers of wind-pollinated plants are small. They aren't scented or brightly coloured because they don't have to attract insects. Wind-pollinated flowers have:

- anthers that hang outside the flower to catch the wind
- lots of small, light pollen grains that are easily blown by the wind
- feathery stigmas, like nets, to catch pollen grains out of the air.

grass flowers

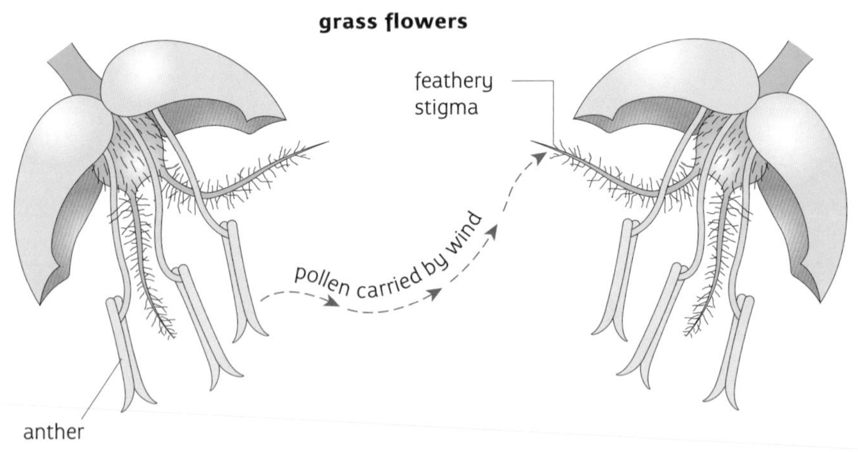

Pollination by insects

Insects, like bees, carry pollen from flower to flower. The pollen sticks on the fur on the insect's body as it collects nectar from the flower. Insect-pollinated flowers have:

- big, coloured, scented petals to attract insects
- large, spiky pollen grains that stick to the insect's fur
- anthers and stigmas inside the flower so they touch the insect when it collects nectar.

Questions

1. What is pollination?
2. Explain the difference between self pollination and cross pollination.
3. Explain how wind-pollinated flowers catch pollen grains carried by the wind.

A bee collects nectar and gets dusted with pollen.

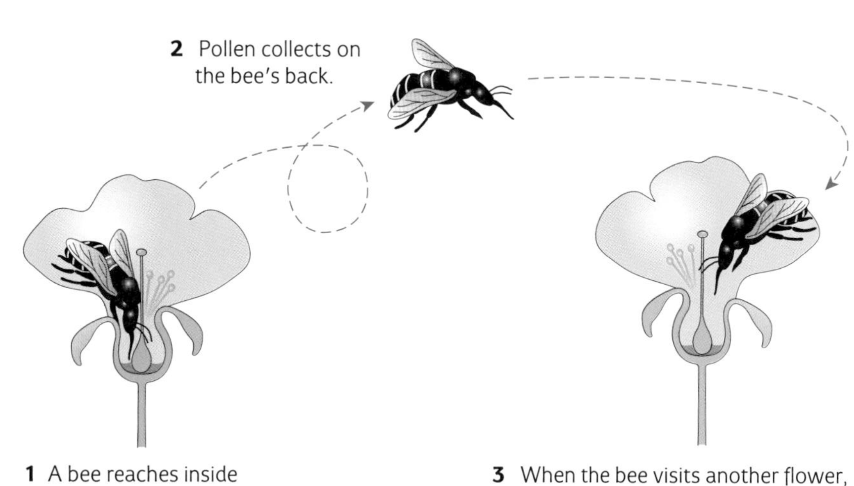

Fertilization

Fertilization happens when the nucleus of a male sex cell joins up with the nucleus of a female sex cell. It happens like this.

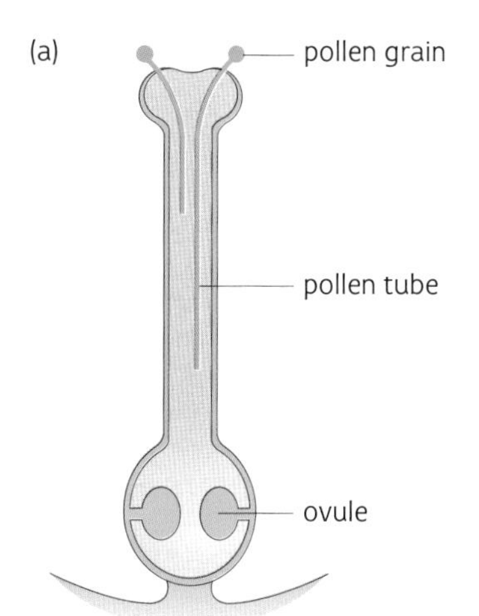

*A **pollen tube** grows from the pollen grain. It grows down through the wall of the ovary and goes into an ovule.*

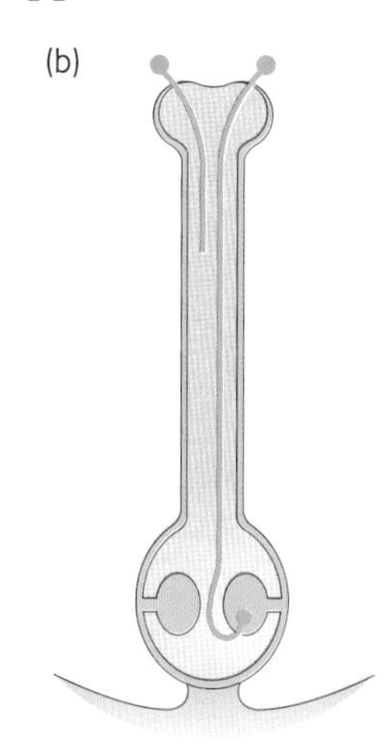

The nucleus from the male sex cell moves down the pollen tube. The end of the pollen tube bursts open, letting the nucleus out. The male nucleus joins up with the female nucleus in the ovule.

(c)

*The fertilized ovule becomes a **seed**. The ovary becomes a **fruit** with the seed inside it. The flower's job has been done – the petals die and drop off.*

Questions

4 Describe three differences between a wind-pollinated flower and an insect-pollinated flower.

5 **a** What is fertilization?

b Describe how fertilization happens in a flowering plant.

6 Kelvin was using a microscope to look at some pollen grains taken from two different flowers. Some pollen grains were small and very smooth. The others were larger and had a spiky, sticky surface.

a Which pollen grains came from hazel catkins?

b Explain your answer.

4.13 Fruits and seeds

Objectives

This spread should help you to

- explain what fruits and seeds are
- explain why dispersal of seeds is important
- describe how fruits and seeds can be dispersed

Fruits can be soft and juicy like these tomatoes.

Seeds inside fruits

Each seed contains a tiny **embryo plant**. It also contains a store of food. Round the seed is a tough protective coat. The ovary grows bigger as the seeds are formed. When fully grown, the ovary with seeds inside is called a fruit.

Pea pods are fruits. The peas inside are seeds.

Nuts are fruits. The seed inside is the bit we eat.

Scattering seeds

Seeds are scattered or **dispersed** to new areas so that the new plants won't be overcrowded when they grow. Scattering gives the plants a better chance of survival. If they were dumped all together in one place the seeds would have to **compete** for water, minerals, and light.

Wind dispersal

Some seeds are dispersed by wind.

Dandelion seeds have 'parachutes' which help them float in the wind.

Sycamore seeds have wings which help them fly in the moving air.

Poppy heads are like pepper pots, shaking the seeds out as they move in the breeze.

Animal dispersal

Some seeds are dispersed by animals.

Burdock fruits have hooks which catch on animals' fur and get carried long distances before falling off.

Birds and other animals eat fruits like these blackberries. The seeds are unharmed as they pass through their bodies.

Nuts like these acorns are carried away by squirrels who usually bury them and forget where they are.

Exploding fruits

Some seeds are dispersed by explosions. On a hot day when the pod is dry, it bursts open scattering the seeds.

Lupin fruits suddenly flick back, throwing seeds everywhere.

Labernum pods twist quickly back to throw the seeds away.

Geranium fruits burst open. The sides spring upwards, scattering the seeds.

Questions

1 What is in a seed?

2 What is a fruit?

3 **a** What does 'dispersed' mean?

b Explain why seeds are dispersed.

4 Describe one example of dispersal by:

a wind **b** animals **c** explosion.

5 Olivia goes to a shop, buys some lettuce seeds and sows them in her garden.

a What sort of seed dispersal is this?

b Explain your answer.

4.14 How seeds grow into plants

Objectives

This spread should help you to

- describe the structure of a seed
- describe the germination of a seed
- explain why seeds are a good source of food for humans

What's in a seed?

A seed is made up of an embryo plant and a food store wrapped in a protective coat. The embryo has a young root called the **radicle**, and a young shoot called the **plumule**. Seeds also have a store for food. The young plant needs this stored food so it can grow or **germinate**. Later it will have leaves to make food by photosynthesis. In some seeds this stored food is in seed leaves called **cotyledons**.

Structure of a bean seed

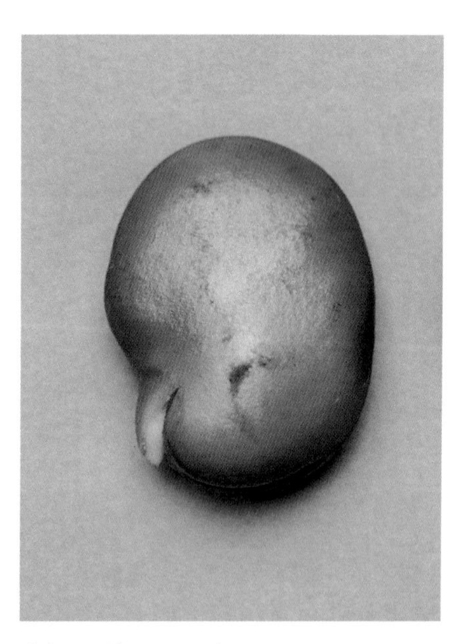

A broad bean seed.

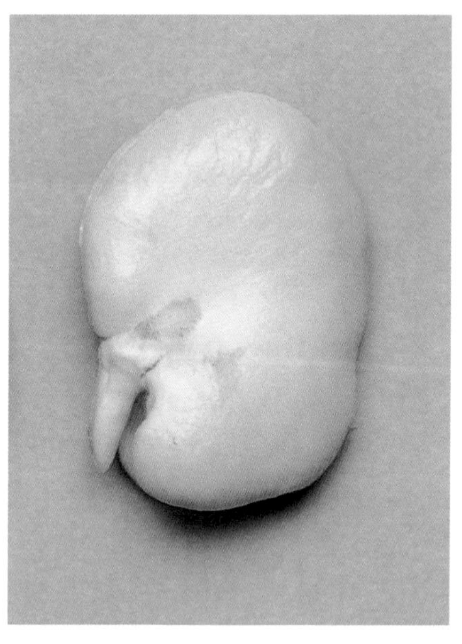

A broad bean seed with its protective coat removed.

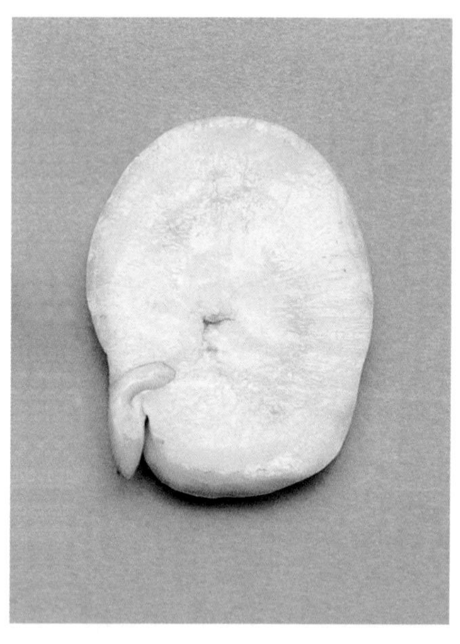

A broad been seed cut in half to show the embryo and cotyledon.

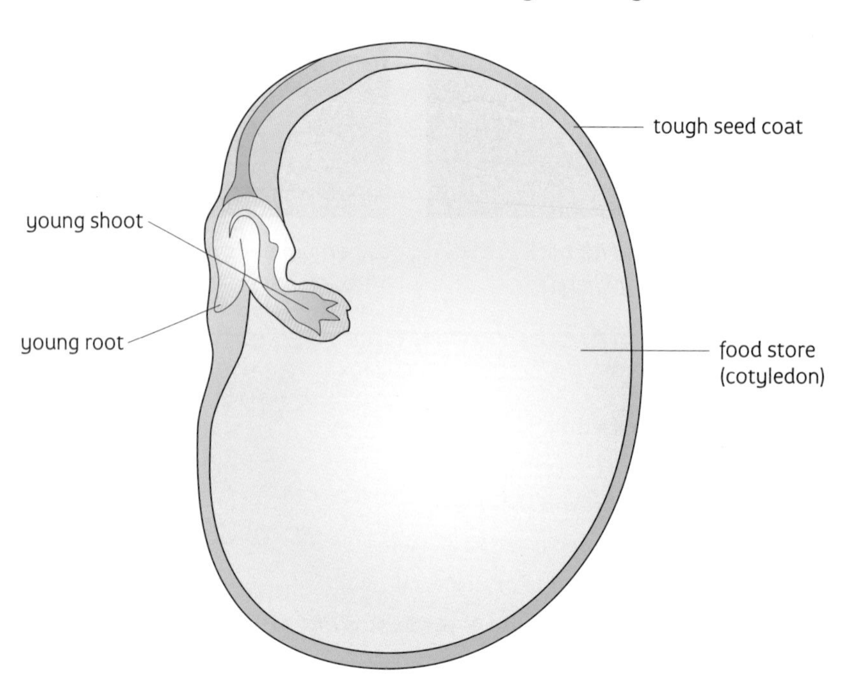

Questions

1 In a seed, what is the scientific name for:

a the young root

b the young shoot

c the food store?

2 **a** What is the biggest part of a seed?

b Why do you think this is?

Germination

If a seed is given the right conditions it will germinate. A seed dispersed from a plant is quite dry. It needs to take in water before it can germinate. Oxygen and warmth are also needed by a germinating seed.

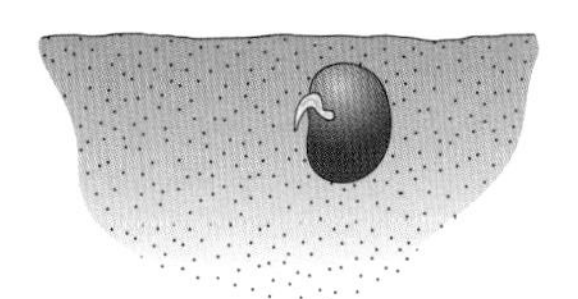

The seed absorbs water and swells up. The protective coat softens and the young root (radicle) emerges. Slowly it pushes down into the soil. The tip of the root is protected by a **root cap**.

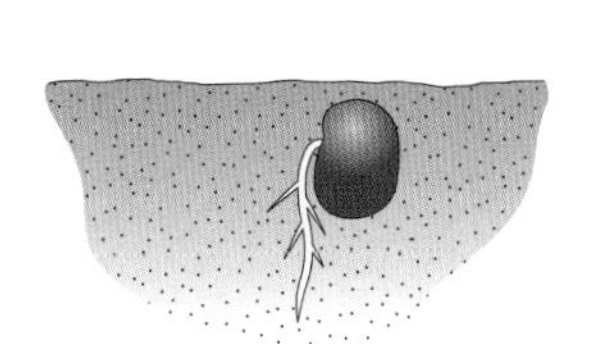

Root hairs grow so that more water and minerals can be taken into the growing seed. This water will help the seed use the food stores in the cotyledons.

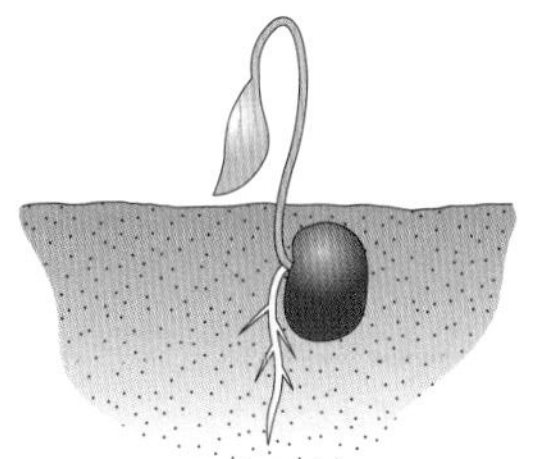

The young shoot (plumule) grows upwards through the soil. Leaves soon grow and the young plant begins to make its own food by photosynthesis.

Wheat seeds are grown on a large scale.

Seeds for food

Many seeds never get the chance to germinate – they get eaten by humans and other animals. Wheat seeds are an important source of food.

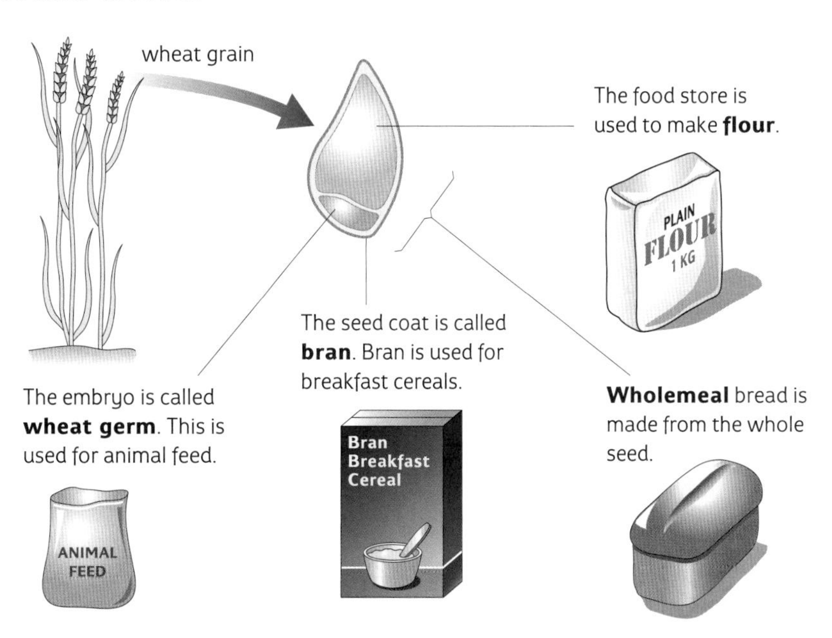

Questions

3 What does 'germinate' mean?

4 What three things does a seed need to germinate?

5 Explain why a germinating seed needs lots of water.

6 Which part of a wheat seed is used for:
a flour **b** animal feed **c** wholemeal bread?

7 **a** What is bran?
b Name a food that contains bran.

4.15 Cloning in plants

Objectives

This spread should help you to

- explain what a clone is
- describe some natural clones
- describe how new plants can be grown by using micropropagation and tissue culture

Clones are genetically identical organisms. Clones can happen naturally in plants and some animals. Today cloning is a 'high tech' business. A lot of commercial plant growers use the technique.

Natural clones

Many plants, as well as being able to produce seeds, are able to reproduce by growing new parts which can live as separate plants. This is called asexual reproduction because no gametes are involved. The new plants are exact genetic copies of the 'parent' plant.

*Strawberries send out special stems called **runners** that spread over the ground. New strawberry plants grow at the tips of these stems.*

*Some stems of the potato plant grow underground and become swollen with stored food. The swelling is a potato or **tuber**. If a tuber is planted, new plants grow from buds – the 'eyes'.*

***Bulbs** such as daffodils are large underground buds with swollen leaves full of food. Each year new bulbs grow inside the old one.*

Humans have grown plants asexually for hundreds of years by cutting pieces off plants they like, and growing them into new plants. These new plants are called **cuttings**.

These geranium cuttings were cut from the stem of a larger plant.

Some plants can be grown from leaf cuttings.

Apple tree 'cuttings' grow by being grafted onto the stem of a healthy plant.

Micropropagation and tissue culture

Micropropagation means growing new plants from very small (microscopic) pieces. Some plants can be grown from a tiny piece of plant containing one bud.

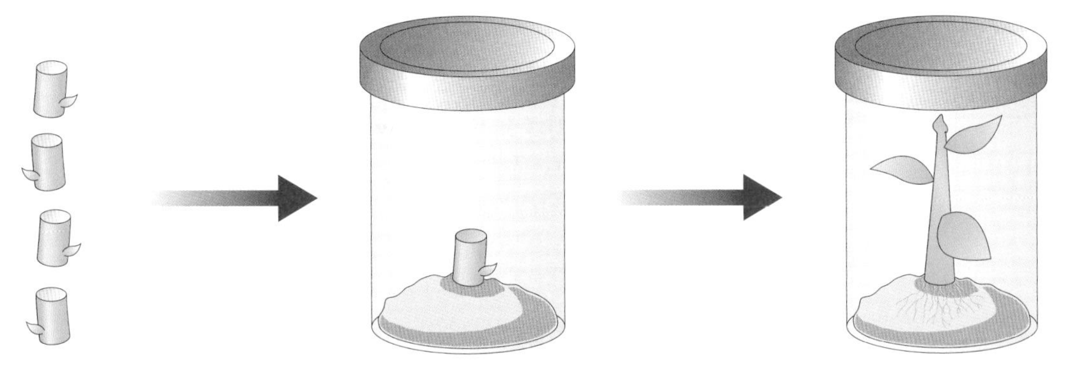

A stem is chopped up into lots of tiny pieces, each with a bud on it. These pieces are sterilized to prevent contamination and put into containers.

The container has a ***growth medium*** *containing everything a plant needs for normal healthy growth. It also contains rooting hormone to encourage the growth of new roots.*

Each piece of stem grows into a new plant with roots. The new plants are kept in a greenhouse until they are big enough to survive.

In **tissue culture** new plants are grown from only a few cells (from tissue) instead of from a bud. The technique is very quick and doesn't take up much space. Plant growers can use it all year round in a greenhouse to produce healthy disease-free plants. Genetically engineered plants are grown using this technique.

The only problem is…

There is no **genetic variation** in cloning. So if one plant gets a disease it is likely that all the other clones will get it as well. That's why, in nature, plants reproduce sexually as well as asexually.

These plants were grown by micropropagation.

Questions

1 What is a clone?

2 Explain how asexual reproduction is different from sexual reproduction.

3 What is: **a** a runner **b** a tuber **c** a bulb?

4 **a** What is a 'cutting'?
b Where are cuttings taken from?

5 What is micropropagation?

6 Explain why conditions for micropropagation must be sterile.

7 **a** What is a growth medium?
b Why does a growth medium contain hormones?

8 Give three advantages of tissue culture.

9 Give one big disadvantage of cloning.

4.16 Selective breeding

Objectives

This spread should help you to

- describe how selective breeding works
- describe some advantages and disadvantages of selective breeding

Choosing the best

For thousands of years humans have been selectively breeding farm animals and food crops. Selective breeding is also called **artificial selection**. This is because humans artificially select only the most productive animals and plants to breed from.

Selective breeding works like this:

1. Choose the animals or plants that have the best characteristics.
2. Breed them with each other.
3. Choose the best offspring and breed them with each other.
4. Do this over and over again to improve the characteristics.

Selective breeding in sheep

Sheep are reared for their meat and their wool. A farmer wants big, meaty sheep that have thick, woolly coats.

Questions

1. Why is selective breeding also called artificial selection?
2. Describe how selective breeding works.
3. A modern cow produces several gallons of milk each day. Explain how farmers have made cows that produce so much milk.

Selective breeding can be very useful. It has given us...

... beef cattle with more meat

... milking cows that give more milk

... dwarf wheat that has lots of seeds

... more offspring in sheep.

This Jacob sheep is a 'rare breed'. Their alleles may be needed some day if diseases wipe out modern varieties.

Problems with selective breeding

In selective breeding, closely related animals or plants are bred together. This is called **inbreeding**. Inbreeding reduces the variety of alleles in a population. This is because only the best animals or plants are chosen each time – there is no variety. This can cause problems if a new disease appears. None of the animals or plants will be resistant to the disease because they are all the same. They could all die, then their alleles would be lost forever.

Questions

4 Champion male racehorses are often 'put to stud'. Racehorse owners pay thousands of pounds to have their female horses mated with the ex-champion. Explain why.

5 a What is inbreeding?

b Explain why inbreeding is a problem.

6 What is the point of keeping 'rare breeds'?

4.17 Genetic engineering

Objectives

This spread should help you to

- explain what genetic engineering is
- describe how human insulin is made by genetic engineering
- describe some other examples of genetic engineering

Many diabetics need insulin to live.

Genetic engineering involves taking genes from one type of cell and putting them into a completely different cell. The idea is that we get a good supply of useful products at a lower cost.

The genes controlling these products are usually put into bacteria cells. The genes then 'tell' the bacteria cells what to do. Bacteria reproduce very quickly, about once every 20 minutes. This means they make the products much faster than the original animal or plant cell.

Making human insulin

Diabetics used to use insulin from pigs or cattle. This is not the same as human insulin and sometimes produces side effects. With genetic engineering, bacteria are used to make human insulin.

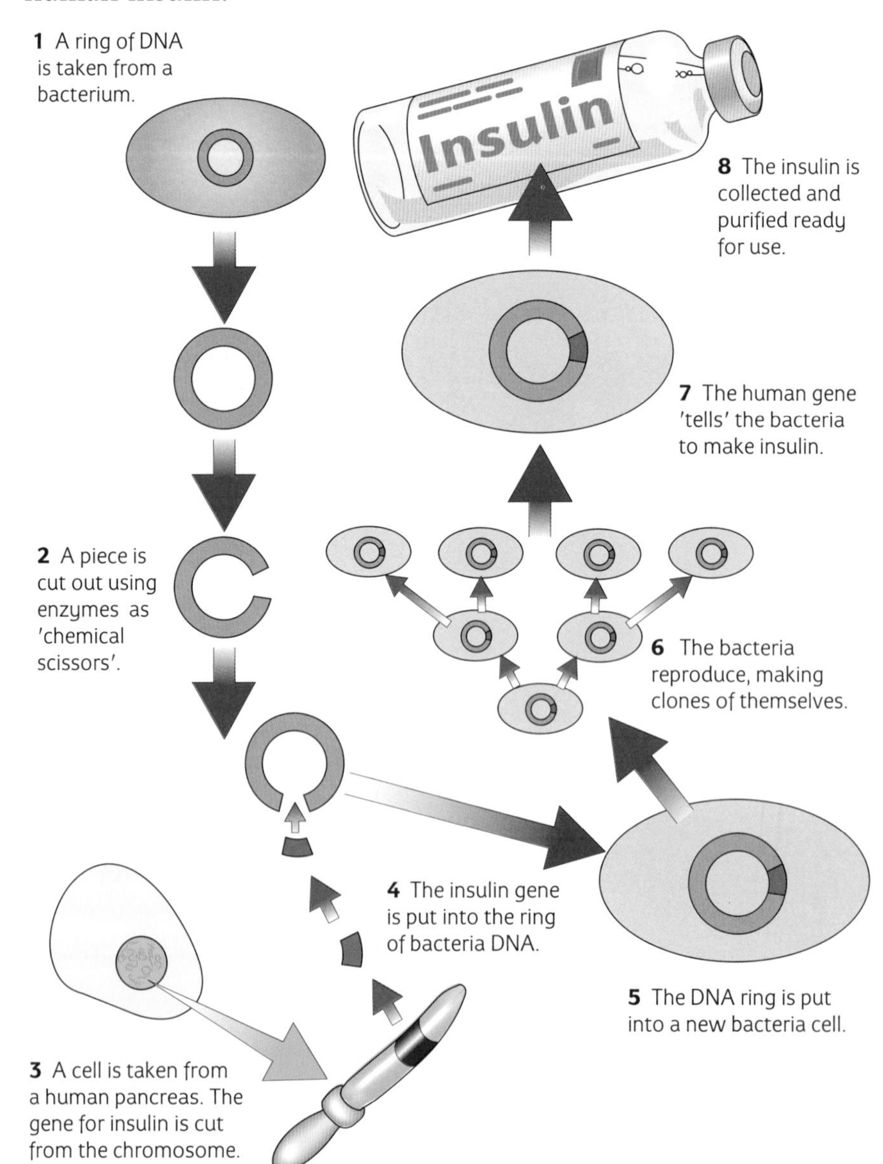

Questions

1 **a** What is genetic engineering?

b What is the point of it?

2 Explain why bacteria are used in genetic engineering.

3 Why would a diabetic prefer to use genetically engineered insulin?

4 Where is the gene for human insulin taken from?

5 Why do you think the enzymes are called 'chemical scissors'?

A human gene which codes for the production of a drug that helps kill viruses can be put into sheep cells. The drug can be collected in the milk from the sheep.

Putting genes into other living things

Useful genes can also be put into animal embryos. Cattle and sheep have been genetically engineered so they make useful drugs in their milk. The drugs are easily collected and purified.

Tomatoes have been produced which are resistant to frost and stay fresh longer. This makes them easier to transport. The genes came from fish! Plants have been engineered to make their own **insecticide** – a chemical that kills insects. Better crops are grown and there is not as much pollution caused by spraying insecticide.

Gene therapy

Gene therapy tries to help those people who are born with genetic diseases like cystic fibrosis. If the correct 'gene' can be put into their cells, the person would be cured of the disease. The problem is how to put enough new genes into enough cells. But it's only a matter of time before it works well.

Embryo transplants in cows

A cow only has one calf each year. To get more than one calf from his best cow a farmer can use **embryo transplants**. It works like this:

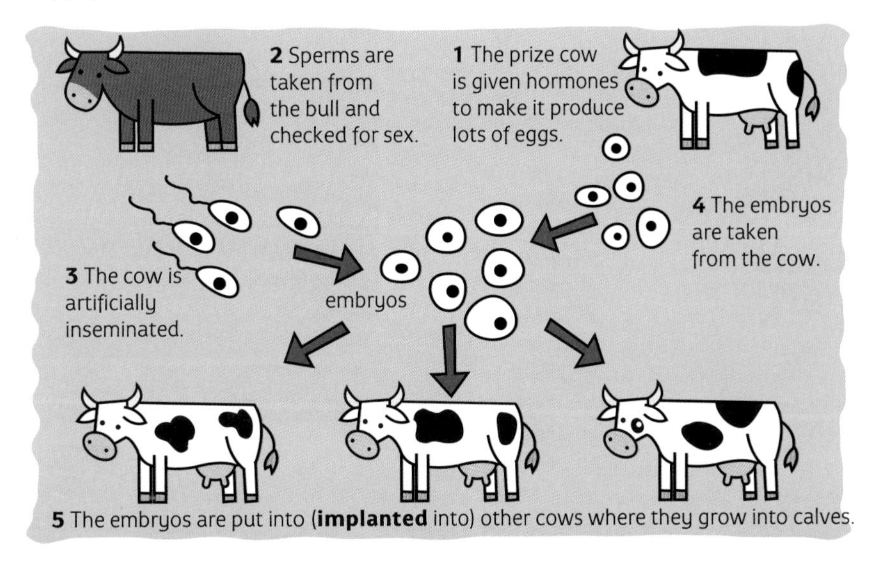

Questions

6 Explain why genes are put into sheep embryos.

7 What is the advantage of having genetically engineered tomatoes?

8 'Foreign genes' put into insect-pollinated plants could quickly spread to wild plants. How could this harm the environment?

9 Cystic fibrosis is a genetic disease.

a What is a genetic disease?

b What are the problems in finding a cure for genetic diseases?

10 Explain why embryo transplants are used in cattle breeding.

4.18 The GM debate

Plants replacing oil

Petroleum oil has been the main source of oils for industry for many years. But its increasing cost, worries about future supply, and problems of pollution have made scientists look for alternative, renewable sources of oils. Scientists have been looking closely at plants that produce oil.

Many plants store oil in their seeds. So long as we grow the plants, we will have a supply of oil. Unlike petroleum oil, plant oil is biodegradable – bacteria can rot it easily and quickly.

Today, genetic engineering is making it possible to produce varieties of oil-seed rape which will make oils for other jobs.

A tree called California bay makes an oil which can be used to make detergent. In 1993 scientists found the gene that 'instructs' the California bay to make its oil. The gene was put into bacteria cells which reproduced, making clones of themselves. Each new bacterium carried the 'new gene'. The bacteria were allowed to infect rapeseed cells which were then grown into whole rape plants using tissue culture. Every cell in the new rape plants carried the 'new gene'. Lots of these plants were grown and field trials carried out.

The modification of oil-seed rape is only one example of **genetic modification**.

Many people are concerned about the possible effects of GM crops on wild plant populations.

Talking point

1 What do you think about 'genetic modification'?

4.19 What is a species?

Objectives

This spread should help you to
- explain what a species is
- describe how new species are formed

Breeding together

A **species** is a population of living things that can breed together and produce **fertile** young. This means the young can also breed. Different species will not usually breed with each other, but if they do, the offspring will be **sterile** (unable to breed).

The horse and the donkey belong to different species.

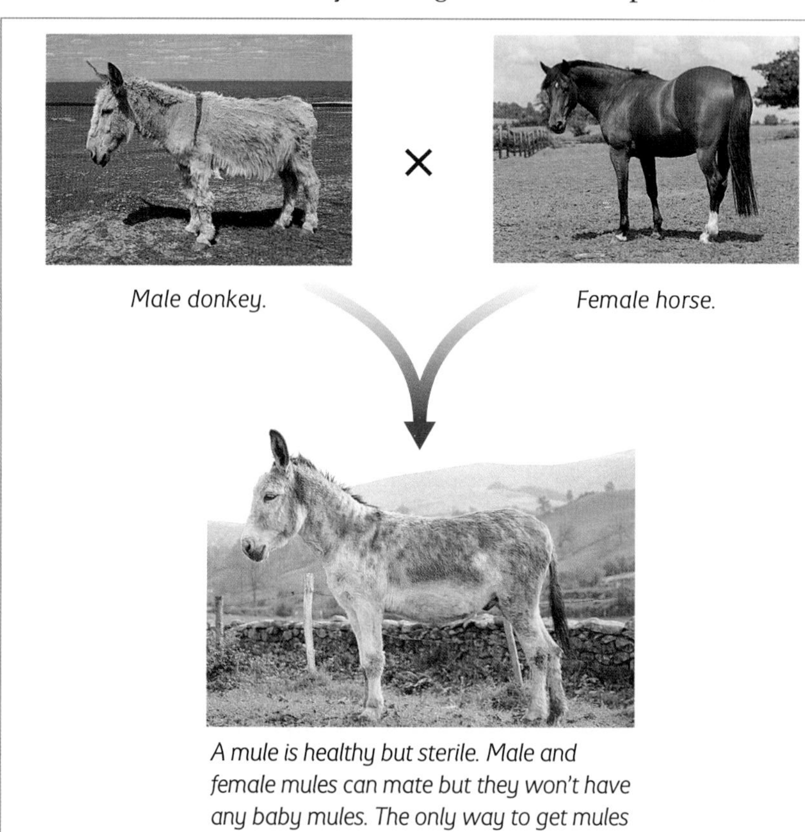

A mule is healthy but sterile. Male and female mules can mate but they won't have any baby mules. The only way to get mules is to cross a horse with a donkey.

Other 'mules'

Tiger × lion (a tigon).

Zebra × donkey (a zedonk).

Questions

1 What is a species?

2 If animals or plants of different species reproduce, what happens to their offspring?

3 During the First World War, mules were used for transport. They were bred on mule farms. Which of these would a mule farmer have on his farm? Explain your answer.

A lots of female horses and a few male donkeys

B lots of male horses and a few female donkeys

C lots of female mules and a few male mules

D lots of male mules and a few female mules

How species are formed

Before a new species can be formed, a large group of animals or plants must be split up into groups. These groups can be separated by mountains, rivers, and seas. This is called **isolation**. When they are isolated there is no chance of them breeding with each other. Breeding continues in each group. Over thousands of years mutations and variations will change the group completely.

This is how a new species of butterfly is formed.

A population of butterflies lives in a large grassy area.

The climate gets wetter and a river cuts the area in half.
Two groups of butterflies are isolated.
The butterflies can't cross the river because it is too wide.

Mutations cause changes to both groups of butterflies.
They both develop patterns on their wings.

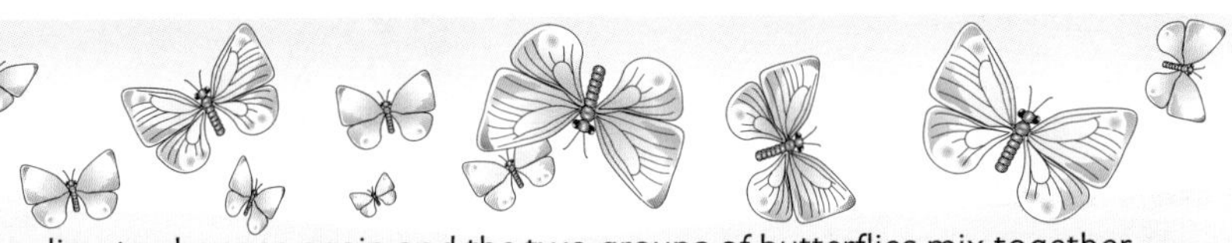

The climate changes again and the two groups of butterflies mix together.
The groups do not recognize each other so do not mate.
Two new species are formed.

Questions

4 What must happen before a new species can be formed?

5 **a** What is isolation?

b Give two ways in which animals or plants can become isolated.

6 How long does it usually take for a new species to appear?

7 Bacteria reproduce every 20 minutes. Elephant reproduce every $2\frac{1}{2}$ years. Which is likely to happen sooner, a new species of bacteria or a new species of elephant? Explain your answer.

4.20 Evolution

Objectives

This spread should help you to

- explain the theory of evolution
- list Darwin's observations that led to his theory of evolution
- describe an example of natural selection in action

Evolution is one way of explaining why there are so many different living things on Earth. When something **evolves** it changes and improves on something that went before.

Modern aircraft evolved by changing and improving the first, simple designs.

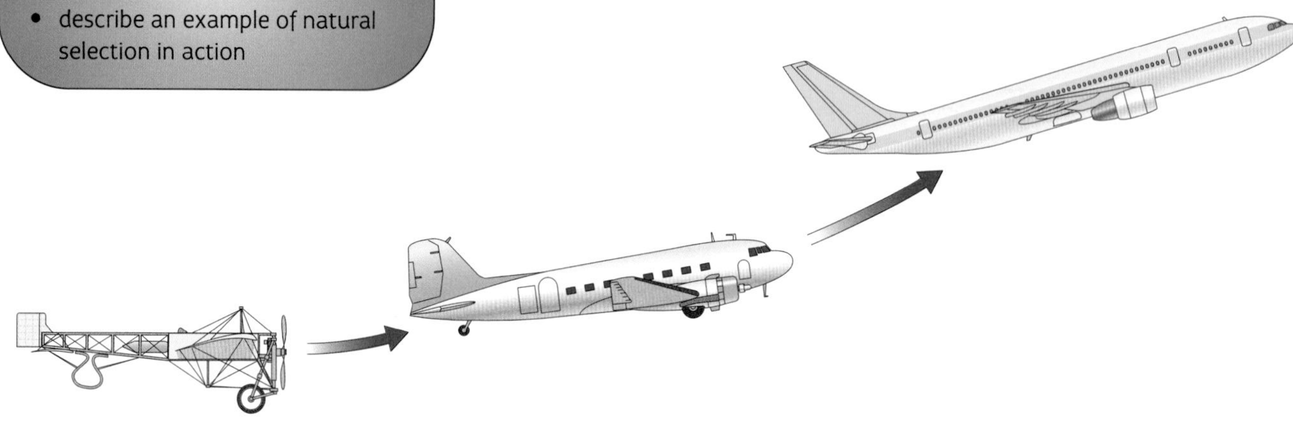

Evolution of living things

The first living things appeared on Earth more than three billion (3000 million) years ago. They were very simple but they could reproduce. Some of the young were different from their parents. These differences meant they could survive better and breed, passing on their differences to their offspring. Over billions of years more changes and improvements have led to all the different animals and plants alive today.

This diagram shows how some animals could have evolved.

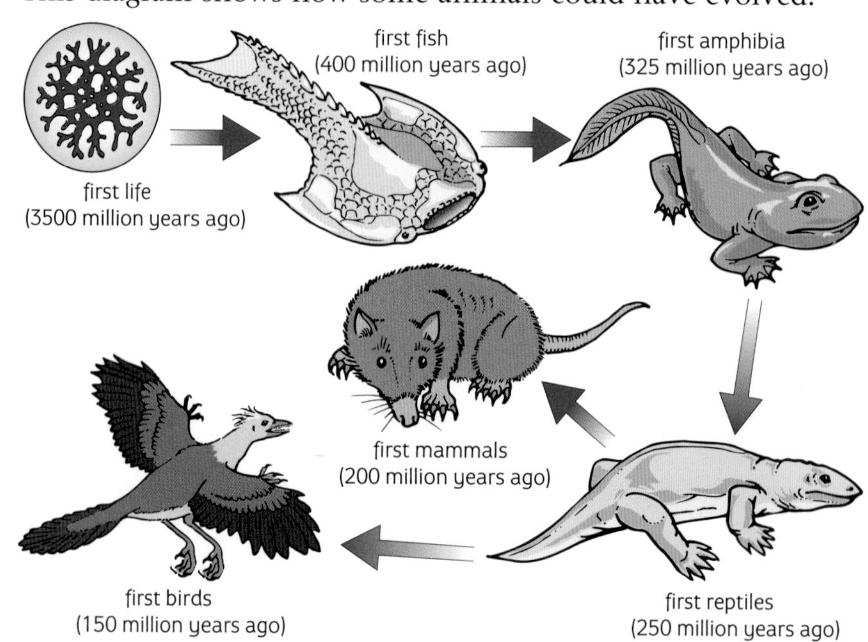

Questions

1. What does 'evolution' mean?
2. When did the first life forms live on Earth?
3. What is one billion?
4. What did amphibians evolve from?
5. Humans are mammals. What have we evolved from?

Charles Darwin 1809–82. Darwin faced strong opposition from religious groups. Gradually his ideas have become widely accepted.

Natural selection

Natural selection means that those animals and plants that are best suited to their surroundings will survive and pass on their advantage to their young. Charles Darwin published his theory of natural selection in 1859. It was a simple idea based on these four things he saw:

- All living things can have lots of offspring.
- Numbers of animals and plants stay roughly the same so there must be **competition** for survival.
- There is variation in populations. Some variations help an animal or plant survive better, and breed; others die.
- Animals and plants inherit changes from their parents.

Natural selection in action

During the day peppered moths rest on the trunks of trees. There are two varieties of peppered moth, a light form and a dark form.

In a clean country area, the trees are clean. The light peppered moth is difficult to see and survives to breed. The dark form will be spotted by birds and eaten. The light form has been naturally selected to live.

In city areas the trees are blackened by dirt. Here the dark peppered moths are more common because they are better camouflaged. This time the dark form has been naturally selected to live.

Questions

6 What is natural selection?

7 What did Charles Darwin base his theory of natural selection on?

8 Where did Darwin get his evidence for his theory?

9 **a** Explain why most peppered moths that live on dirty trees are the dark form.

b How could you show that the light form and dark form of peppered moth are the same species?

4.21 Evidence for evolution

Objectives

This spread should help you to

- describe how fossils are made
- explain how the fossil record gives evidence for the theory of evolution

Fossils

Most of the evidence for evolution comes from **fossils.** Fossils are any sort of preserved remains of an animal or plant. Fossils can be made in three ways.

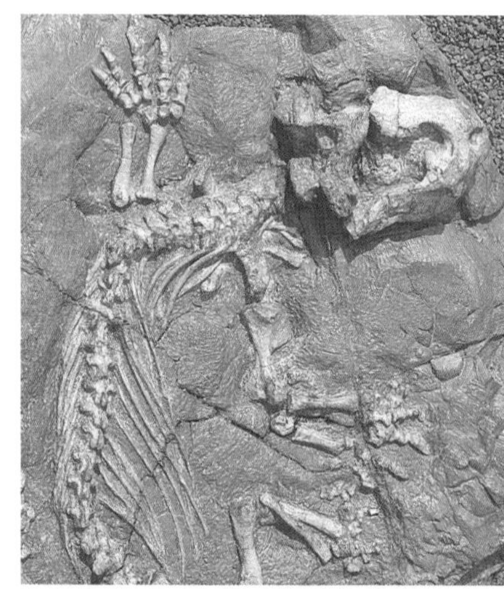

1 Hard parts like bones, shells, or plant fibres don't decay very quickly. They become covered with ***sediment*** *and buried deep underground. Over thousands of years the sediment becomes rock. The animal or plant remains are replaced by minerals which turn to stone. A fossil is made. These dinosaur fossils are perfect copies of the original.*

2 Soft parts like leaves usually decay too fast to fossilize. But sometimes, in certain conditions, decay is very slow. The leaf is buried under sediment and is replaced by minerals to form a fossil.
Coal sometimes contains fossils of leaves. This fern lived around 300 million years ago.

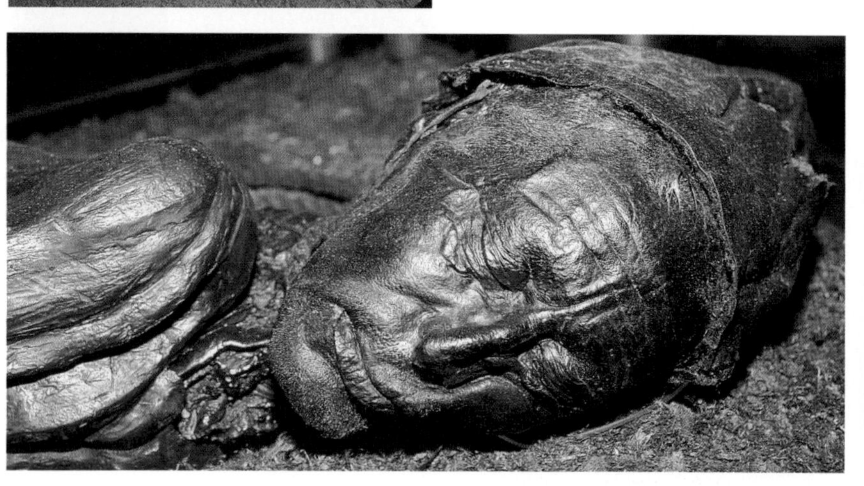

3 Whole animals or plants can be fossilized when no decay happens. Insects are found fossilized in amber. The amber was tree sap that the insect got stuck in millions of years ago. The bodies of woolly mammoths have been found in Siberia. Their meat was still edible after thousands of years in the deep freeze! The body of this man was found preserved in a peat bog. He dates from around 220 BC. Bogs are too acidic for decay bacteria to work.

Questions

1. What is a fossil?
2. Which part of an animal doesn't decay very quickly?
3. Explain why fossils of the soft parts of animals and plants are rare.
4. **a** What is amber?
   b Explain why the body of an insect can be perfectly preserved in amber.
5. Explain why decay bacteria can't work in a peat bog.

The fossil record

Fossils tell us what animals and plants looked like and how long ago they lived. Generally the deeper a fossil is found, the older it is. Sometimes rocks are pushed up by earth movements so it is important to know the age of the rocks where a fossil is found.

The Grand Canyon, USA.

The Grand Canyon in Arizona is the deepest crack in the Earth's surface. It is about a mile deep with young rocks at the top and the oldest rocks at the bottom. The canyon has lots of fossils. They are found in the order in which the animals and plants lived on Earth. This is called a **fossil record**. Near the top, dinosaur fossils have been found. Halfway down, the fossils of fish appear. At the bottom there are no fossils at all.

The fossil record tells us a lot about how animals and plants have evolved, but it is not complete. There are gaps in the fossil record called 'missing links'. These are probably because most animals and plants don't turn to fossils – they just rot away completely.

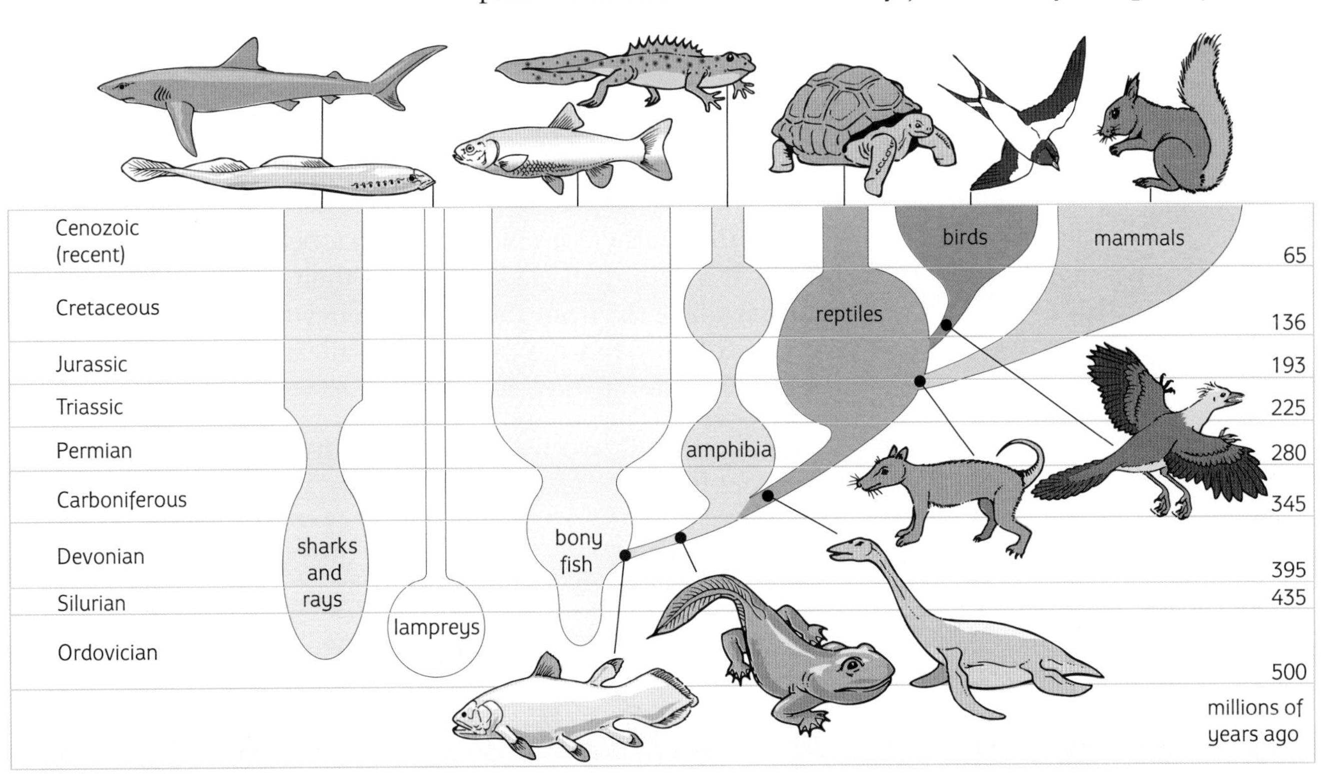

Questions

6 Explain why it is important to know the age of rocks when studying fossils.

7 What is a fossil record?

8 When did the first reptiles appear on Earth?

9 **a** What is a missing link?

b Explain why there are missing links.

4.22 Patterns of inheritance

Objectives

This spread should help you to

- use a genetic diagram to show how characteristics are inherited
- explain the difference between phenotype and genotype

Genetic diagrams

We can show the results of crosses between two people in a **genetic diagram**. Genes are shown by letters. Capital letters are used for dominant alleles and small letters for recessive alleles. In the diagram below, **H** is the allele for black hair and **h** is the allele for blond hair.

Father has blond hair. His cells contain two alleles for blond hair. He is **hh**.

The alleles separate. One goes to each sperm.

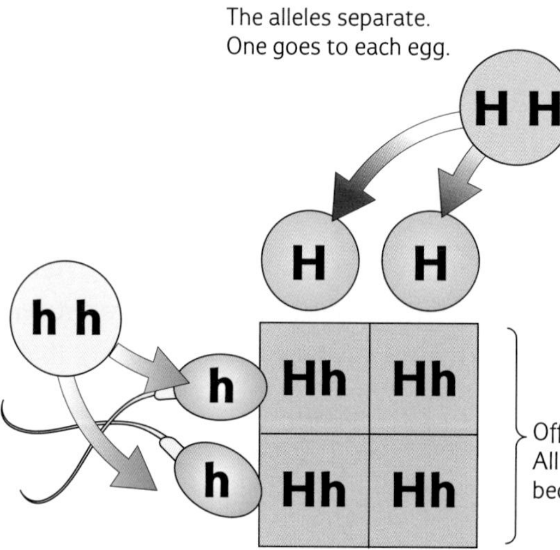

Mother has black hair. Her cells contain two alleles for black hair. She is **HH**.

Offspring – the alleles come together. All the children have black hair because **H** is the dominant allele.

In the example above, all the children had black hair even though their father was blond. This happened because they inherited a dominant allele (**H**) from their mother and a recessive allele (**h**) from their father. They are heterozygous (**Hh**) for hair colour. They are carrying the allele for blond hair even though they don't show it.

What happens if two heterozygous (**Hh**) people have children together?

The alleles separate. One goes to each egg.

Mother has black hair. Her cells contain one dominant allele and one recessive allele. She is **Hh**.

Father has black hair. His cells contain one dominant allele and one recessive allele. He is **Hh**.

The alleles separate. One goes to each sperm.

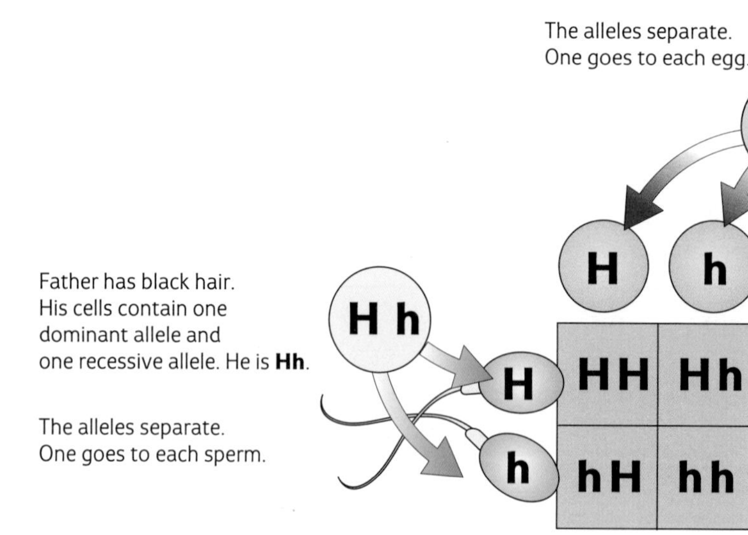

Offspring – the alleles can come together in four ways. Children with **HH**, **Hh** and **hH** have black hair. The child with **hh** has blond hair.

The child with blond hair is possible because the two recessive alleles (**hh**) can come together. But the children of these parents are three times more likely to have black hair than blond. It's all down to luck!

Phenotype and genotype

A **phenotype** is the outward appearance of a person, such as blond hair or blue eyes. A **genotype** is the kind of allele a person has.

This woman has the phenotype 'blond hair'. Her genotype is **hh**.

This man has the phenotype 'black hair'. His genotype could be **HH** *or* **Hh**.

Questions

1 What are used to show alleles on a genetic diagram?

2 How are:

a dominant **b** recessive alleles shown on a genetic diagram?

3 If **E** is the allele for brown eyes, and **e** is the allele for blue eyes, what eye colours will these pairs of alleles give?

a **EE** **b** **Ee** **c** **ee**

4 In question 3, which of the pairs of alleles is:

a homozygous **b** heterozygous?

5 **a** Draw a genetic diagram to show the cross between a man heterozygous for brown eyes (**Bb**) and a woman with blue eyes (**bb**).

b What are the phenotypes of the offspring?

6 In mice, **D** gives dark fur and **d** gives light fur. What are the genotypes for dark-haired mice?

7 Bianca has brown hair. She dyes it red.

a Has Bianca changed her phenotype? Explain your answer.

b Has Bianca changed her genotype? Explain your answer.

4.23 Mendel and inheritance

One of Mendel's experiments

1 Mendel crossed tall plants with short plants. He brushed pollen from the anthers of one plant onto the stigma of another plant.

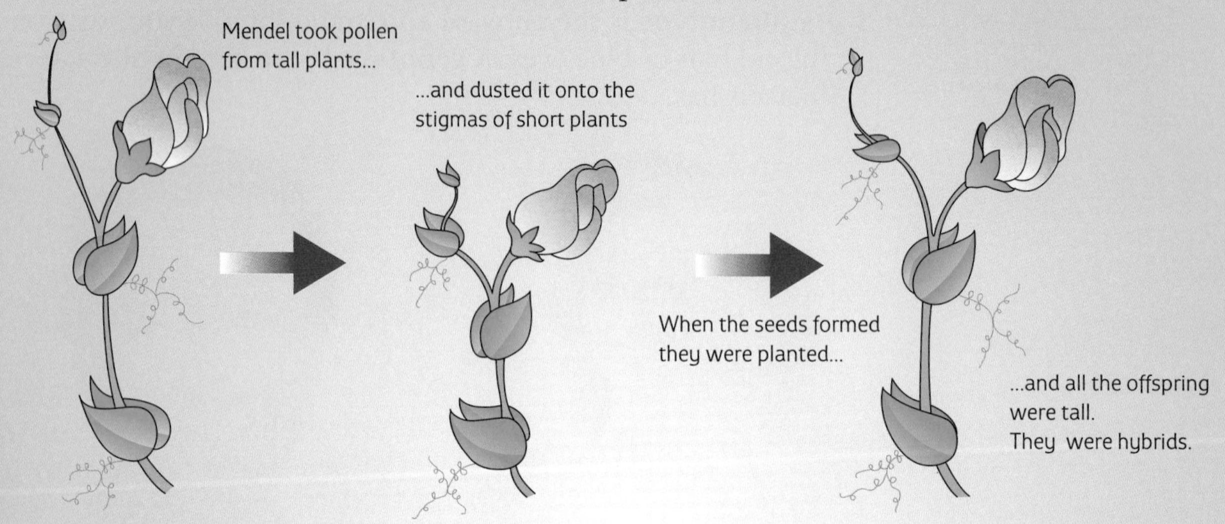

Clearly there was something in the tall plants that dominated something in the short plants, making all the offspring tall.

2 Mendel allowed his hybrid plants to flower and pollinate themselves.

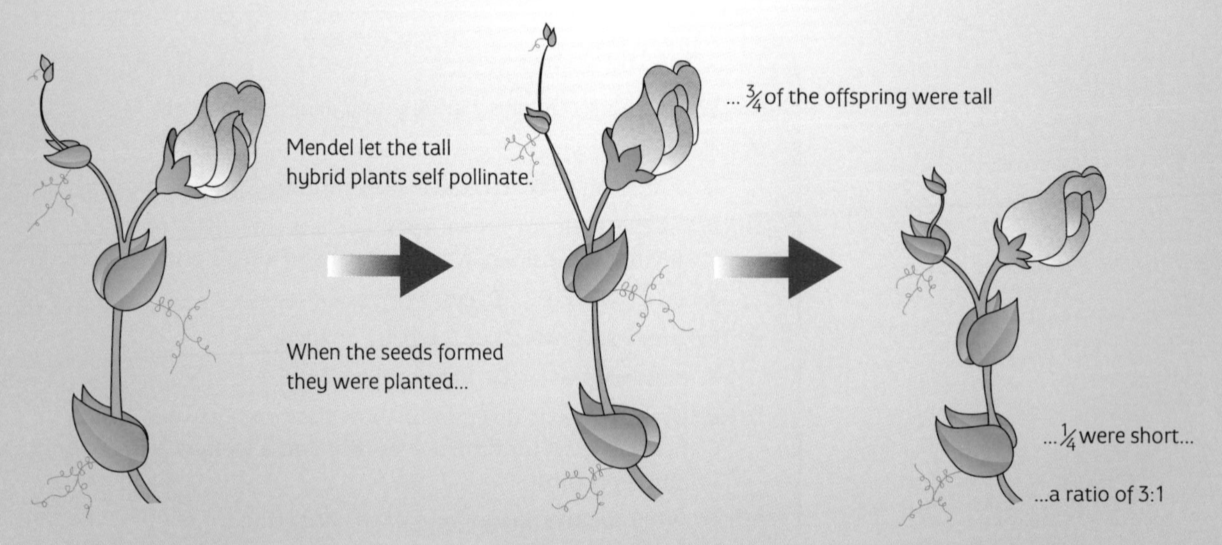

The reappearance of the short plants in the second cross told Mendel that the instruction for short stems must be carried by all of the tall plants from the first cross. Short plants were produced when two of these instructions got together.

Gregor Mendel was an Austrian monk. He had always been interested in how characteristics were inherited so he did lots of breeding experiments on garden peas.

Mendel was a proper scientist!

There are three main reasons why Mendel's work was so successful.

1 He only studied the inheritance of one characteristic at a time.

2 He counted and recorded his results accurately.

3 He interpreted his results and drew conclusions from them.

It's amazing...

...that Mendel did all this work and discovered so much without any knowledge of chromosomes and genes – they hadn't been discovered then!

Mendel's work wasn't appreciated...

...when he published his results in 1866. It seems that other scientists just couldn't handle the new ideas – it was all too much for them to understand. Mendel used statistics to support his theory of inheritance, and statistics was a very new subject, especially to the biologists of the day. It wasn't until 1900, 34 years after Mendel published his work and 16 years after he died, that the importance of his work was realized.

Talking points

1 Why didn't Mendel know about chromosomes and genes?

2 What did the reappearance of the short plants in the second cross tell Mendel?

3 Explain why Mendel's work was so successful.

4 Explain why Mendel's work wasn't appreciated until after his death.

4.24 Boy or girl?

Objectives

This spread should help you to
- describe the sex chromosomes
- describe how sex is inherited

Sex chromosomes

Humans have 46 chromosomes. There are 22 pairs that match exactly. But the last pair doesn't always match. These are the **sex chromosomes**. Sex chromosomes control whether you are a boy or a girl.

There are two types of sex chromosome, called X and Y. Females have two X chromosomes (XX). Males have one X chromosome and one Y chromosome (XY). The diagram shows how sex is inherited.

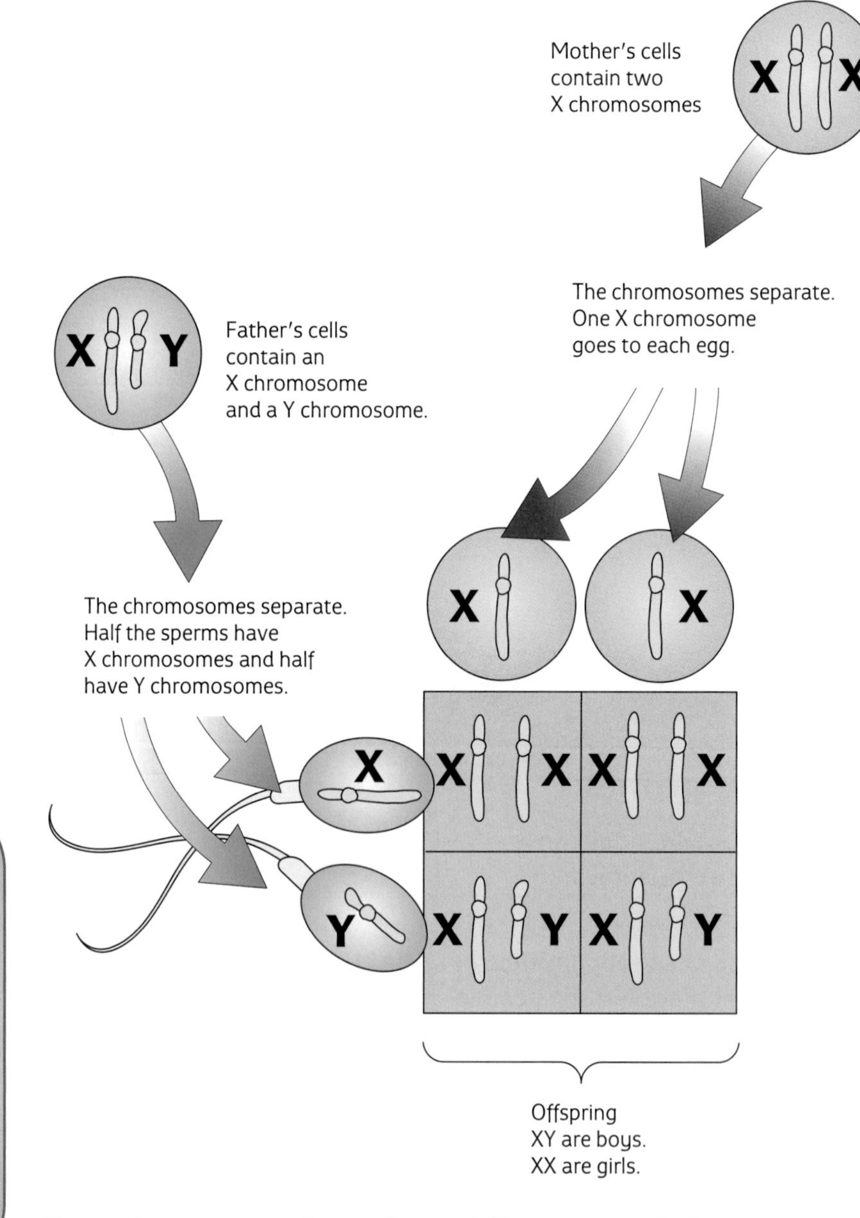

Since there are equal numbers of X sperms and Y sperms, a child has an equal chance of being a boy or a girl.

Questions

1. What do sex chromosomes do?
2. **a** How many types of sex chromosome are there?
   b What are they called?
3. Explain why half the population is male and half is female.
4. A man and a woman want two children. Explain why they can't be certain of having a boy and a girl.

This is a photograph of the 46 chromosomes of a woman. They were photographed though a microscope, then cut up and sorted into 23 matching pairs.

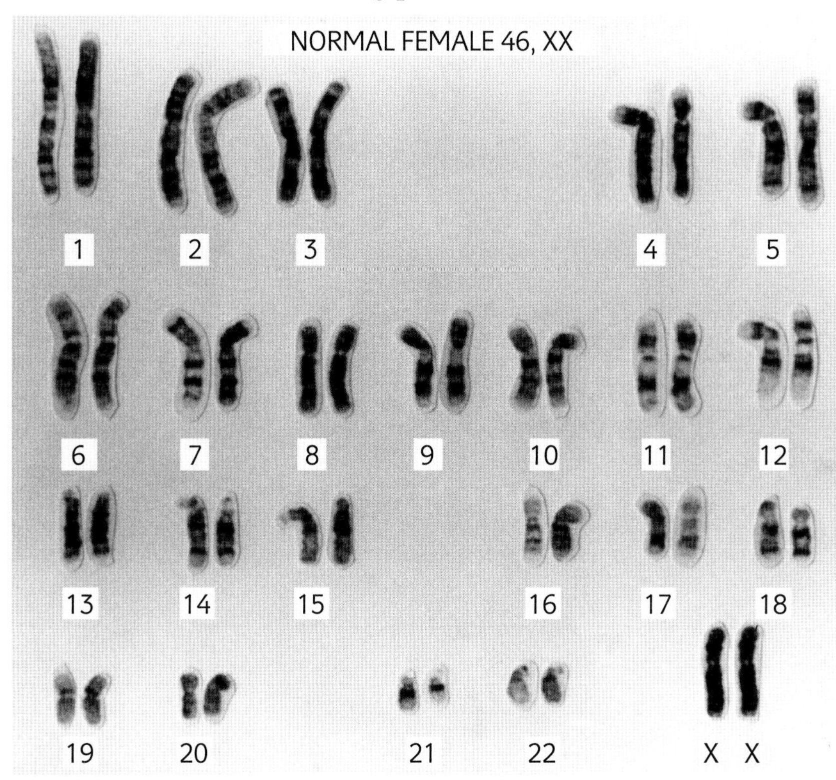

This is a photograph of the 46 chromosomes of a man. These have been sorted into 22 matching pairs and a pair of X and Y chromosomes.

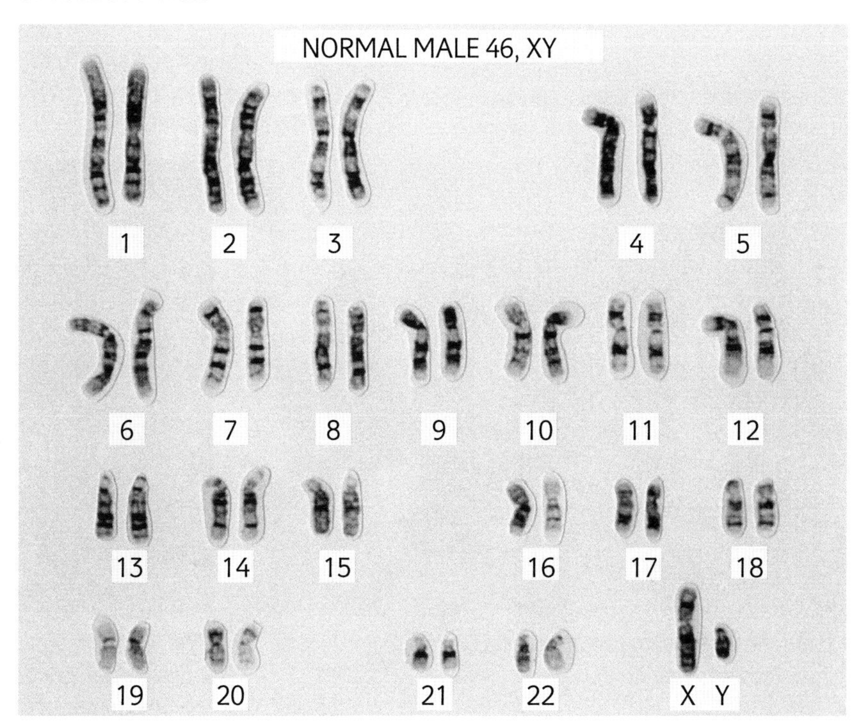

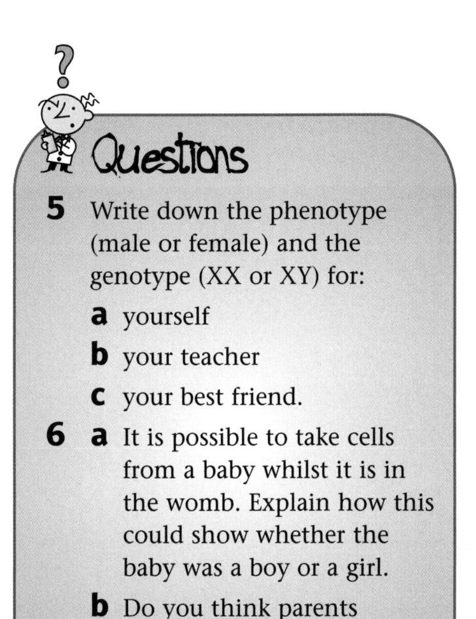

Questions

5 Write down the phenotype (male or female) and the genotype (XX or XY) for:

a yourself

b your teacher

c your best friend.

6 **a** It is possible to take cells from a baby whilst it is in the womb. Explain how this could show whether the baby was a boy or a girl.

b Do you think parents should be able to tell the sex of their baby before it is born? Explain your answer.

4.25 Mutations and inherited diseas

Objectives

This spread should help you to

- explain what a mutation is
- describe how Down's syndrome and Huntington's disease are caused

What is a mutation?

A **mutation** is a change in a chromosome or a gene which alters the way an organism develops. Strange, new characteristics appear which make an individual different from the rest of the population. This may lead to a **genetic disease** (inherited disease). Mutations usually happen when cells are dividing and the DNA is copying itself. If a mutation happens in sex cells (gametes), the effects will be inherited by the offspring.

Down's syndrome

Sometimes pairs of chromosomes don't separate properly when sex cells are made. Both chromosomes go into the same sex cell. This kind of chromosome mutation causes **Down's syndrome**. A woman produces an egg containing 24 chromosomes instead of 23. When the egg is fertilized, the baby has 47 chromosomes instead of 46. The 'extra' chromosome is at pair 21. Having an extra chromosome causes physical and mental problems for the child.

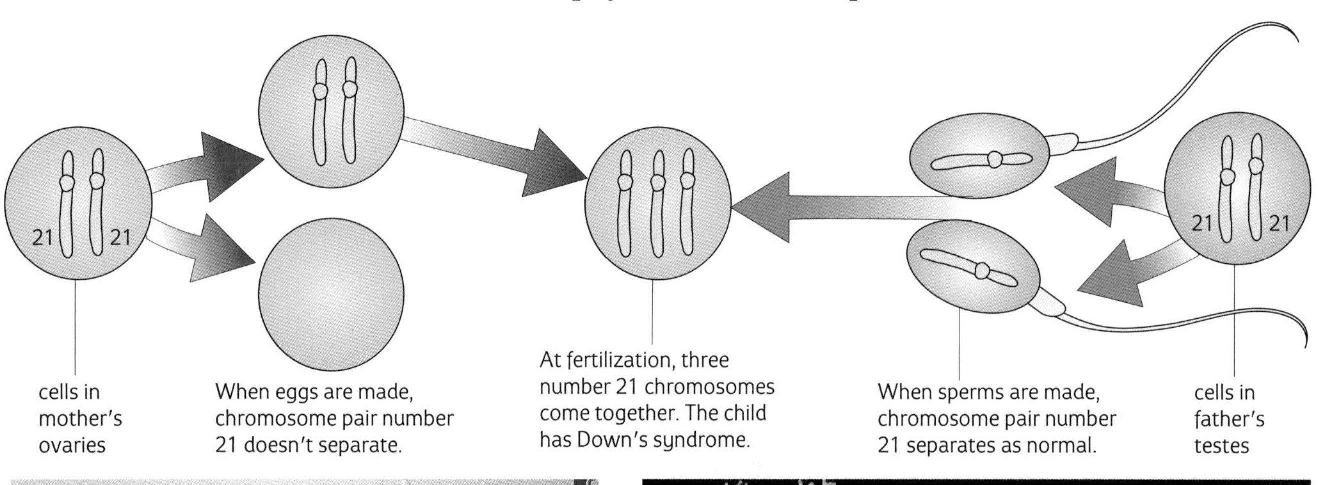

Someone with Down's syndrome has three 'twenty-first' chromosomes instead of two.

The 47 chromosomes of a girl with Down's syndrome. They have been sorted into 22 matching pairs and three number 21 chromosomes.

Huntington's disease

Huntington's disease affects the nervous system. It causes shaking, sudden body movements, and mental illness. This disease is caused by a dominant allele, so it can be passed on by one parent who has the disease. The genetic diagram below shows how it happens. **N** stands for the dominant Huntington's allele and **n** stands for the recessive allele.

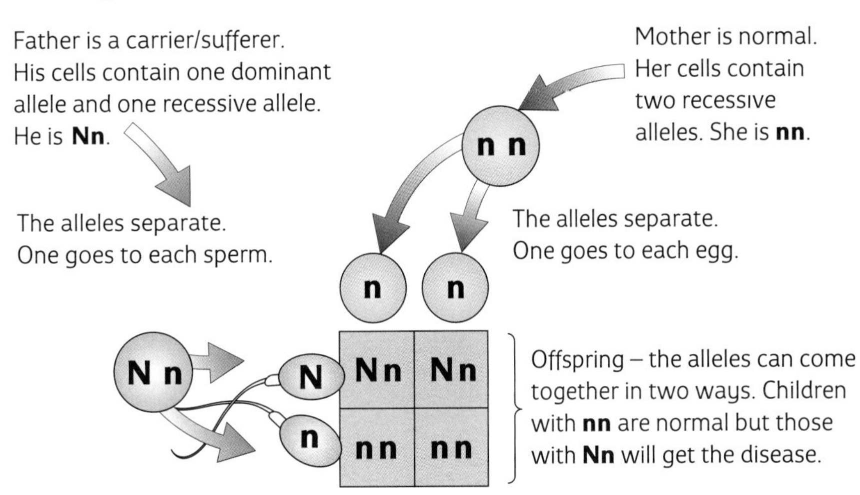

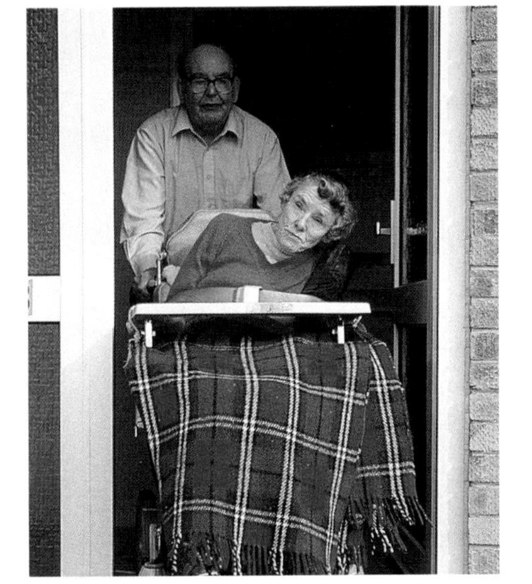

Huntington's disease is an inherited genetic disease that affects people as they get older.

People don't know if they've got the disease until about age 40. By that time they could have had children and passed the disease on.

Questions

1 **a** What is a mutation?
b When do mutations happen?
c What happens if a mutation happens in a sex cell?

2 **a** How many chromosomes does a person with Down's syndrome have in each of their cells?
b Explain how this happens.
c Describe the effect this has on the sufferer.

3 What is Huntington's disease?

4 **a** Is Huntington's disease carried on a recessive or dominant allele?
b Explain why Huntington's disease can be passed on even if one parent is perfectly healthy.

5 Roughly at what age does someone know if they've got Huntington's disease?

6 Explain why it would be useful to for someone know if Huntington's disease was in their family.

More mutations

Objectives

This spread should help you to

- describe how sickle cell anaemia and cystic fibrosis are caused
- describe how mutations are caused
- explain how mutations can lead to evolution

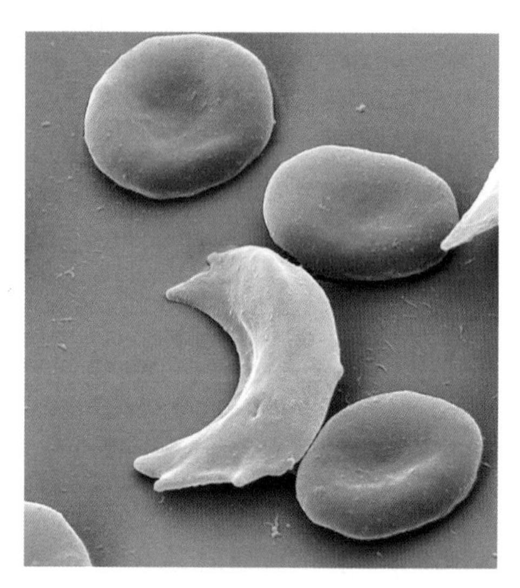

Blood with sickle cells looked at through a microscope.

Sickle cells

Sickle cell anaemia is a genetic disease that causes the red blood cells to become sickle shaped instead of the usual round shape. These sickle cells get stuck in the blood vessels and the body doesn't get enough oxygen. People with the disease die when they are very young. Unlike Huntington's disease, sickle cell anaemia is caused by a recessive gene allele. To get the disease a person must inherit a recessive allele from *both* parents. This genetic diagram shows how it happens. **S** stands for the normal dominant gene allele and **s** stands for the recessive sickle cell allele.

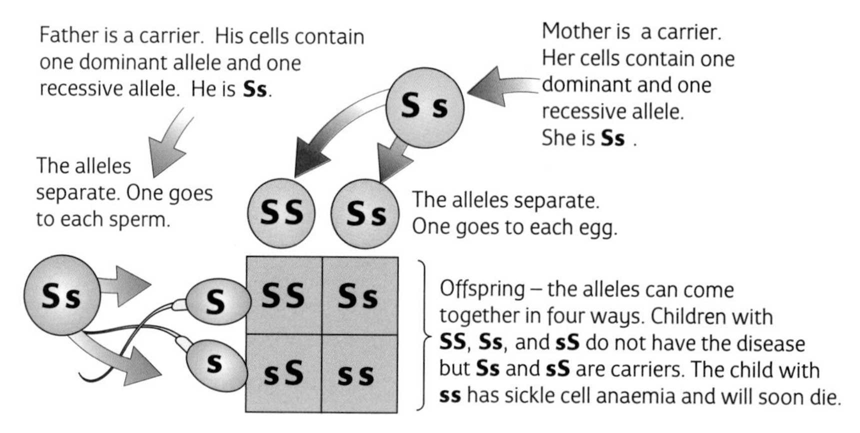

Most people with sickle cell anaemia die before they can have children. So why hasn't the disease died out? The answer is that carriers of the disease (**Ss**) are more resistant to malaria than non-carriers (**SS**).

Malaria is a disease that can kill. It is common in Africa. Being a sickle cell carrier increases someone's chances of survival. If they live longer they are likely to have more children, even though some of the children may die of sickle cell anaemia.

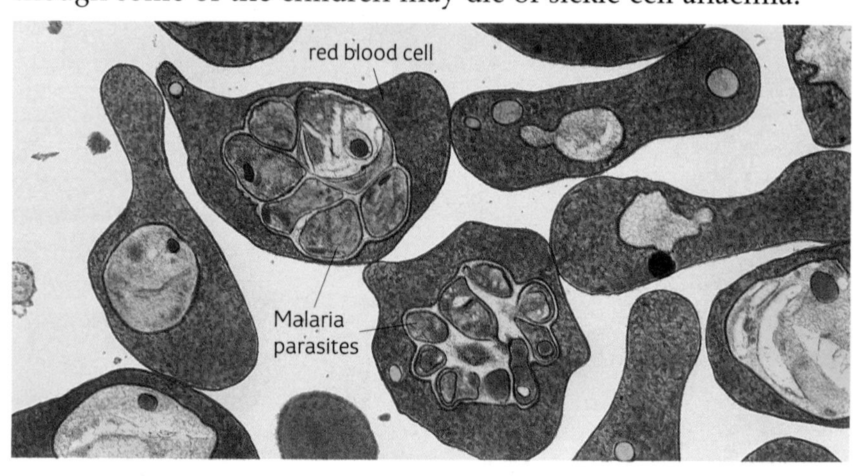

The parasite that causes malaria gets inside red blood cells. It can't get into sickle cells as easily so a carrier of sickle cell anaemia is less likely to get malaria.

Questions

1. **a** What is sickle cell anaemia?
   b Explain why sickle cell anaemia kills people.
2. Explain why the gene allele for sickle cell anaemia is common in Africa.
3. **a** Is sickle cell anaemia carried on a recessive or dominant allele?
   b Explain why both parents have to be carriers of the disease before it can be passed on to their children.

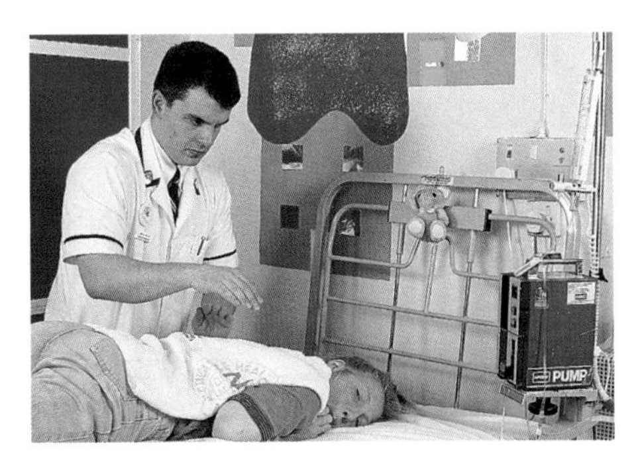

Physiotherapy helps to clear the lungs of thick, sticky mucus.

Cystic fibrosis

Cystic fibrosis is inherited in the same way as sickle cell anaemia. So a person will only get cystic fibrosis if they inherit the recessive allele from both parents. The disease causes the mucus in the lungs to be thicker than normal. This thicker mucus can't be easily cleared from the airways and bacteria get stuck, causing chest infections.

What causes mutations?

Mutations happen naturally. However, the chances of a mutation are increased if you expose yourself to...

*...nuclear radiation or **ionizing radiation** such as alpha, beta, or gamma*

...high frequency waves like X-rays or ultraviolet light

... chemicals that cause cancer like those found in cigarette smoke.

Mutations and evolution

Most mutations are harmful to the organism that has them, but some are useful – they lead to **evolution**.

Occasionally a mutation gives an individual an advantage over the others in the population. For example, bacteria can mutate and become resistant to antibiotics. The mutant lives and the 'normal' bacteria die. A new 'resistant strain' of bacteria is the result. This causes big problems in hospitals.

Mutations have made some wild rats resistant to rat poison. These rats survive and pass on their mutant alleles. So now, more and more rats are resistant to poison. The wild rat population is increasing rapidly as more resistant rats are born.

Questions

4 Why are people with cystic fibrosis likely to get chest infections?

5 Explain how the inheritance of Huntington's disease is different from the inheritance of sickle cell anaemia and cystic fibrosis.

6 What is ionizing radiation?

7 Give two examples of high frequency waves.

8 What do the chemicals in cigarette smoke do?

9 Why are most mutations harmful?

10 Explain how mutations can lead to evolution.

4.27 The human genome project

The human genome project began in 1990 and was completed in 2000. It is a project to map all of the human genome. This has provided scientists all over the world with information about most of the genes on our chromosomes.

What is a genome?

A **genome** is all the genetic material in an organism. It contains the master plan for the organism and how it will behave. The master plan is carried in DNA molecules.

The human genome is in every nucleus of a person's cells. It is estimated to contain 30 000 genes arranged in order on 23 chromosomes.

The human genome project has benefited among others, medicine, forensic science, and agriculture.

Medicine

As a result of the human genome project, there have been great leaps forward in medical research. The genes associated with many genetic diseases have been found. These diseases include cystic fibrosis, muscular dystrophy, inherited breast and colon cancer, and Alzheimer's disease.

Scientists are now looking at new methods of treating diseases. One of these is gene therapy, in which defective genes are replaced by normal ones. Gene therapy is still experimental and there are many problems to overcome. The main problem is how to get the 'new' gene into the patient's cells.

Scientists have studied the genetic makeup of bacteria like this Escherichia coli. *This bacterium is used in modern genetic engineering to make products such as human insulin.*

DNA forensics

Forensic scientists identify individual humans by looking at part of the human genome. They use the data to create a **DNA fingerprint** or genetic fingerprint. It is very unlikely that two people have exactly the same genetic fingerprint.

This forensic scientist is looking at a person's genetic fingerprint. No two people have exactly the same genetic fingerprint.

Agriculture

By studying plant and animal genomes, scientists will be able to breed stronger, disease-resistant plants and animals. Farmers are able to increase production and reduce waste because their crops and herds are healthier.

Alternative uses for crops such as tobacco have been found. One researcher has genetically engineered tobacco plants to produce a bacterial enzyme that breaks down explosives such as TNT and dinitroglycerine.

Genetically engineered plants like these tobacco plants can be used to clean up the environment.

Talking points

1 Roughly how many genes are there in the human genome?

2 Name three genetic diseases.

3 a What is gene therapy?
  b Explain why gene therapy is not used yet.

4 a What is a DNA fingerprint?
  b Give two uses for DNA fingerprinting.

5 What is the advantage of producing stronger and disease-resistant plants and animals?

Practice questions

1 The diagrams show some butterflies.

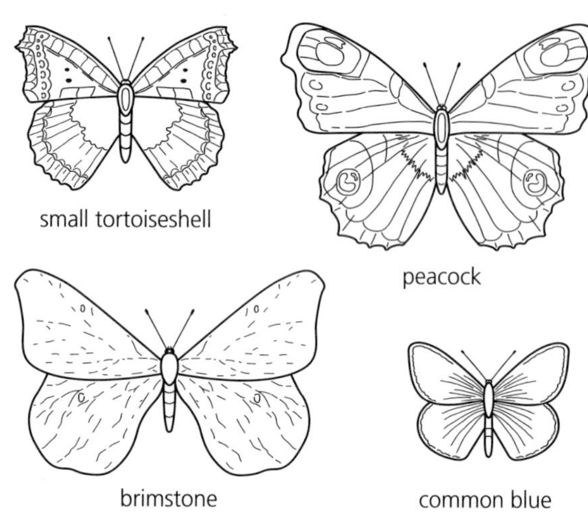

a Give three ways in which the butterflies are different from each other.

b Common blue butterflies vary in length. There is a full range from small to large. What kind of variation is this?

c Some brimstone butterflies have spots missing. What kind of variation is this?

d The peacock butterfly looks as if it has eyes on its wings. The brimstone butterfly looks like a dead leaf when it lands on a plant. Suggest how these features might help the butterflies survive.

2 The graph shows the number of students in a class by height range.

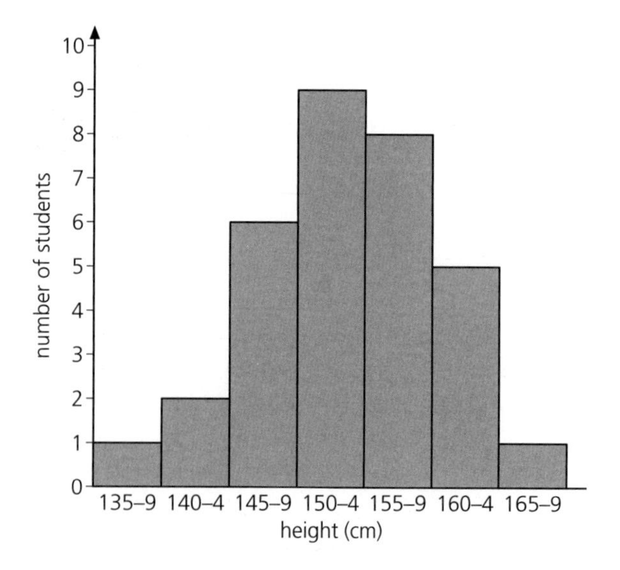

a How many students are in the class?

b What height would a student have to be in order to be in the:

i tallest five students

ii shortest ten students?

c What is the most common height in the class?

d What do we call this kind of graph shape?

e Suggest two other things the students could measure and still get this shape of graph.

3 The diagram shows the menstrual cycle for a human female.

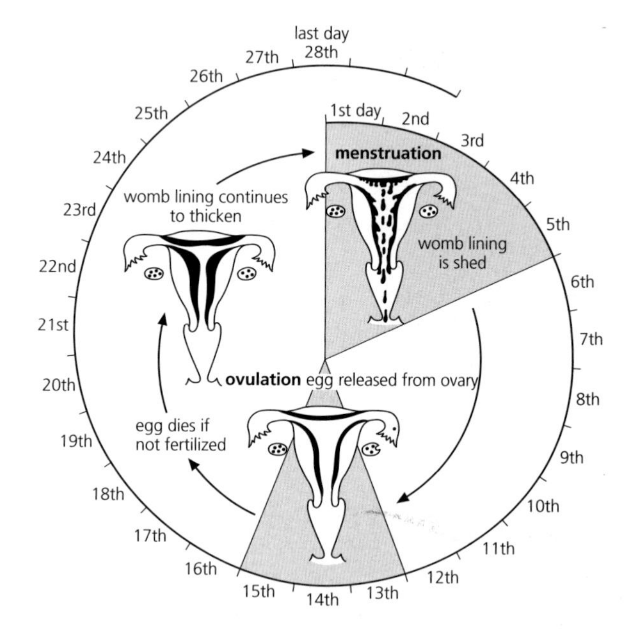

a How long is the menstrual cycle?

b If a woman begins to produce eggs at the age of 13 and stops at 48, roughly how many eggs will she produce during this time?

c During one cycle a woman's period started on 16 March. On what date would an egg be released from an ovary? (Use the diagram to help you.)

d Explain why the lining of the womb needs to thicken.

e What is another name for menstruation?

f Why do you think menstruation stops when a woman has a developing baby in her womb?

g The menstrual cycle is controlled by hormones. One of these causes ovulation. Another hormone keeps the lining of the womb thick, stopping it breaking down.

These hormones can be used as fertility drugs. Give one example of when a woman might want to use a fertility drug.

4 A pregnant woman can have a test known as amniocentisis. This test is done to see if the baby is developing normally. A needle is used to take some fluid from the sac holding the baby. This fluid contains cells which can be looked at under the microscope.

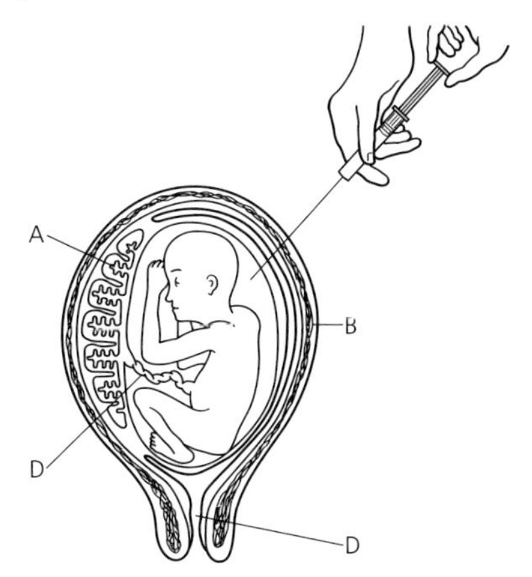

a i List the letters A–D and write the correct label beside each one from the words below.

placenta, umbilical cord, vagina, womb wall

ii Suggest why it is important for the doctor to know the exact position of the baby before the needle goes in.

iii Suggest what might happen if the needle damages the sac holding the baby.

b Amniocentisis is used to see if a baby has Down's syndrome. Someone with Down's syndrome has an extra chromosome in each of their cells.

The diagram shows a set of chromosomes from a person with Down's syndrome.

1 2 3 4 5
6 7 8 9 10 11 12 X
13 14 15 16 17 18
19 20 21 22 Y

i How many chromosomes are there in the body cells of a normal person?

ii How many chromosomes are there in the body cells of someone with Down's syndrome?

iii At which pair of chromosomes is the extra chromosome?

iv When sex cells are made, the chromosome pairs usually separate. Suggest how an extra chromosome could get into the cells of someone with Down's syndrome.

5 Genes can change by having a mutation.

a What is a mutation?

The chances of a mutation happening are increased by exposure to ionizing radiation or certain chemicals.

b Give two examples of ionizing radiation.

c Give one example of a chemical that can cause a mutation to happen.

d Some mutations have been quite useful. They can lead to evolution.

i Suggest how a mutation that produced a slightly longer neck might help an animal that eats leaves.

ii Suggest one way that the long neck of the giraffe might have evolved.

6 Sex cells or gametes contain chromosomes which pass information on from one generation to the next. In humans, the male gamete is a sperm.

a What is the human female gamete called?

b This diagram shows a sperm.

i What is the tail for?

ii Where are the chromosomes?

c The sex of humans is controlled by X and Y chromosomes. Women have XX chromosomes and men have XY chromosomes.

i What type of sex chromosome could you find in a sperm?

ii What type of sex chromosome could you find in a female gamete?

d In birds, the female has XY chromosomes and the male has XX chromosomes. Draw a diagram showing how sex is inherited in birds.

Practice questions

7 The drawings show a carthorse, used for pulling heavy loads, and a racehorse, used for sprinting.

Both horses are varieties of the same species. They have been bred using selective breeding to enable them to do their different jobs.

a How is the carthorse built to do its job?

b Describe how selective breeding has been used to produce the carthorse.

c Modern racehorses have been produced by breeding mares (females) and stallions (males) that can run fast. Give two reasons why a person should check the pedigree of a racehorse before buying it.

d **i** What is a species?

ii How could you prove that the carthorse and the racehorse are varieties of the same species?

8 The following diagrams show how the horse has evolved over millions of years.

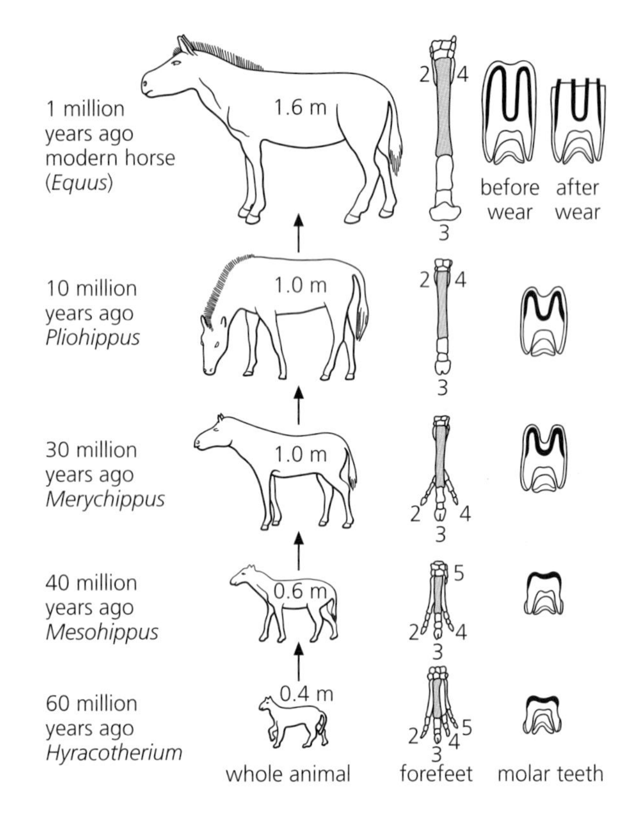

a The horse's oldest ancestor is *Hyracotherium*.

i How long ago did *Hyracotherium* live on Earth?

ii How tall was *Hyracotherium*?

iii What evidence is there to suggest that *Hyracotherium* fed on soft food?

b Name two types of fossil shown in the diagram.

c **i** Describe what has happened to the number of toes as the horse has evolved.

ii Suggest why this might have something to do with horses running fast.

d Suggest one reason why the horse had to evolve.

9 Copy the diagram below. It shows a cross between a woman with brown eyes and a man with brown eyes. **B** stands for the allele for brown eyes and **b** stands for the allele for blue eyes.

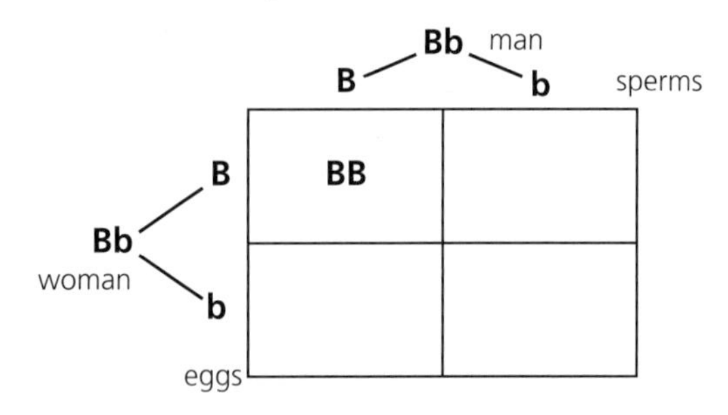

a Finish off the diagram by filling in the boxes.

b Which allele is dominant?

c What do we call the other allele?

d Which of the offspring are:

i homozygous for eye colour

ii heterozygous for eye colour?

e Which offspring could eventually produce children with:

i brown eyes **ii** blue eyes?

f Explain why you can't be sure about the genotype of a brown-eyed person.

10 New plants can be produced by tissue culture or micropropagation. A tiny piece of tissue is taken from a donor plant and put on a growth medium. The tissue grows and is then split up into small bits which grow into clones of the original plant.

a What do these words mean?

i tissue

ii donor

iii clone

b Suggest why only a tiny piece of tissue is needed.

c Explain why tissue culture requires sterile conditions.

d Why do you think the process is sometimes called micropropagation?

e Give one advantage and one disadvantage of tissue culture.

f Suggest what the growth medium should contain.

11 The diagram shows a cross between a woman with normal blood and a man who is carrying the allele for sickle cell anaemia.

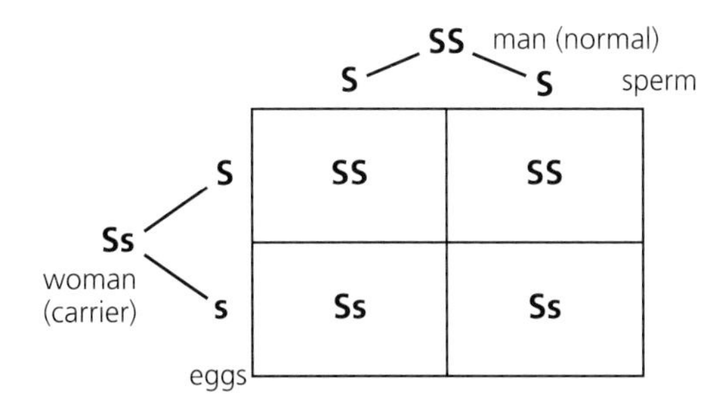

a What is a carrier?

b What are the chances of the couple having a child who:

i has normal blood

ii is a carrier of the disease

iii has sickle cell anaemia?

c Draw a diagram to show what might result from a cross between two people who both carry the allele for sickle cell anaemia.

d Explain why it is an advantage to be a carrier of the sickle cell allele if you live in an area where there is a lot of malaria.

12 Read this information about cystic fibrosis, then answer the questions.

Cystic fibrosis is a genetic disease. Around one in 20 people in Britain are heterozygous for the condition. A person will only get cystic fibrosis if they inherit the recessive allele from both parents. At present about one in 2000 babies is born with cystic fibrosis. The disease causes the mucus secretions of the lungs to be more viscous, or thicker, than normal. This thicker mucus cannot be easily cleared from the airways and bacteria often remain to cause chest infections.

a What is a genetic disease?

b What do these words mean?

i heterozygous

ii recessive

iii viscous

c Explain why someone with cystic fibrosis is likely to suffer from chest infections.

d Describe a treatment that can be given to a person with cystic fibrosis to help them lead as near normal a life as possible.

13 The diagrams show different stages in the development of a baby in the womb. They are in the wrong order.

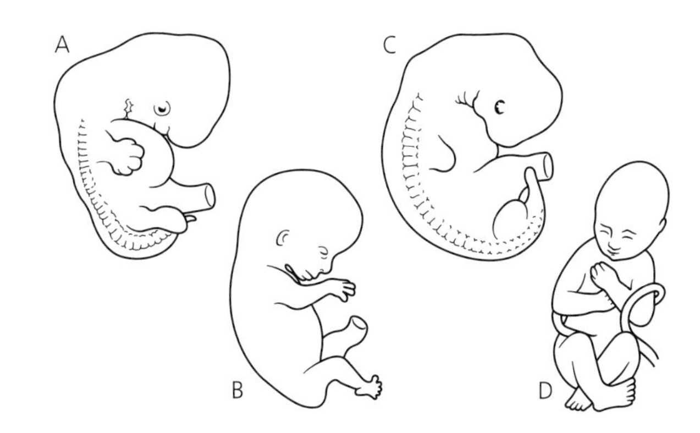

a Write the letters in the correct order.

b What word is used to describe a woman who is carrying a baby in her womb?

c How many months after fertilization is the birth of the baby?

d Give two advantages of having a baby develop inside its mother's body.

AIDS acquired immune deficiency syndrome: a disease that stops the body's natural defence system from working

alleles different forms of the same gene

artificial selection choosing and breeding the best animals or plants to produce a desired characteristic

asexual reproduction reproduction involving no gametes

chromosomes thread-like structures in the nucleus of every cell; they carry genetic information

clones genetically identical organisms

continuous variation characteristics that are not easily separated; there are lots of possibilities

contraception methods of birth control

cross pollination the transfer of pollen from the anther on one plant to the stigma on another plant

deoxyribonucleic acid the chemical that chromosomes and genes are made of

discontinuous variation distinct characteristics; you either have them or you haven't

DNA the common name for deoxyribonucleic acid

dominant the 'stronger' allele in a pair of alleles; the dominant character shows in the appearance of a heterozygous organism

double helix the twisted ladder shape of a DNA molecule

ejaculation the release of sperm from the testes through the erect penis

embryo plant a partly developed plant with a tiny root and shoot

embryo transplant implanting the embryos from a chosen animal into other animals

environmental variation variations caused by what an organism does, not its genetics

evolution the production of new species from existing ones by small changes over a long time

fertility treatment giving a woman oestrogen to make her ovaries produce eggs

fertilization in humans: the joining of the nucleus of a sperm with the nucleus of an egg; in plants: the joining of the nucleus of a pollen grain with the nucleus of an ovule

gametes sex cells

gene therapy putting the 'correct' gene into the body of someone with an inherited disease

genes the parts of chromosomes that determine the characteristics of an organism

genetic disease a disease caused by faulty genes, that can be passed on from parents to children

genetic engineering taking genes from one organism and putting them into another

genetic variation differences in the kind of genes that organisms of the same species have, e.g. different heights

genome all the genetic material in an organism

genotype the kind of genes an organism has

grafted a cutting grown on another plant is grafted, e.g. rose

heterozygous describes an organism that has two different gene alleles for one character

HIV human immuno-deficiency virus: the virus that causes AIDS

homologous pair two chromosomes with the same shape and size as each other

homozygous describes an organism that has two identical gene alleles for one character

hybrid another name for heterozygous

inbreeding breeding from closely related individuals, reducing genetic variety

inherit to pass on from parents to children

meiosis cell division that produces gametes

menopause when the menstrual cycle stops in women aged around 45–50

menstruation (or period) a monthly loss of blood from the womb lining which passes out of the vagina

micropro-pagation growing new plants from microscopic pieces

mitosis cell division that happens when living things grow

mutation a mistake in the copying of a DNA molecule

natural selection organisms best suited to their surroundings survive and breed

nectar sweet liquid at the base of the petals in a flower, which attracts insects

oestrogen a female sex hormone that causes an egg to be released from an ovary

ovary female sex organ; in humans: the organ where eggs are produced; in plants: part of the carpel, which surrounds and protects the ovules

ovulation in humans: the release of an egg from an ovary

ovule the female sex cell in plants

penis part of the male reproductive system used to pass sperms into the female during mating

phenotype the outward appearance of an individual

pollen grains containing the male sex cells in plants

pollination the transfer of pollen from anthers to stigmas

puberty a time when boys and girls become sexually mature

recessive the 'weaker' allele in a pair of alleles; the recessive character is not shown in a heterozygous organism

runners stems that spread over the ground to help in asexual reproduction of plants like strawberries

selective breeding another name for artificial selection

self pollination the transfer of pollen from anther to stigma on the same plant

sexual intercourse when the male moves his erect penis up and down the female's vagina

sexually transmitted diseases (STDs) diseases passed on during sexual activity

species a population of living things that can breed together to produce fertile offspring

sperm male sex cells

sterile not able to produce offspring

tissue culture new plants grown from a few cells (tissue)

vagina opening of the female reproductive system; the penis is placed there during mating

venereal diseases (VD) another name for sexually transmitted diseases

womb female reproductive organ, where the baby develops

Exam-style questions

1 a Write down the following organs. For each one, choose the correct organ system and write it alongside.

Organ	Organ system
artery	excretory
intestine	respiratory
kidney	reproductive
lung	circulatory
nerve	digestive
penis	nervous

b i What does the word 'organism' mean?

ii Give one example of an organism.

2 The diagram shows a farmer spreading manure onto a field.

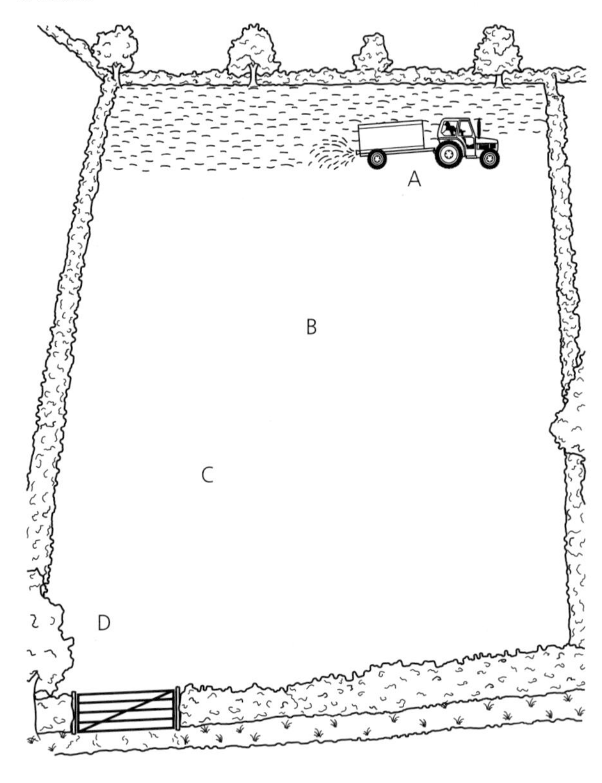

a Which letter, A, B, C, or D, shows where the smell will be strongest?

b Which letter, A, B, C, or D, shows where the smell will be weakest?

c Which of the following represents the overall direction the 'smell' particles are moving in?

A → D

B → A

C → B

D → A

d What do we call this spreading of particles through the air?

e Give one example where the same process happens in the body.

3 The diagram shows a piece of food moving along the large intestine.

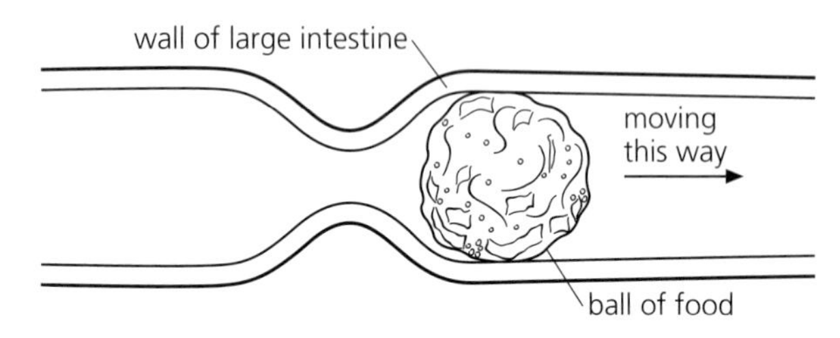

a i Describe how the food is moved.

ii Explain why someone who doesn't eat much dietary fibre can suffer from constipation.

b The small intestine produces enzymes. One of these is called lipase.

i What is an enzyme?

ii What food type does lipase act on?

iii Explain why enzymes like lipase are important.

4 The table tells you how much carbohydrate, fat, and protein there is in the milk produced by some female mammals.

Mammal	Carbohydrate g/100 ml	Fat g/100 ml	Protein g/100 ml
cow	5.0	3.5	3.5
human	6.8	4.0	1.5
polar bear	0.5	31.0	12.6
whale	1.3	42.0	10.0

a Which milk has the most:

i carbohydrate

ii fat

iii protein?

b Describe the habitat of a:

i cow

ii polar bear

iii whale.

c i Why is fat important in the diet?

ii Explain why the milk from a polar bear contains more fat than the milk from a cow.

d Protein is needed for growth.

i Suggest why polar bear cubs grow faster than human babies.

ii Why is it important for polar bear cubs to grow fast?

5 Two students set up this experiment to investigate respiration.

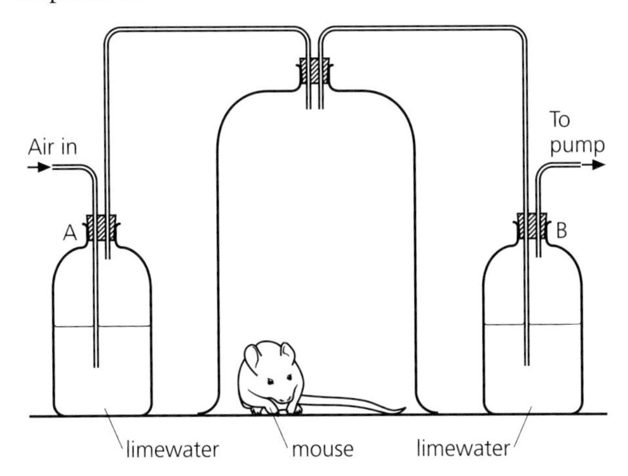

a What gas does limewater test for?

b The limewater in bottle B went cloudy before the limewater in bottle A. What does this show?

c Condensation appeared on the inside of the mouse container. Suggest where this might have come from.

d What do the letters X and Y stand for in this word equation for respiration?

glucose + oxygen → ___X___ + ___Y___

e Explain why it is important that the students stop the experiment and release the mouse as soon as they have got their results.

6 Bread is made by mixing flour and water with some yeast and sugar. After being mixed thoroughly, the dough is left in a warm place to swell up. Once the dough has risen, it is baked in a hot oven.

a What is yeast?

b In bread-making, what does thc yeast feed on?

c Name the gas produced by the yeast that makes the dough rise.

d The yeast is respiring anaerobically. Explain what this means.

e i What else is produced during anaerobic respiration in yeast?

ii What happens to this product during baking in a hot oven?

7 The diagrams show a healthy artery (A) and an artery lined with cholesterol (B).

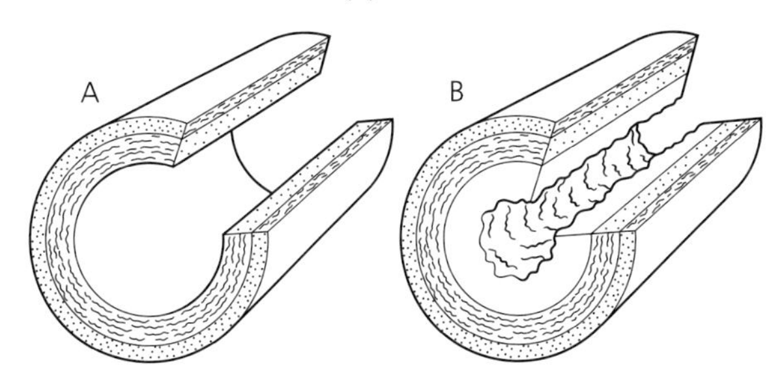

a What is an artery?

b Describe the difference between the two arteries.

c What effect will the cholesterol have on the flow of blood along artery B?

d i Name one disease resulting from cholesterol lining arteries.

ii Explain how it is caused.

iii Suggest one way in which this disease might be avoided.

8 The diagrams show light passing through a fat lens (A) and a thin lens (B).

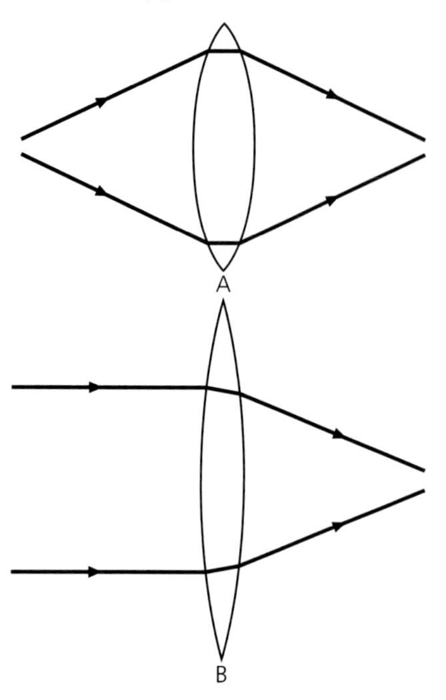

a What do the arrows show?

b Which diagram shows light coming from a near object?

c Explain how the shape of the lens is changed in the human eye.

d Copy the diagram below.

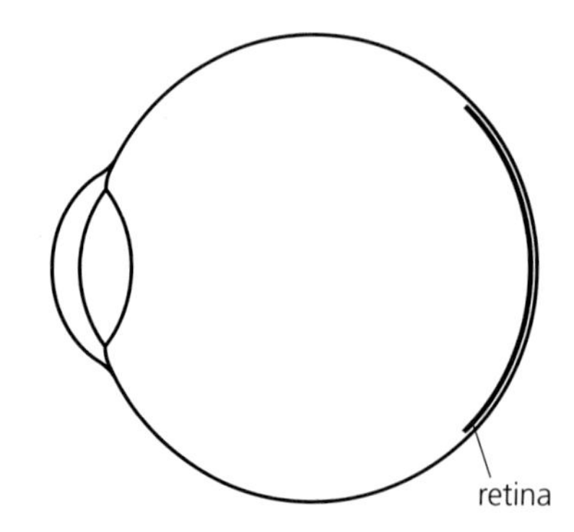

i Draw two lines, one from the top and one from the bottom of the object, to show how an image is focused on the retina.

ii Which ray of light from an object doesn't bend as it goes through the lens?

9 Copy the diagram.

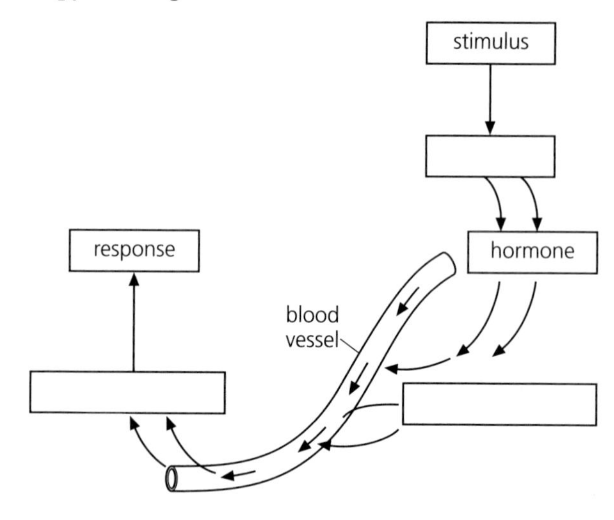

a Use these words to fill the gaps.

bloodstream, gland, target organ

b Diabetes is a disease caused when the body doesn't produce enough insulin.

i Name the gland that produces insulin.

ii Name the target organ for insulin.

iii How does the target organ respond when it receives insulin?

iv What happens to the blood sugar level if not enough insulin is produced?

v Give one way that sufferers can overcome the problems caused by the disease.

10 The diagram shows the human female reproductive organs.

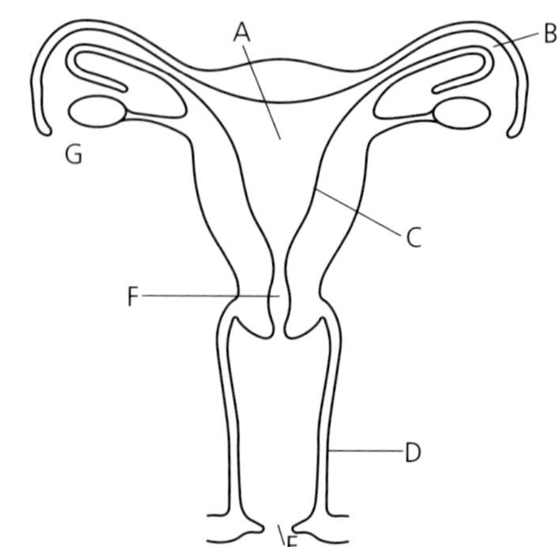

a i Which letter shows where the penis is placed during mating?

ii Which letter shows where an egg might be produced?

iii Which letter shows where fertilization usually happens?

iv Which letter shows where the fertilized egg grows into a baby?

b The graphs show what happens in the menstrual cycle.

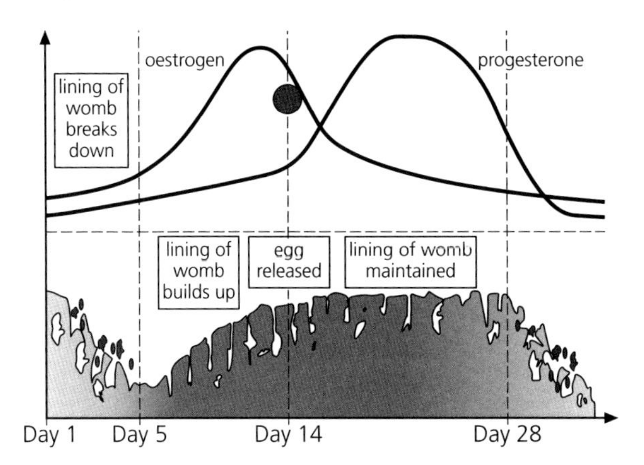

i Roughly, how long is the menstrual cycle?

ii Roughly, how long does a woman's period last?

iii Suggest what the hormone oestrogen does to the womb lining.

iv Suggest what happens when levels of progesterone fall.

v Explain why the level of progesterone must stay high during pregnancy.

11 Plants need minerals for normal, healthy growth. Magnesium is needed to make chlorophyll. Nitrates and phosphates are used to make protein and new cells.

a i Where do plants get these minerals from?

ii Describe how minerals get into a plant.

b Suggest how a shortage of magnesium might affect photosynthesis.

c Most artificial fertilizers contain nitrates and phosphates. Suggest why a farmer uses these fertilizers.

d i Sometimes fertilizers run off into rivers and streams. Explain why this causes a rapid growth of algae in rivers and streams.

ii What effect does this have on other living things in these habitats?

12 The diagram shows the route water takes through a plant.

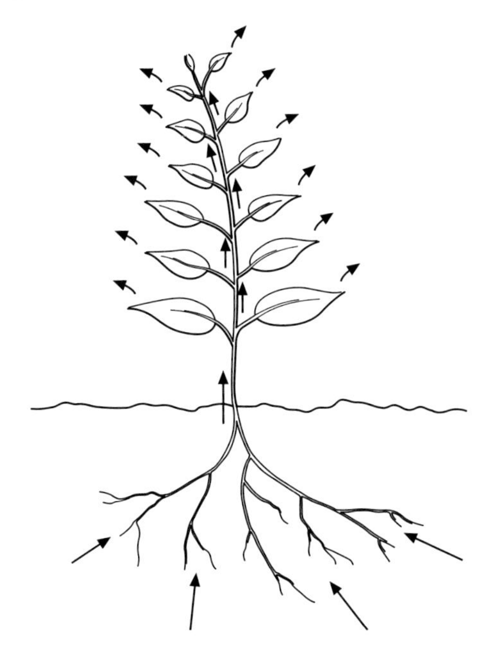

a Explain how water gets into the roots from the soil.

b How are the roots adapted to get water from the soil?

c Name the tubes that carry water through the plant.

d Name the holes in the leaves through which water escapes.

e i What is transpiration?

ii Where does transpiration happen?

iii What is meant by the 'transpiration stream'?

iv Explain why a cactus plant needs to have a slow transpiration stream.

Exam-style questions

13 The diagram shows hair colour in a family.

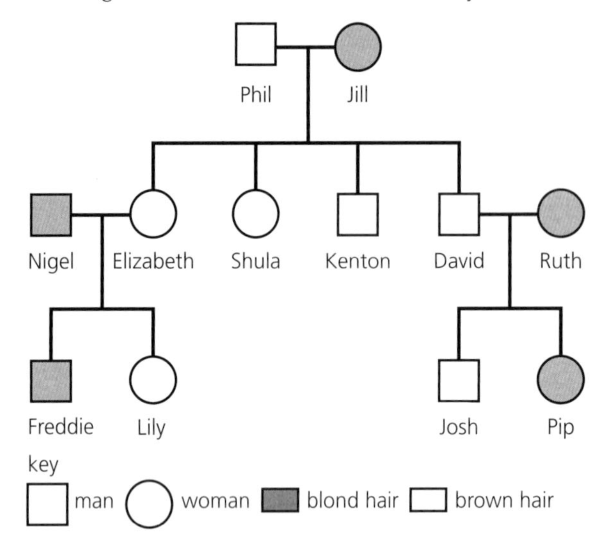

a **i** How many women in this family have blond hair?

ii How many men in this family have brown hair?

b Who has Pip inherited blond hair from?

c **i** Which allele for hair colour is dominant?

ii Explain your answer.

d If Freddie marries a blond woman, what are the chances of them having a blond child?

14 Copy the paragraph. Use these words to fill the gaps.

conditions, hard, millions, plants, preserved, sedimentary, warmth

Fossils are the remains of animals and _____ which lived _____ of years ago. They are usually found in _____ rocks. Most fossils are formed from _____ parts of the body that do not decay easily. Sometimes the whole body is _____. This is because the _____ for decay are missing. For bacteria to decay a dead body, they need oxygen, moisture, and _____.

15 Read this article from a magazine.

WILDLIFE THREATENED BY POLLUTION

WILDLIFE LIVING around the coast of Britain is being threatened by pollution. Sewage, farm waste, and toxic waste from industry are being carried out to the sea by rivers in ever increasing amounts.

In a recent report, scientists studying habitats of wading birds have identified high levels of zinc, chromium, and nickel. These have been traced to industrial factories inland.

There are also high levels of phosphates. Undoubtedly these have come from sewage and fertilizer that has run off into rivers and streams. Algae grows rapidly in waters where phosphate levels are high. In a process known as eutrophication, the algae reduces the amount of oxygen available for other life forms in the water.

Research has shown that plankton in particular is threatened. Plankton is at the bottom of every food chain in the sea. Once that goes, then everything else will follow.

Action is needed now. The Government must bring in legislation to fight this most serious issue before it is too late.

a **i** Name three pollutants mentioned in the article.

ii What is eutrophication?

iii Explain why the loss of plankton is a serious matter.

iv Suggest a law that the Government could make that might help reduce the threat to wildlife.

b One food chain in the sea is:

plant plankton → animal plankton → herring → cod

The biomass of each member of this food chain is:

plant plankton	100 kg
animal plankton	80 kg
herring	10 kg
cod	2 kg

i What is biomass?

ii Draw a pyramid of biomass for this food chain.

16 Jenny's pet dog, Skip, became ill so she took him to the vet. Skip had a bacterial infection. The vet prescribed antibiotics (antibiotics kill bacteria). She told Jenny to give Skip two tablets every day after meals for the next ten days.

After four days Skip seemed much better so Jenny stopped giving him the tablets. Unfortunately Skip became very ill again a few days later.

Jenny started giving Skip the tablets again but this time they did not make Skip better.

Once again Skip was taken to the vet. The vet was very annoyed with Jenny for not continuing with the antibiotics for the full ten days. She gave Jenny some different antibiotics for Skip. This time Jenny made sure she gave Skip all the tablets as instructed. Skip got better.

a Explain why Skip got better after the first four days of treatment.

b Were all of the bacteria killed after four days? Explain your answer.

c Suggest why the antibiotics had no effect the second time they were used.

d What is this an example of?

e Why is it important to take a full course of antibiotics?

Multiple-choice questions

Write the letter for the correct answer in each question.

There is one mark for each correct answer.

1 The part of a plant cell that controls what the cell does is the:

A membrane
B chloroplast
C cytoplasm
D nucleus

2 The part of a plant cell where all chemical reactions happen is the:

A chloroplast
B cytoplasm
C nucleus
D vacuole

3 Which words are in the correct order?

A tissue, organ, organism, cell
B organ, cell, organism, tissue
C cell, tissue, organ, organism
D organism, organ, cell, tissue

4 All animal cells have a nucleus *except*:

A muscle cells
B nerve cells
C red blood cells
D white blood cells

5 The pH of the stomach contents is usually:

A 1–2
B 3–5
C 6–8
D 9–11

6 The job of bile is to:

A digest protein
B dissolve starch
C emulsify fats
D lower the pH

7 Oxygen goes from the air in the lungs into the blood by:

A absorption
B diffusion
C filtration
D osmosis

8 Heat is carried round the body in the:

A blood plasma
B blood platelets
C red blood cells
D white blood cells

9 Antibodies are made by:

A blood plasma
B blood platelets
C red blood cells
D white blood cells

10 Most photosynthesis happens in:

A flowers
B leaves
C roots
D stems

11 Most stomata are found underneath:

A flowers
B leaves
C roots
D stems

12 The word equation for photosynthesis is:

A carbon dioxide + oxygen $\xrightarrow{\text{sunlight}}$ water + glucose
B glucose + oxygen $\xrightarrow{\text{sunlight}}$ water carbon + dioxide
C oxygen + water $\xrightarrow{\text{sunlight}}$ carbon dioxide + glucose
D water + carbon dioxide $\xrightarrow{\text{sunlight}}$ glucose + oxygen

13 Osmosis is the way that water gets:

A into the air
B into roots
C out of leaves
D out of flowers

14 Which way does sugar usually go through a plant?

A flowers → leaves → stem
B leaves → stem → roots
C roots → stem → leaves
D stem → leaves → roots

15 Phototropism is a plant's response to:

A gravity
B light
C touch
D water

16 Diabetes is a disease caused by too little:

A adrenaline
B insulin
C oestrogen
D testosterone

17 Barbiturates are drugs which are:

A depressants
B stimulants
C anaesthetics
D hallucinogens

18 The place where a plant or animal lives is called:

A an ecosystem
B a community
C a habitat
D a population

19 The diagram shows a pyramid of numbers

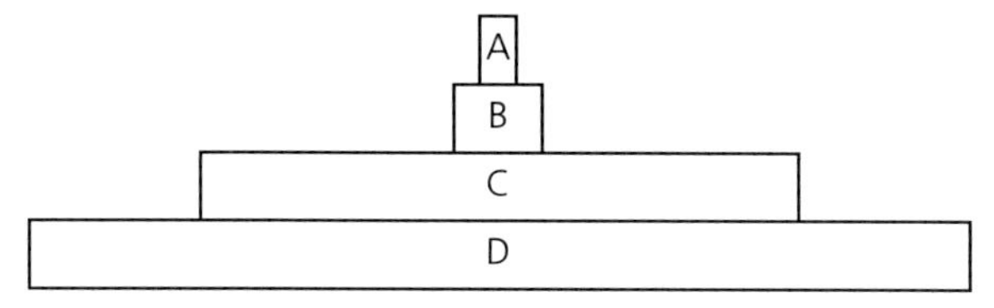

Herbivores are at level:

A
B
C
D

20 The diagram shows a pyramid of biomass.

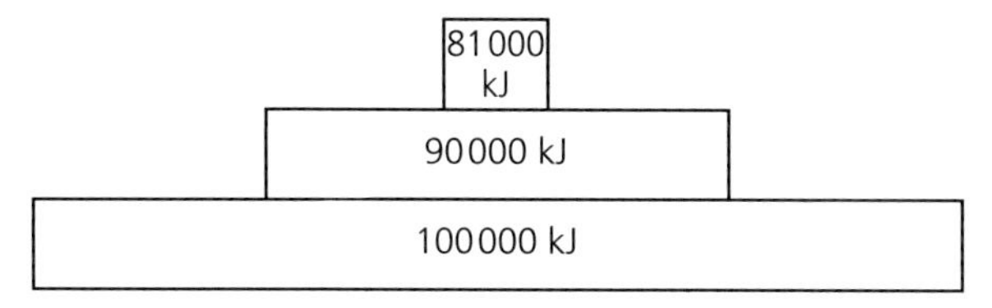

How much energy is lost at each stage?

A 1%
B 10%
C 90%
D 99%

21 If raw (untreated) sewage is put into a river:

A bacteria numbers fall and oxygen levels rise
B bacteria numbers rise and oxygen levels fall
C bacteria numbers fall and oxygen levels fall
D bacterial numbers rise and oxygen levels fall

22 Lichens grow on roof tiles. They are killed by air pollution. Which of these houses is likely to have most lichens on its roof tiles?

A a house near a chemical works
B a house near a city centre
C a house near a coal-fired power station
D a house near a sewage works

23 Which of these is *not* a 'greenhouse gas'?

A carbon dioxide
B methane
C oxygen
D water vapour

24 Herbicides are used to kill:

A unwanted animals
B unwanted fungi
C unwanted pests
D unwanted plants

25 A bull's sperm can be injected into a cow's vagina. The production of calves in this way is called:

A cloning
B embryo transplantation
C sexual reproduction
D tissue culture

26 In humans, fertilization usually happens in the:

A egg tube
B ovary
C womb
D vagina

27 This question is about the order of events in producing a new plant. Which words are in the correct order?

A fertilization, pollination, germination
B germination, pollination, fertilization
C pollination, germination, fertilization
D pollination, fertilization, germination

28 A pure bred tall pea plant is crossed with a pure bred short pea plant. The allele for tallness is dominant.

The result of the cross are likely to be:

A all medium
B all short
C all tall
D half tall and half short

29 Which of these is an example of continuous variation?

A eye colour
B finger length
C shape of nose
D tongue rolling

30 Which of the following is an inherited disease?

A German measles
B Huntington's chorea
C scarlet fever
D whooping cough

Revision and exam guidance

How to revise

There is no one method of revising that works for everyone. It is therefore important to discover the approach that suits you best. These guidelines may help you.

Give yourself plenty of time There are very few people who can revise everything 'the night before' and still do well in an examination the next day. You need to plan your revision to begin several weeks before the examinations start.

Plan your revision timetable Draw up a revision timetable well before the examinations start. Once you have done this, follow it – don't be sidetracked. Stick your timetable somewhere prominent where you will keep seeing it – or better still put several around your home!

Relax You will be working very hard revising. It is as important to give yourself some free time to relax as it is to work. So build some leisure time into your revision timetable.

Ask others Friends, relatives, and teachers will be happy to help if you ask them. Don't just give up on something that is causing you a problem. And don't forget your parents too!

Find a quiet corner Find the conditions in which you can revise most efficiently. Many people think they can revise in a noisy, busy atmosphere – most cannot! And don't try to revise in front of the television. Revision in a distracting environment is very inefficient.

Use routemaps or checklists Use routemaps, checklists, or other listing devices to help you work your way logically through the material. When you have completed a topic, tick it off. Tick off topics you already feel confident about. That way you won't waste time revising unnecessarily.

Make short notes and use colours As you read through your work or your textbooks, make brief notes of the key ideas and facts as you go along. But be sure to concentrate on understanding the ideas rather than just memorizing the facts. Use colours and highlighters to help you.

Practise answering questions As you finish revising each topic, try answering some questions. At first you may need to refer to your notes or textbooks. As you gain confidence you will be able to attempt questions unaided, just as you will in the exam.

Give yourself a break When you are revising, work for perhaps an hour, then reward yourself with a short break of 10 to 15 minutes while you do something different. Look out the window, stretch your legs, have a soft or hot drink. But when your 10 or 15 minutes are up, get back to work!

Success in examinations

Most people become a bit nervous about an important examination. If you have done most of your work consistently for two years and revised effectively, the following steps should help you to minimize anxiety and ensure that your examination results reflect all your hard work.

Be prepared Make sure you have everything you need ready the night before, including pens, pencils, ruler, and calculator. Check that you have anything else required well in advance.

Read carefully Before you start, spend a few minutes reading the paper all the way through. Make sure you know exactly what you have to do.

Plan your time Work out how much time you should spend on each question, based on how many marks it has. Allow yourself a few minutes at the end of the exam to check through your work.

Answer the question! When you are ready to start a question, read through it again carefully to make sure it really does say what you think it says. Follow the instructions to the letter: you will get marks for answering the question but not for giving other information about the subject.

Present your work clearly Write as clearly as you possibly can in the time available and think through what you are going to write before you begin writing. Draw diagrams clearly and simply, using single lines where appropriate. Label your diagrams and make sure the label lines point exactly to the relevant places. The examiner will be trying to award you marks – make it easy for him or her to do so!

Keep calm!!! If you find a question you have no idea about, don't panic! Breathe slowly and deeply and have another look. It's likely that if you stay calm and think clearly, the question will start to make more sense, or at least you may be able to answer part of it. If not, then don't agonize about it – concentrate first on the questions you can answer.

A1 Sorting and naming

Sorting into groups

Biologists have devised a system of **classification** for all known living things. The smallest classification group is the **species**. Members of a species can breed together to produce fertile offspring.

Similar species are grouped into **genera** (singular: **genus**). All members of the same genus have common features. For example, the domestic cat, *Felis domestica*, and the lynx, *Felis lynx*, belong to the same genus. This genus is called *Felis*.

Similar genera are grouped into **families**. The cat family is called Felidae. It includes the domestic cat, the lion (*Panthera leo*), and the cheetah (*Acinonyx jubatus*).

Similar families are grouped into an **order**. Cats, dogs (Canidae), bears (Ursidae), and weasels (Mustelidae) are all in the order Carnivora.

Similar orders are grouped into **classes**. The order Carnivora belongs to a most important class, the Mammalia. Class Mammalia includes bats, monkeys, horses, whales, kangaroos, apes, and humans.

Similar classes are grouped into **phyla** (singular: **phylum**). Mammals are members of the phylum Chordata (animals with a spinal cord). Phylum Chordata also includes fish, amphibians, reptiles, and birds.

The biggest groups of all are the kingdoms. The chordates, along with other animal phyla, belong to the **Animal Kingdom**. Other kingdoms are the **Kingdom Prokaryotae**, **Kingdom Protoctista**, **Kingdom Fungi**, and the **Plant Kingdom**.

Scientific names

Every living thing is given a scientific name. This is because each organism must have a name which refers to only it and to nothing else. An organism's scientific name is understood all over the world. It avoids confusion, especially if an organism has more than one common name. For example, in North America, the puma, cougar, and mountain lion all refer to the same animal. By using the scientific name, *Felis concolor*, scientists know which animal they are talking about.

Examples of classification

Group		
kingdom	Animal	Plant
phylum	Chordata	Angiosperms
class	Mammalia	Dicotyledons
order	Carnivora	Ranales
family	Canidae	Ranunculaceae
genus	Canis	Ranunculus
species	familiaris	bulbosus
scientific name	*Canis familiaris* (domestic dog)	*Ranunculus bulbosus* (buttercup)

Questions

1. Which is the:
   a smallest **b** largest
   classification group?
2. To what order do bears belong?
3. Name seven classes in the Animal Kingdom.
4. Explain why living things are given scientific names.
5. Write a classification table like the one on this page for:
   a the brown bear (*Ursus arctos*)
   b the stoat (*Mustela erminea*).

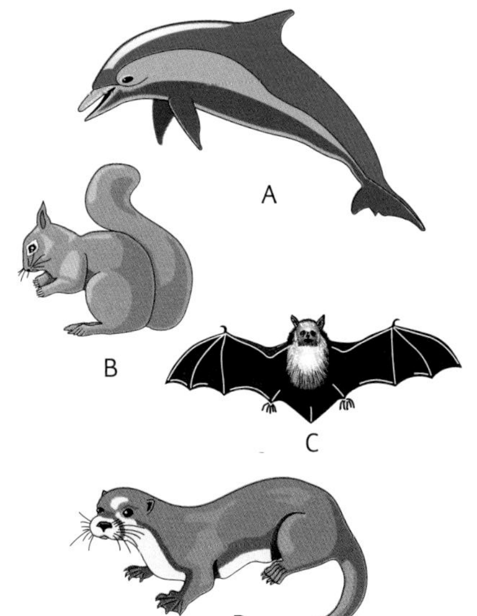

Naming living things

What do we do when we want to find the name of an organism we don't recognize? The answer is to use a **key**. A key is a series of questions which we ask ourselves. Each answer leads to another question. This goes on until eventually the name of the organism is found.

Here is a simple key which will help you to name the four 'unknown' animals shown on the left.

Question 1	Does the animal have flippers? Answer:	Yes **dolphin**	No Go to question 2
Question 2	Does the animal have wings? Answer:	Yes **bat**	No Go to question 3
Question 3	Does the animal have a bushy tail? Answer:	Yes **squirrel**	No **otter**

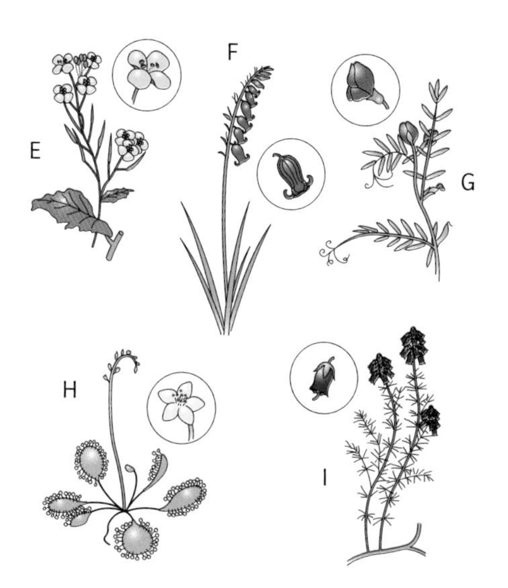

Here is a different kind of key. It works in just the same way as the one above, but is written in the form of a flow chart. Use it to identify the wild flowers shown on the left.

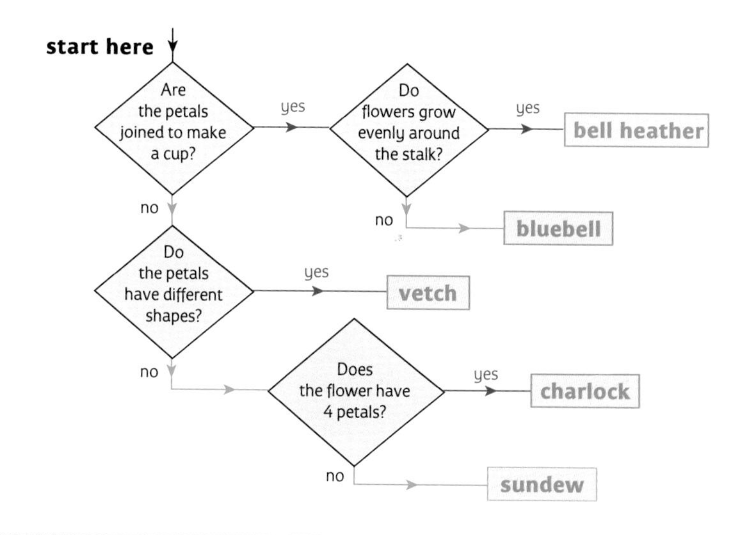

Questions

6 What are the names of the animals A–D?

7 What are the names of the wild flowers E–I?

8 Why are keys useful?

9 Explain how to use a key.

10 Make up a key to help someone identify these leaves:

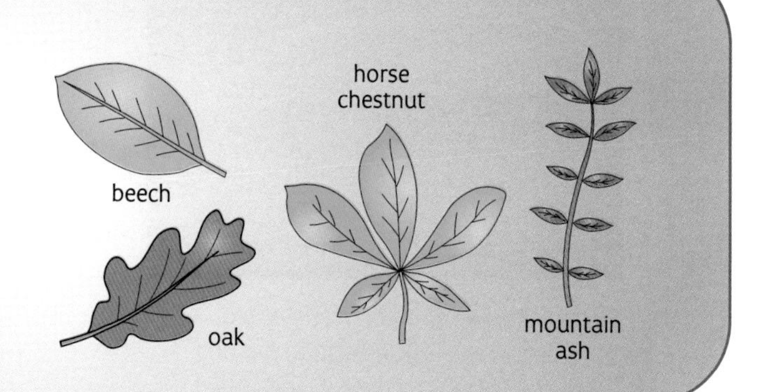

Kingdom Prokaryotae

This kingdom contains single-celled organisms whose cells do not have a nucleus.

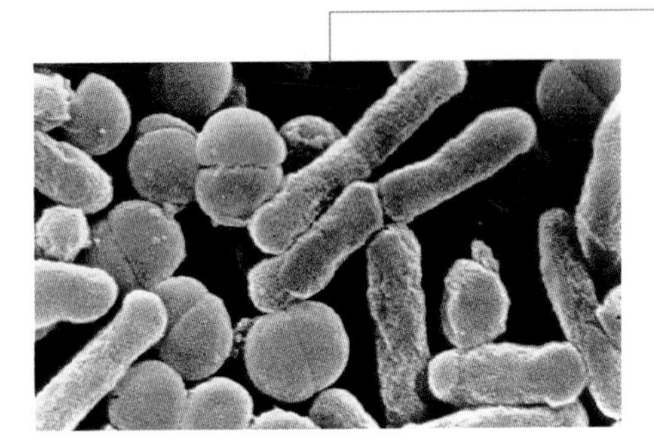

Bacteria *like these live in our intestines. Some are rod shaped, others are round or twisted in a spiral.*

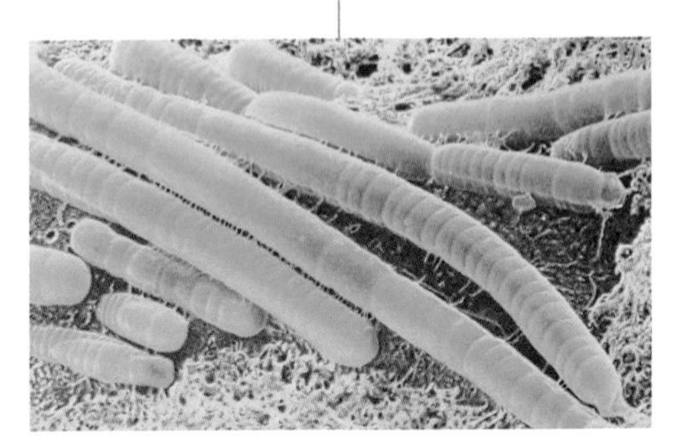

Blue–green algae *make their food by photosynthesis. They live in the sea, near the surface, giving the blue–green colour.*

Kingdom Protoctista

This kingdom has single-celled or very simple, plant-like organisms whose cells have a nucleus.

Protozoa (single-celled organisms)

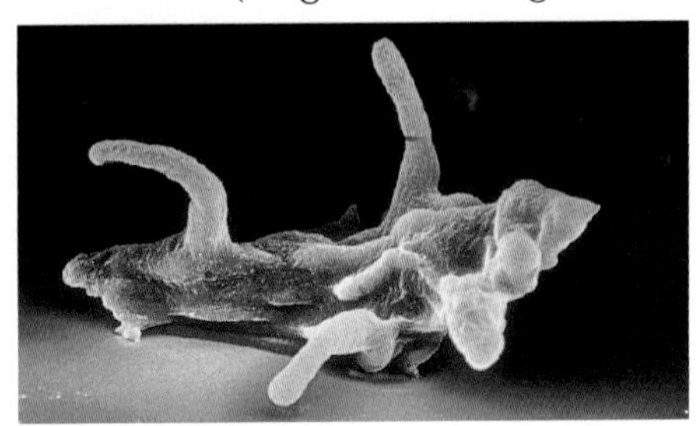

Amoeba *is a single-celled organism. It moves by changing its shape. It feeds on bacteria and other tiny creatures.*

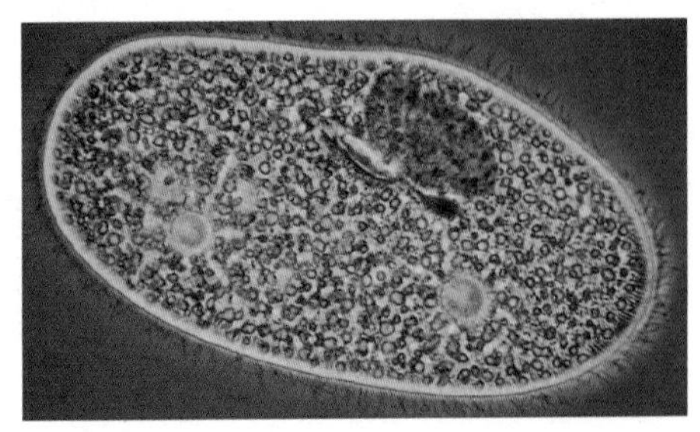

Paramecium *is another single-celled organism. It moves by beating the tiny hair-like cilia which cover it. These cilia also help it to trap other tiny creatures.*

Algae (simple plant-like organisms)

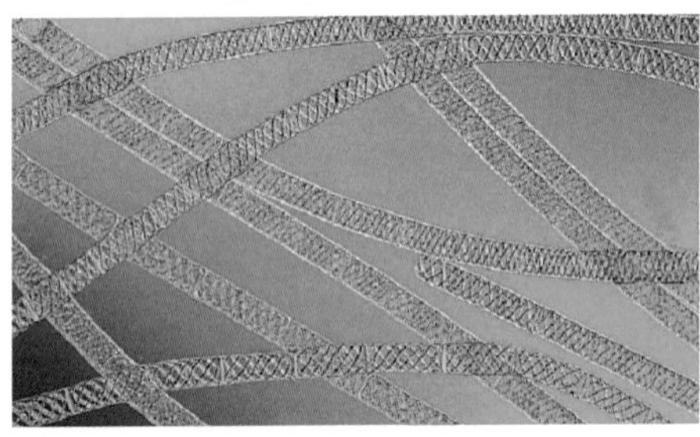

Under a microscope Spirogyra *looks like tiny threads. It lives in ponds.*

Fucus, *or bladderwrack, is a brown algae (seaweed). It lives on rocky sea shores.*

Kingdom Fungi

Fungi are made up of threads called **hyphae**. Fungi look a bit like plants, but they can't make their own food. Most fungi feed on dead things – they are decomposers.

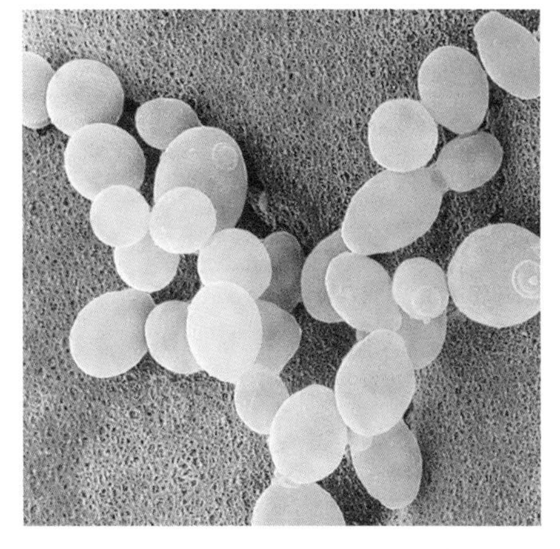

Yeast is a single-celled fungus. It lives on the surface of fruit like grapes. Yeast feeds on sugar.

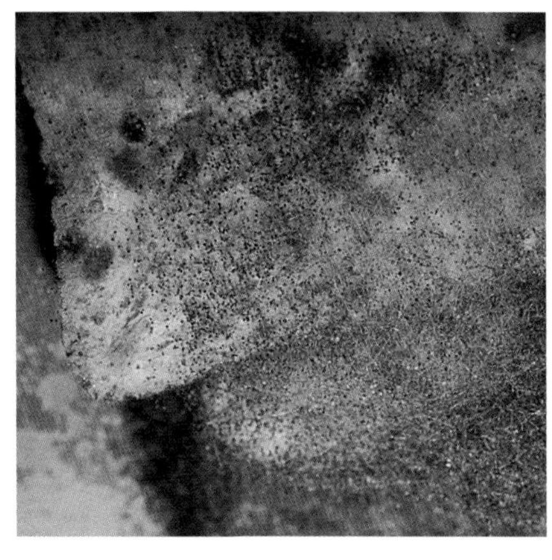

Mould grows on stale food such as bread. The white threads are hyphae. Each black dot contains thousands of spores. These are released for reproduction.

Puff-ball fungus. When a raindrop hits the puff-ball a cloud of spores 'puffs' out.

Mushrooms are fungi you can eat. These parasol mushrooms grow in fields and woods.

Many fungi are poisonous. This fly agaric grows in woods. Its bright colour makes it easy to see.

Bracket fungi are found on dead trees. They feed on the dead wood.

Questions

1 What shape are the bacteria that live in our intestines?

2 Explain why the sea is a blue–green colour.

3 What is different about the cells of living things in the Kingdom Prokaryotae?

4 Give one way that *Amoeba* and *Paramecium* are:
a the same **b** different.

5 To what group of Protoctista do seaweeds belong?

6 Explain why fungi can't be classified as plants.

7 Name one fungus:
a you can safely eat
b that is poisonous.

8 Bread mould is sometimes called 'pin mould'. Can you suggest a reason why?

Groups of living things (2)

Animal Kingdom

Animals are multicellular organisms whose cells have a nucleus and a cell membrane but no cell wall. Their cells are specialized into tissues, organs, and systems. Animals usually move to get their food, which is swallowed and digested inside their bodies. There are more than 1.5 million animal species. Over 1 million of these are insects.

Coelenterates have a body like a bag.

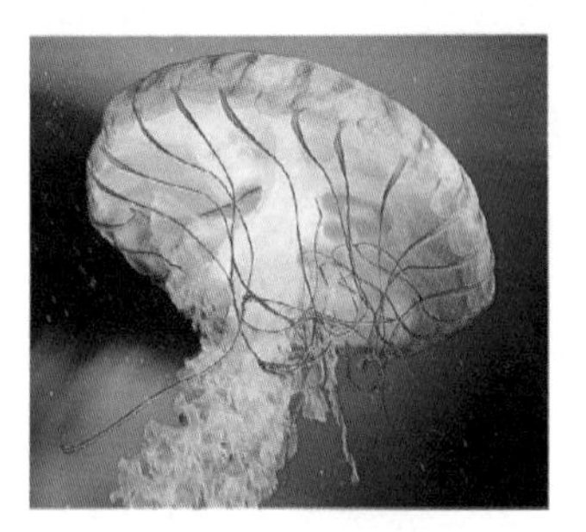

Jellyfish live in the sea. They move by opening and closing like an umbrella. They have sting cells to catch their prey.

Flatworms have flat bodies.

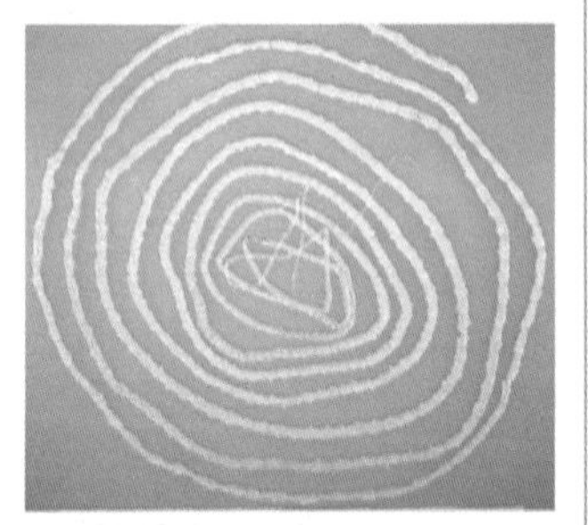

Tapeworms live in the intestines of other animals and eat their digested food. This one lives in cattle. It can grow to 8 metres long!

Segmented worms have bodies made of rings or segments.

Earthworms 'eat' soil and digest the tiny creatures and dead leaves in it. The rest passes through their bodies.

Molluscs have soft bodies with one or two shells.

Snails have coiled shells. They move about on a slimy foot. They have rough tongues for scraping up food.

Arthropods have jointed limbs, at least one pair of antennae (feelers), and a tough outer skeleton.

Insects

This dragonfly has six legs and two pairs of wings. Its body is divided into three parts – head, thorax, and abdomen.

Arachnids

Spiders are the most common arachnids. They have eight legs and no wings. Spiders spin webs from silk to catch their prey.

Crustaceans

Crabs have a hard outer skeleton or shell. They have two pairs of antennae and five pairs of legs. One pair of legs is used to catch prey and put food into the mouth.

Myriapods

Millipedes have long thin bodies and lots of legs. They eat plants.

Chordata are animals that usually have a backbone.

Fish

This trout spends all of its life in water. Its body is covered in soft scales which come off easily. Its eggs are fertilized outside the body in the water.

Amphibians

Frogs can live both on land and in water. On land they use their lungs to breathe. In water they breathe through their smooth, moist skin. Frogs usually go into water to breed because their eggs are fertilized outside the body.

Reptiles

Lizards have a tough, dry, scaly skin. The scales don't come off easily. All reptiles breathe with lungs. They reproduce by laying eggs and burying them in sand.

Birds

All birds have wings and a body covered in feathers. Birds reproduce by laying eggs which they incubate until they hatch. Birds have light, hollow bones to help them fly.

Mammals

All mammals have fur (even humans have a bit!). They give birth to live young and feed their babies on milk produced in mammary glands.

Questions

1 Which group of animals has a body like a bag?

2 What do tapeworms eat?

3 Give one feature of molluscs.

4 Describe how snails eat.

5 What group of arthropods has:

a three pairs of legs **b** four pairs of legs

c two pairs of antennae?

6 Name the three parts of an insect's body.

7 Explain how amphibians such as frogs can live on land and in water.

8 Why must frogs breed in water?

9 How is reproduction different in reptiles and in birds?

10 What is the difference between fish scales and reptile scales?

11 What two features help birds to fly?

12 How would you recognize a mammal?

Plant Kingdom

Plants have more than one cell. In fact they usually have many thousands of cells specialized into tissues, organs, and systems. Plants have cells that contain chlorophyll. This means they can make their own food by photosynthesis.

Over 300 000 plant species have been classified so far – but there are many more still to be discovered.

The Plant Kingdom is divided into two large groups depending on whether or not the plant produces seeds.

Plants without seeds

Mosses and **liverworts** grow in damp places. They make **spores** in capsules which are held up in the air by stalks.

Moss plants have stems and tiny leaves. Lots of plants grow together for support and to stop them drying out.

When the spores are ripe, the lid drops off each moss capsule letting the spores out. Spores are light and are easily scattered by the wind.

Liverworts don't have a stem. They look like leaves growing flat on the ground.

Ferns are larger and more developed than mosses. Their stems, leaves, and roots are similar to those of the flowering plants.

The stem of this fern supports the leaves for photosynthesis. The leaves have a cuticle to stop water loss. This is why some ferns can live on open moorland.

Like mosses, ferns reproduce by spores. Spores are made in capsules on the backs of fern leaves. The capsules are protected by a small brown cover.

Ferns don't have buds like flowering plants. Instead the leaves are tightly coiled and unwind as they grow.

Plants with seeds

Conifers are trees or shrubs with needle-like leaves. They reproduce by seeds rather than spores, but the seeds are made in cones not flowers and they are not enclosed in an ovary.

Male cones produce lots of pollen. This is carried by the wind to the female cone.

The fertilized female cone grows seeds under its flaps. When the seeds are ripe, the flaps open to let them float to the ground.

Flowering plants reproduce by seeds which are made inside flowers. The seeds are enclosed in an ovary. There are two kinds of flowering plant – **monocotyledons** and **dicotyledons**.

A horse chestnut tree in flower. Its seeds are used for playing 'conkers'. Is it a monocotyledon or a dicotyledon?

Monocotyledons, like these daffodils, have long thin leaves with parallel veins.

Dicotyledons, like this water lily, have broad leaves with a network of veins.

Questions

1 What two large groups is the Plant Kingdom divided into?

2 Give one way the mosses and liverworts are:

a the same **b** different.

3 Explain why mosses grow together in clumps.

4 Why are some ferns able to survive in open areas such as moorland?

5 **a** Describe how young fern leaves grow.

b How is this different from flowering plants?

6 Give two features of conifers.

7 How are conifers pollinated?

8 **a** Name the two groups of flowering plants.

b Describe how each group is different.

c Name one example of each group.

Note: The answers are arranged in three sections:
- end-of-module practice questions
- exam-style questions
- multiple-choice questions

There are no answers to end-of-spread questions.

End-of-Module practice questions

Module 1

1 **a** movement – changing position; respiration – releasing energy from food; sensitivity – responding to something; feeding – taking in food or raw materials; excretion – getting rid of waste; reproduction – producing more of the same kind; growth – increasing in size **b** No. The car doesn't grow, respond, or reproduce.

2 **a i** controls the cell's activities **ii** controls what enters and leaves the cell **b** chemical reactions **c i** to carry messages **ii** long and thin to cover distances

3 **a** stomach – digests food; kidney – gets rid of urea; lung – exchanges oxygen and carbon dioxide; brain – controls what the body does; eye – gets information about the surroundings **b i** A breathing/respiratory system **ii** B digestive system **iii** C excretory system **iv** D circulatory system

4 **a i** 7 **ii** to help chemical reactions **b i** to give intestine muscles something to push on **ii** it contains 14% fibre **c i** 701.5 kJ **ii** 5.1 g

5 **a** A mouth; B liver; C gall bladder; D large intestine; E gullet; F stomach; G pancreas; H small intestine; I rectum; J anus **b i** D **ii** C **iii** F **iv** H **v** A, F, G, or H **vi** A **vii** I

6 **a** glucose **b** amino acid **c** fatty acid, glycerol **d** amylase **e** lipase **f** protease

7 **a i** add iodine; blue/black colour shows starch present **ii** heat with Benedict's solution; orange colour (brick red) shows glucose present **b** glucose; they are small **c** starch molecules are too big **d** small intestine (villi) **e** blood

8 **a** to absorb digested food **b i** easy for food to get through **ii** absorbs more food **iii** food carried quickly to other body parts

9 **a** A windpipe or trachea; B muscle; C rib; D lung; E bronchiole; F alveoli; G diaphragm; H bronchi **b i** oxygen and carbon dioxide **ii** blood and air **c i** easy for gases to get through **ii** gases dissolve and pass through walls in solution **iii** gets gases to and from lungs

10

Breathing in	Breathing out
rib muscles contract	rib muscles relax
diaphragm contracts	diaphragm relaxes
chest gets bigger	chest gets smaller
air sucked into lungs	air pushed out of lungs

11 **a i** 6 **ii** 9 **b i** $3.4 - 2.9 = 0.5\ dm^3$ **ii** $5.0 - 1.8 = 3.2\ dm^3$ **c** after race, runner needs to repay oxygen debt **d** $3.2 \times 9 = 28.8\ dm^3$

12 glucose, energy, oxygen, carbon dioxide, water, emergency, lactic acid, cramp, muscles, oxygen debt

13 **a** A artery taking blood to lungs; B artery taking blood to body; C vein bringing blood from lungs; D left atrium; E left ventricle; F right ventricle; G valve strings; H valve; I right atrium; J vein bringing blood from body **b i** E **ii** F **c i** B, C, D, and E **ii** A, F, I, and J

14 **a** antibodies – chemicals that kill microorganisms; antitoxins – break down poisons; haemoglobin – joins with oxygen; plasma – a straw-coloured liquid; platelets – help the blood to clot; red cells – made in the bone marrow; white cells – have a big nucleus **b** two from: digested food; urea; carbon dioxide; hormones **c** white cells eat microorganisms

15 **a i** very tiny living things **ii** bacteria; viruses **b i** stops microorganisms going into the air **ii** kills any microorganisms that are in the food **iii** removes any microorganisms from the skin **iv** stops growth of microorganisms on body **v** the stranger may have an infection **c** three from: nose; mouth; digestive system; breathing system; reproductive system; cuts in skin; etc. **d** natural immunity, white cells, antibodies, antitoxins, vaccine, artificial immunity **e** Jack already has antibodies against chickenpox because he had the disease. He is naturally immune.

Module 2

1 **a** A flower; B bud; C stem; D root; E leaf **b** bud – where new growth takes place; flower – where reproduction takes place; leaf – where photosynthesis takes place; root – anchors the plant in the soil; stem – supports the plant

2 **a** A vascular bundle; B xylem; C phloem; D root hair; E root cap; F growing tip **b** xylem; phloem **c i** xylem **ii** phloem **iii** root hairs **iv** root cap

3 **a** A waxy cuticle; B leaf tissue; C stoma; D upper skin; E chloroplast; F guard cell **b** cells that make food by photosynthesis – leaf tissue; makes the leaf waterproof – waxy cuticle; hole that lets gases in and out of the leaf – stoma; has chlorophyll inside – chloroplast

4 **a** A energy; B glucose; C oxygen; D water; E carbon dioxide **b** Sun **c** water; carbon dioxide **d** glucose; oxygen **e** water + carbon dioxide $\xrightarrow{\text{energy}}$ glucose + oxygen

5 **a** photosynthesis is slow to start with; it speeds up and levels off at a steady rate **b** lack of carbon dioxide **c** speed of photosynthesis would go up **d** photosynthesis speeds up as temperature rises **e** temperature is higher on warm days and plenty of light is available so speed of photosynthesis is faster; more food = more growth

6 **a i** normal size; yellow leaves **ii** small size; yellow leaves **b** size; plant without nitrogen is smaller **c** food crops need leaves to photosynthesize and make food; bigger leaves means more food made

7 **a i** it has got longer **ii** it has got shorter **b** osmosis (or definition) **c** water enters by osmosis; volume inside cells gets bigger; cell contents push out against cell wall, making cells rigid; in sugar solution, water leaves by osmosis and opposite occurs **d** water enters cells by osmosis making them rigid (as above); rigid cells in stem will keep stem upright; with lack of water, the opposite happens and the plant wilts

8 **a** the loss of water vapour from a plant's leaves **b** air would break the continuous line of water **c** air could get into the twig and break the line of water **d** put apparatus in different places and measure speed of transpiration **e** open and close stomata; waxy cuticle; falling leaves in winter **f i** helps transport water and minerals up to the plant from roots **ii** loss of water in hot, dry conditions leads to wilting and death

9 **a i** they have bent towards the hole **ii** they have grown towards/responded to the light **b** phototropism **c** put foil caps on some tips or remove the tips from some of the seedlings **d i** auxin **ii** it would grow quicker **iii** selective weed killer; hormone rooting powder

10 **a** A ciliary muscle; B retina; C sclera; D optic nerve; E jelly; F lens; G liquid; H cornea; I pupil; J iris; K suspensory ligament **b i** gets fatter **ii** gets thinner **c i** gets bigger **ii** gets smaller **d** carries messages to brain

11 **a** ears – a fire alarm; eyes – colours of the rainbow; nose – a girl's perfume; skin – a hot plate; tongue – a bitter drink **b** ear **c** smell gives flavour to food, cold deadens sense of smell

12 **a** A adrenal glands; B ovaries; C testes; D pancreas; E pituitary gland **b** adrenaline – speeds up heartbeat and breathing; insulin – controls the amount of glucose in the blood; oestrogen – gives girls their female features; testosterone – gives boys their male features **c** adrenaline **d** insulin **e** in the blood **f** nervous messages travel along nerve cells directly to where they are going; hormones depend on speed and direction of blood flow

13 **a i** alcohol **ii** caffeine **iii** aspirin **iv** ecstasy **b i** the body can't do without it **ii** it can damage the body in other ways

14 **a** A sweat gland; B dead skin; C blood vessels; D oil gland; E hair; F fat layer **b i** air trapped between hairs; air is a good insulator **ii** heat from body evaporates sweat, so you feel cooler **c i** contract or get narrower **ii** less blood near surface of skin so less heat is lost

15 a A cells with cilia; B cells that make mucus; C windpipe wall; D mucus; E cilia **b** in a smoker: no/fewer cilia; thicker/more mucus **c** coughing **d** bronchitis **e i** lung cancer or emphysema **ii** lung cancer – tar collects and irritates alveoli, causing tumour; emphysema – chemicals in smoke destroy alveoli, reducing surface area for absorbing oxygen

Module 3

1 a i roots; stem; leaves **ii** roots – anchorage; stem – supports leaves; leaves – photosynthesis **b** the leaves need to be in a position to get light for photosynthesis **c** there is no water around it for support **d** helps keep the leaves up near the water surface where there is most light

2 a

Examples of competition	Examples of predation
two stags fighting for control over a herd of deer	farm cats killing mice and stopping them eating cereal crops
trees in a forest growing upwards to get light	lions hunting a zebra for food
a seagull chasing off other birds from food in a garden	a spider catching a fly in its web

b two from: mice; zebra; fly

3 a i strong beak and claws **ii** juicy stem with thick waxy cuticle; leaves reduced to spines **iii** long thin mouthparts act like a drinking straw **iv** strong jaws for crushing bones and sharp teeth for tearing flesh **b i** no urine so no water loss; at night it is cool so no water is lost in sweat when they are active **ii** living in holes during the day means it can't be eaten; it is more difficult for a predator to see them at night

4 a i both populations grow and then the yeast population starts to go down **ii** *Paramecium* are eating the yeast **b** there is less yeast/food for the *Paramecium* so they die **c** it takes time for *Paramecium* to die after yeast die; yeast reproduce faster than *Paramecium* **d** it would continue to increase quickly; then level off; then eventually fall **e** two from: lion–zebra; cat–mouse; spider–fly; etc.

5 shortage of light – plants can't photosynthesize; overcrowding – pigs fight if kept together in large numbers; forest fires – squirrels' homes destroyed; sudden fall in temperatures – birds die in bad winters; floods – rabbits' burrows washed away

6 a 500 000–10 000 years ago **b** the population fell **c** because of the effect of disease; low numbers so reproduction rate low **d** better living conditions; medical treatment; people live longer, etc.

7 a natural ecosystems – woodland, lake, garden (if 'wild' and neglected); artificial ecosystems – tropical fish aquarium, field of wheat, garden (if looked after), grazing meadow **b i** nothing (other than energy) is added to it or taken away **ii** if we create too much waste and pollution or use up resources before they can be renewed, the Earth ecosystem will be destroyed – once gone, it's gone for good

8 a i no pollution at A **ii** pesticides get into bodies of fish; pesticides build up in bodies of fish-eating birds to higher levels **iii** sewage is decomposed by bacteria; more sewage means more bacteria; more bacteria means less oxygen as they use it for respiration **iv** chemical pollution from the factory is blown over plantation; rain washes pollutants onto ground/trees; tries die **v** fertilizers get into water via drainage ditch; water weed grows owing to increase in nutrients

9 a carnivore – animal that eats other animals; omnivore – animal that eats both plants and animals; decomposer – animal that feeds on dead plants and animals; herbivore – animal that eats plants; producer – plant that makes its own food **b i** grass **ii** grasshopper **iii** lizard or hawk **c** it shows the direction in which the food goes

10 a any plant → snail → water beetle; any plant → tadpole → water beetle **b i** duck weed or *Elodea* or milfoil or algae **ii** caddis fly or snail or tadpole **iii** leech or dragonfly nymph or beetle larvae or water beetle or stickleback **iv** water beetle or leech or fish louse or dragonfly nymph or beetle larva **v** fungi or bacteria or flatworm or water skater or crayfish **c** increased numbers of animals water beetles feed on; reduced numbers of plants **d** any plant → tadpole → stickleback → beetle larva (or dragonfly nymph) → leech

11 a a diagram showing the number of plants or animals at each stage/level of a food chain **b** grass → antelopes → lion C; oak tree → caterpillars → sparrows A; rabbits → fox → fleas D; tomatoes → whitefly → parasites (or microscopic grubs) B **c** a diagram showing the mass of plants and animals at each stage/level of a food chain **d** all will be pyramid shaped

12 a i light energy **ii** chemical energy **b** it gets less **c** to any other herbivore e.g. rabbits, snails; respiration of grass, etc.; faeces, urine, respiration of cow, etc. **d** 125 kJ **e** they compete with the cows for energy in the grass

13 a A respiration; B burning; C decay; D photosynthesis **b** oxygen **c** carbon dioxide **d i** glucose + oxygen → carbon dioxide + water + energy **ii** carbon dioxide + water + energy → glucose + oxygen **iii** they are the exact opposite of each other

14 a decay or decomposition **b** bacteria or fungi **c** oxygen, water, and warmth (suitable temperature) **d i** something that has come from a living thing (including its body) **ii** it takes too long to decay; the river would become polluted before decay was complete **e** make a compost heap or put it in the 'green' bin at the local tip

15 a to get money for timber; for space to grow crops to sell **b** three from: less photosynthesis going on; destroys habitats for (rare) animals and plants; no protection against erosion of soil by wind and rain; burning scrub/branches creates more carbon dioxide/greenhouse gases **c** share the world's wealth and provide more financial aid from richer countries; look for other ways of making money, e.g. conservation of rare plants for useful things

Module 4

1 a size; colour; markings **b** continuous variation **c** discontinuous variation **d** peacock – predators think it is a larger animal (carnivore); brimstone – camouflage/hides it from predators

2 a 32 **b i** in range 160–9 cm **ii** in range 135–54 cm **c** in range 150–4 cm **d** a normal distribution cure **e** any continuous variable, e.g. weight; finger length; arm length; shoe size; etc.

3 a 28 days **b** 12 eggs a year for 35 years = 420 eggs **c** 30 March **d** womb lining prepares to receive a fertilized egg **e** period **f** if womb lining is shed then baby is lost as well **g** when she is not producing eggs of her own

4 a i A placenta; B womb wall; C vagina; D umbilical cord; **ii** it could harm the baby **iii** loss of fluid; no protection for baby **b i** 46 **ii** 47 **iii** pair 21 **iv** pair 21 of one parent didn't separate

5 a a change in the chemical structure of a gene **b** two from: alpha, beta, gamma radiation, X-rays, etc. **c** tar from cigarettes, mustard gas, benzene, phenol, etc. **d i** the animal can reach leaves higher up the tree so gets more food and has a better chance of survival **ii** some giraffes have longer necks than others (variation); they win the competition for food; they are more likely to survive and breed, passing on the genes for a longer neck

6 a an egg or ovum **b i** swimming **ii** in the nucleus **c i** X *or* Y **ii** X

d

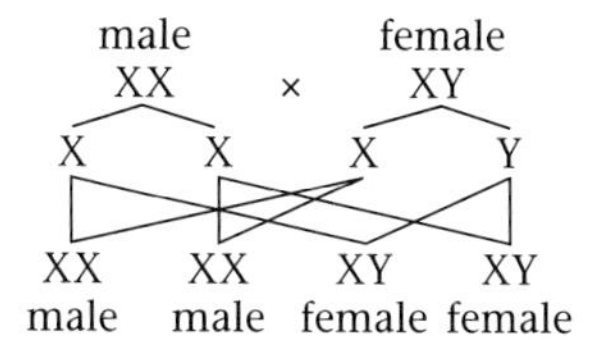

7 a it is big, muscular, heavy, strong **b** a male and a female with the desired characteristics are selected and bred together; this is continued for further generations to increase the chance of breeding true **c** it is possible that an animal with undesired characteristics was used in the breeding line; there is always a chance that those undesired characteristics will show themselves in later generations (a 'throwback') **d i** a group of living things that can breed to produce fertile offspring **ii** breed from them; their offspring should be fertile

8 a i 60 million years ago **ii** 0.4 m **iii** it has flat/blunt teeth **b** bones and teeth **c i** they have got fewer as the animal has evolved **ii** only the middle toe(s) touch the ground when running, the others become useless **d** changing habitat/food source/to avoid predators, etc.

9 a

(BB)	Bb
Bb	bb

b B **c** recessive **d i** **BB** and **bb** **ii** **Bb** **e i** **BB** or **Bb** **ii** **Bb** or **bb** **f** there are two possible genotypes **BB** and **Bb**; both have the dominant **B** gene allele so will have brown eyes

10 a i a group of similar cells **ii** the organism the tissue comes from (that donates the tissue) **iii** a cell that is an identical copy of the original cell **b** all the genetic information is contained in one nucleus of one cell **c** to avoid contamination/keep the culture disease free **d** cells/tissue used is microscopic; propagation means to grow **e** advantage – one of: fast, disease free, high yield, little space needed, etc.; disadvantage – all plants produced are identical; no variation **f** sugar for energy; mineral elements; amino acids for protein/growth; growth hormones e.g. auxin to encourage growth

11 a someone who has the gene but doesn't show the condition **b i** 50:50 (50% or 0.5) **ii** 50:50 (50% or 0.5) **iii** no chance (0% or 0)

c

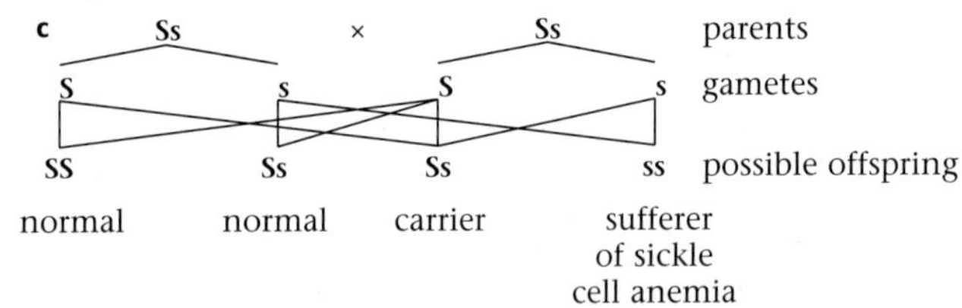

d carriers are more resistant to malaria than 'normal' people; they survive

12 a a disease passed from parent to children/an inherited disease **b i** carries two different alleles for a particular character **ii** an allele whose effects are hidden if a dominant allele is also present **iii** thick/not so runny **c** mucus usually keeps lungs free of bacteria by carrying them up to the throat; if the mucus can't flow easily the bacteria stay in the lungs and cause infection **d** physiotherapy to keep air passages clear of sticky mucus; inhale spray containing normal genes for runny mucus (gene therapy)

13 a C A B D **b** pregnant **c** 9 **d** the baby is protected; it has a good food supply

Exam-style questions

1 a artery – circulatory; intestine – digestive; kidney – excretory; lung – respiratory; nerve – nervous; penis – reproductive **b i** something able to exist on its own **ii** name of any living thing

2 a A **b** D **c** A → D **d** diffusion **e** examples include gas exchange in the lungs, digested food from gut into blood, etc.

3 a i muscles in the wall of the intestine contract behind the food ball, pushing it along (peristalsis) **ii** if there is no fibre the muscles have nothing to push on, so food will not move along the intestines **b i** a chemical that speeds up the digestion of food **ii** fats **iii** food must be digested into small molecules so it can be absorbed into the body

4 a i human **ii** whale **iii** polar bear **b i** field/barn/farm **ii** cold/freezing land **iii** cold seas **c i** it provides insulation/energy store **ii** polar bears need insulation against the cold **d i** they have more protein in their milk **ii** to get body weight/fat insulation quickly; to hunt for food, etc.

5 a carbon dioxide **b** air breathed out has more carbon dioxide than air breathed in **c** a waste product of respiration is water which condenses on the inside of the container **d** carbon dioxide, water **e** prevents the mouse becoming stressed

6 a a microscopic fungus/microorganism **b** sugar **c** carbon dioxide **d** without oxygen **e i** alcohol **ii** evaporates from the dough

7 a a blood vessel that carries blood away from the heart **b** the artery lined with cholesterol has a narrower hole through the centre **c** it slows down because there is less space for the blood to get through **d i** angina or heart attack **ii** angina – blood flow in arteries supplying the heart is reduced so less oxygen reaches the heart muscle; heart attack – blood flow to the heart is cut off by a blood clot getting stuck in an artery supplying the heart so no oxygen reaches the heart muscle **iii** eat less fatty food, take more exercise, no smoking, reduce alcohol intake, avoid stress, etc.

8 a the direction of light **b** A **c** by a ring of muscle (ciliary muscle) around the lens attached to it by ligaments **d i** straight lines correctly drawn with arrows **ii** the one passing straight through the middle

9 a stimulus → gland → hormone → bloodstream → target organ → response **b i** pancreas **ii** liver **iii** converts glucose into glycogen **iv** it rises **v** control the diet or injections of insulin

10 a i D **ii** G **iii** B **iv** C **b i** 28 days **ii** 4/5 days **iii** causes it to build up **iv** the womb lining breaks down **v** to keep the lining of the womb in place to supply baby with nutrients and oxygen

11 a i the soil **ii** minerals dissolve in water; taken into plant roots by diffusion or active transport **b** magnesium is needed for chlorophyll; no chlorophyll means no photosynthesis **c** both are needed for good growth of plants; good growth means good crops **d i** algae, like other plants, use the fertilizer to grow fast **ii** cuts out light to plants below, they cannot photosynthesize, no oxygen is produced, water life dies

12 a osmosis – water moves from high water concentration in the soil to a lower concentration in the root cells **b** they have root hairs to increase the surface area **c** xylem **d** stomata **e i** loss of water vapour from a plant **ii** mainly from the leaves but also from the stems of some plants **iii** a continuous stream of water through the plant from roots to leaves **iv** cacti live in dry, hot places so they need to prevent water loss as much as possible

13 a i three (Jill, Ruth, and Pip) **ii** four (Phil, Kenton, David, and Josh) **b** both parents **c i** brown hair **ii** all of Phil and Jill's children have brown hair even though Jill is blond **d** 100% (they will both carry two recessive 'blond' alleles)

14 plants, millions, sedimentary, hard, preserved, conditions, warmth

15 a i three from: zinc; chromium; nickel; phosphates **ii** rapid growth of algae which reduces the amount of oxygen in the water **iii** plankton is at the bottom of the food chain; without it all other life dies **iv** any sensible suggestion that reduces the polluting effects, e.g. chemical plants having to treat water before letting it enter rivers; they can't just ban effluent going into rivers – they need to consider commercial aspects as well **b i** mass of living material **ii** correct pyramid using a sensible scale, e.g. 1 cm = 10 kg, with plant plankton at bottom rising to cod at the top

16 a the antibiotics killed the bacteria **b** no; stronger/more resistant bacteria in the population were able to survive **c** stronger/more resistant bacteria had reproduced; even stronger/more resistant bacteria would be produced; antibiotic would only kill the few weak ones in the population **d** natural selection **e** to kill off all of the bacteria population first time

Multiple-choice questions

1 D **2** B **3** C **4** C **5** A **6** C **7** B **8** A **9** D **10** B **11** B **12** D **13** B **14** B **15** B **16** B **17** A **18** C **19** C **20** B **21** D **22** D **23** C **24** D **25** C **26** A **27** D **28** C **29** B **30** B

Index

Index

Index